R & R   Ck Reg

George Carvey

Theory and Practice of Pulse Plating

Edited by
Jean-Claude Puippe
Werner Flühmann AG
Dübendorf, Switzerland
Frank Leaman
Electrochemical Company, Inc.
York, PA

Sponsored and Published by
American Electroplaters and Surface Finishers Society

Established 1909

American Electroplaters and Surface Finishers Society
The International Technical and Educational for
Surface Finishing
International Headquarters in Orlando, Florida

PUBLISHED BY THE
AMERICAN ELECTROPLATERS AND SURFACE FINISHERS SOCIETY
12644 Research Parkway, Orlando, Florida 32826

International Standard Book Number 0-936569-02-6
0-936569-01-8

Library of Congress Card Number 86-72342

Manufactured in the United States of America

Theory and Practice of Pulse Plating

DEDICATION

This text is dedicated to the late Professor Dr. Norbert Ibl, who served as head of the Department of Chemical Engineering at the Swiss Federal Institute of Technology, Zurich, Switzerland. Dr. Ibl, through research, numerous lectures and articles in the field of electrochemistry, contributed greatly to the study of the pulsed deposition of metals. His work established much of the present theoretical basis of pulse plating. Several of his students have authored chapters in this text.

EDITORS

J.Cl. Puippe is R&D Manager for Werner Flühmann AG, Dubendorf, Switzerland. He received his PhD in 1978 from ETH in Zurich, where he worked under the direction of Dr. Norbert Ibl on pulse plating research. He worked on laser enhanced electroplating as an invited scientist at IBM. His present fields of interest include alloy plating, difficult-to-plate metals, and plating on plastics. He serves AESF as a member of the Research Board and as 2nd Vice President of the International Branch.

Frank Leaman is president of Electrochemical Company, Inc., York, PA. He received a BA from Eastern Mennonite College and did graduate work in chemistry at Franklin & Marshall College. He holds several patents in electroless plating processes.

CONTRIBUTORS

H.Y. Cheh received a BASc from Ottawa University in 1962 and a PhD in chemical engineering from the University of California at Berkeley in 1967. He came to Columbia University in 1970, and is currently Professor of Electrochemistry and Chairman of the Department of Chemical Engineering and Applied Chemistry. His research deals with electroplating, energy conversion, heterogenerous kinetics and high pressure thermodynamics.

D. Landolt is professor at the Swiss Federal Institute of Technology Lausanne (EPFL) and head of the Laboratoire de Metallurgie Chimique. His research interests include electrochemical metal deposition and dissolution, corrosion and surface chemistry.

O. Dossenbach is a lecturer in the Department of Chemical Engineering and Industrial Chemistry at the Swiss Federal Institute of Technology (ETH), Zurich. He holds a diploma in chemical engineering and a PhD in technical sciences from ETH. His main interests are in electrochemical engineering and in electrolytic metal deposition.

W. Paatsch is a member of the Federal Institute for Testing Materials (BAM), Berlin, Federal Republic of Germany. He received his diploma in physics in 1967 and his PhD in 1971 from the Technical University of Berlin, where he has held a part-time professorship since 1979. His areas of specialization are in surface science and technology, particularly in the fields of electrodeposition and corrosion.

B. Sutter is head of the Corrosion and Surface Treatment group at the Technical Center for Mechanical Engineering (CETIM), France. His laboratory specializes in anticorrosion studies and in the evaluation of new plating techniques, including pulse plating. He graduated from the Paris Conservatoire National des Arts et Métiers in 1977.

A. Knödler studied chemistry and metallurgy at Technische Hochschule Stuttgart, Stuttgart, West Germany. In 1956 he joined the Research Institute for Precious Metals and Metal Chemistry in Schwäbisch Gmünd, where he is engaged in the fields of precious metal plating, alloy plating and metals analysis.

Arthur J. Avila received a BS in Chemical Engineering from New York University and a MS in Metallurgy from the Stevens Institute of Technology. In 1972 he retired as Senior Staff Engineer from the Western Electric Company. Since then he has been serving as a consultant in the fields of electroplating, pollution control, refining and metallurgy.

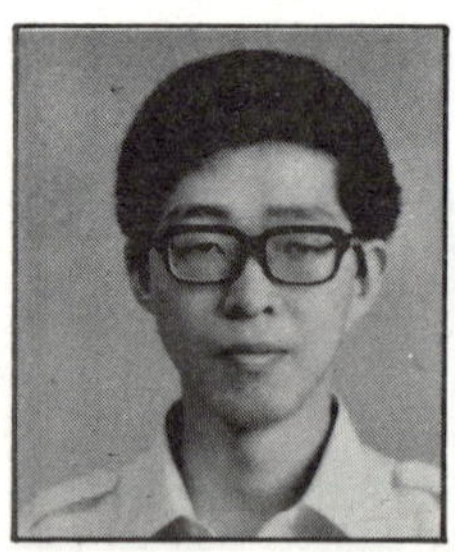

Hidetaka Konno is a Research Associate in the Analytical Chemistry Laboratory, Faculty of Engineering, Hokkaido University. He graduated from Hokkaido in 1969 and received his PhD in 1976. He served as a visiting research scientist at Lehigh University during 1981-83. His present research centers on surface modification of metals and surface analysis.

Norman M. Osero is president of Kraft Dynatronix, Inc., Amery, Wisconsin. He has been active in the development, manufacturing, and marketing of pulse plating equipment since 1971. He is a member of the American Electroplaters and Surface Finishers Society and the American Electrochemical Society. He holds a degree in electronics from Northcentral Technical Institute, Wausau, Wisconsin.

TABLE OF CONTENTS

PREFACE

Research in the electrodeposition of metals by a cyclic variation or interruption of the applied current offers exciting alternatives for the electroplater and electroplating engineer. Optimized pulse plating conditions produce many desirable deposit characteristics, which are enumerated in the text.

Pulse plating is essentially an interrupted direct current plating technique. However, the pulse frequency range for electrodeposition is much narrower than the one of the electrical current in a first order conductor. Limitations occur at high frequencies because of capacity effects at the electrodes, and at low frequencies the electrodeposition is not much different than at DC conditions. In pulse plating, the most sensible pulsed current or pulsed voltage conditions are those for which at least one phenomenon related to the electrodeposition mechanisms occurs under nonstationary conditions during a relevant part of the pulse duration.

The text deals with the theoretical aspects of pulse plating, the results of numerous research studies related to metallic deposition with interrupted current techniques, practical applications for pulse plating and a review of pulse plating equipment that has become commercially available for the electroplater. It is intended to serve as a valuable reference to engineers who select finish specifications for metals and electroplaters engaged in the field of electrolytic metals deposition.

Jean-Claude Puippe and Frank H. Leaman

ACKNOWLEDGEMENTS

The editors gratefully acknowledge the assistance of F.H. Reid, author of numerous articles in the field of precious metal electrodepositon and technical editor of the book *Gold Plating Technology,* who helped edit certain chapters and who prepared the index for this book; D.S. Lashmore, manager of the electrodeposition group at the US National Bureau of Standards, who also provided editorial support; and the Werner Flühmann Company, which provided encouragement and financial support throughout the preparation of the present book.

Chapter 1

Qualitative Approach To Pulse Plating

J.Cl. Puippe

1. Introduction

In electrolysis, in contrast to chemical synthesis, one can easily control the reaction rate of a system by working at a given current density, or easily select the magnitude of the driving force for the reaction by adjustment of the electrode potential. Modern electronics has greatly enhanced this inherent advantage of electrolysis by allowing current or voltage to be applied as almost any function of time. Pulse plating takes full advantage of this possibility. There are quite a number of variations, some of which are schematically represented in Fig. 1.1 and in Fig. 13.1. Typical waveforms include cathodic pulse followed by a period without current and/or by an anodic pulse; DC with superimposed modulations; a train of cathodic pulses followed by a train of anodic pulses; galvanostatic or potentiostatic pulses; square-wave or modified sine-wave pulses. It should be noted, however, that owing to the fundamental nature of the phenomena involved in the deposition process one cannot make full use of the range of conditions that would be allowed by modern electronics. There are two main limiting factors: the charging of the electrical double layer at the metal-electrolyte interface and the mass transport considerations.

2. Capacitance effects

The electrical double layer at the electrode-solution interface can be approximated to a plate capacitor with an interplate distance of a few angstroms and therefore with a high capacitance. Charge must be provided to this double layer in order to raise its potential to the value required for metal deposition at the rate corresponding to the applied pulse current which is supplied by the generator. The electrode behaves like a capacitor with a resistance in parallel as schematically represented in Fig. 1.2, the resistance being a function of the current density. The charging of the double layer requires a certain time which depends on the current density and on other physico-chemical parameters of the system. For practical purposes the charging time should be much shorter than the pulse duration, otherwise the current pulse is strongly distorted. Moreover, the time required for discharge of the double layer should be much shorter than the off-time between two pulses. In an extreme case, where the charging and discharging times of the double layer are much longer than the on-time and off-time of the pulse respectively, the pulsed current is virtually a direct current and the term "pulse plating" is hardly applicable. The current required at the beginning of the pulse for charging the electrical double layer is not lost for the metal deposition, since the charge is recovered at the end of the pulse while the capacitor is discharging.

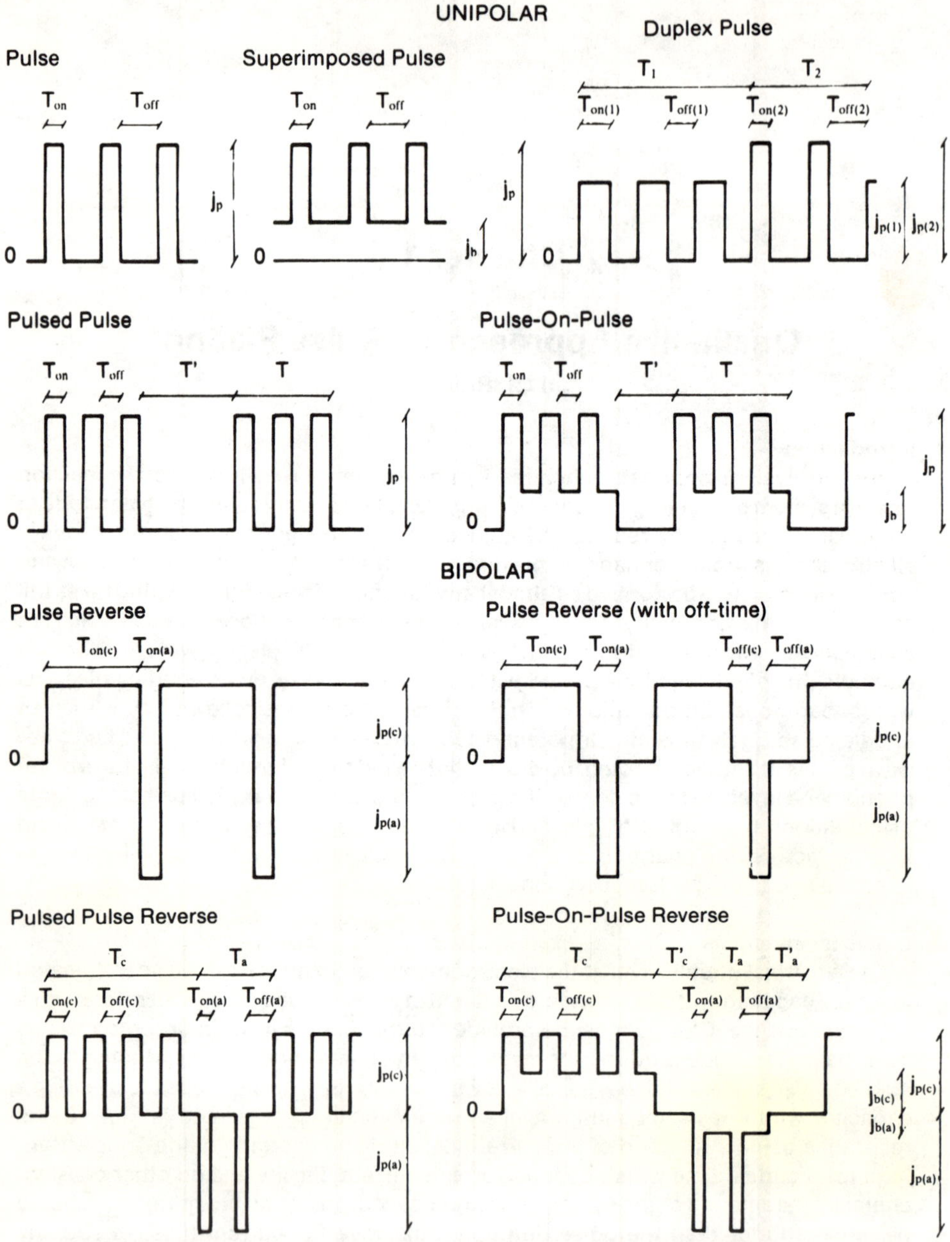

The parameters shown in this figure are independent. The average current density, j_m, dependent parameter, can be calculated from the above mentioned parameters. Pulse reverse is characterized by three different average current densities:

j_m, $j_{m(c)}$, $j_{m(a)}$, with $j_m = j_{m(c)} + j_{m(a)}$

Fig. 1.1 Schematic representation and suggested nomenclature for some square-wave modulated current systems, and definition of related parameters.

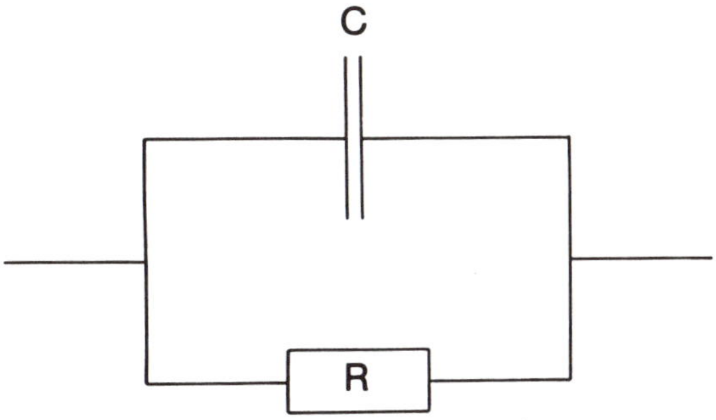

C capacitor
R resistance

Fig. 1.2 Equivalent circuit of an electrode

Pulses in a frequency range where capacitive effects are relevant do not significantly influence the current efficiency, but affect the amplitude of the pulse and hence the energy of the electrodeposition reaction. In fact, the potential benefit of pulse plating, for example, in influencing the structure and therefore the properties of deposits, is lost. In most cases the user of pulse plating should therefore avoid working with on- and off-times shorter or of an order of magnitude comparable with the charging or discharging time of the double layer. Quantitative data on the charging and discharging time of the electrical double layer will be given in Chapter 4.

3. Mass transport effects

The limitations of the useful range of pulse conditions due to mass transport effects arise from the depletion of cations in the diffusion layer (see Fig. 1.3). In pulse plating with short pulse durations, two distinct cathodic diffusion layers can be defined instead of one as in DC. Let us first discuss this concept briefly:

In the immediate vicinity of the cathode the concentration pulsates with the frequency of the pulsating current, decreasing during the pulses and relaxing in the interval between them. Thus a pulsating diffusion layer exists close to the cathode. If the duration of the pulse is short, the diffusion layer does not have time to extend very far into the solution and in particular does not extend to the region where convection takes over the mass transport. Therefore the metal deposited during the pulse must be transported from the bulk of the solution towards the pulsating diffusion layer by diffusion, which means that a concentration gradient also builds up into the bulk of the electrolyte. The thickness of this diffusion layer corresponds essentially to that which would be established under the same hydrodynamic conditions in DC electrolysis. Through this outer diffusion layer cations are also supplied towards the cathode

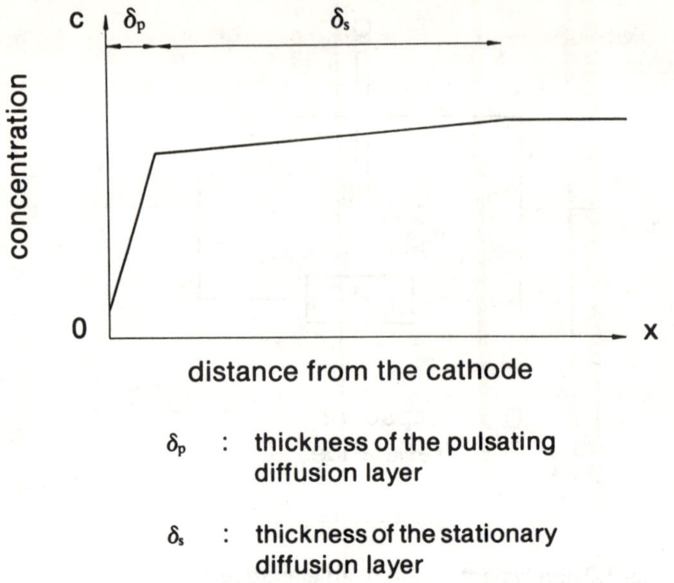

Fig. 1.3 Concentration profiles of the two diffusion layers in pulse electrolysis at the end of a pulse.

during the off-time, and it is this supply that allows the relaxation of the pulsating diffusion layer during the off-time. The outer diffusion layer is essentially stationary.

The two distinct diffusion layers are related to two kinds of limitations. The depletion of the cationic concentration in the pulsating diffusion layer limits the pulse current density, and the depletion of the cationic concentration in the outer diffusion layer limits the average current density. Since the concentration gradient in the pulsating diffusion layer can be very high, increasing with shorter pulse length, the pulse current density can reach extremely high values, e.g., up to 10,000 times the usual DC values, without decrease of current efficiency because of hydrogen evolution. There is a quantitative relation between the maximum pulse duration as far as the current efficiency is concerned and the pulse current density. This maximum pulse duration is called the transition time, and a method for its calculation will be given in Chapter 5. To sum up, the first limitation due to the mass transport effect in pulse plating is that the pulse duration should not exceed the transition time, while the second is that the maximum average current density cannot exceed the DC limiting current density. The latter is directly related to the establishment of the stationary diffusion layer. It is important to emphasize these deposition rate aspects in pulse plating because mistakes can be found in the literature, claiming that the average deposition rate can be very much increased as compared to DC. Cheh (1) and Ibl (2) have, in fact, demonstrated that the average current density in pulse plating cannot exceed the limiting current density in DC plating. However, Cheh (1) pointed out that the practical current density, which is below the diffusion limiting current density, can sometimes be increased in pulse plating as compared to DC while keeping a constant deposit quality. The increase, however, is only modest and does not represent the main advantage of the pulse plating technique, which rather consists in the improvement of deposit properties, which are related to the deposit structure.

4. Crystallization

The crystallization of the electrodeposited metal is a very important step of the electrogrowth since it influences directly the structure of the deposit and therefore its properties which represent the main interest for the user. The crystallization is the process by which the adatoms or adions incorporate in the crystal lattice. Crystallization (as schematically represented in Fig. 1.4) occurs either by the build-up of old crystals or the formation and growth of new ones. These two processes are in competition and can be influenced by different factors. High surface diffusion rates, low population of adatoms, and low overpotentials are factors enhancing the build-up of old crystals, while conversely low surface diffusion rates, high population of adatoms, and high overpotentials on the surface enhance the creation of new nuclei. In pulse plating, since the pulsed current density is usually considerably higher than the corresponding DC density, the population of adatoms on the surface during pulse deposition is higher than during DC deposition, resulting in an increased nucleation rate and therefore in a finer grained structure. As will appear in the chapters describing practical applications, grain refinement has quite often been observed by experimenters. Grain refinement in pulse plating is also favored by the enhancement of nucleation rates due to high overpotentials which result from high pulse current densities.

While adatom population and overpotential can be directly influenced in a predictable direction by external factors, e.g., by adjusting the electrical parameters of the system, this is not the case for surface diffusion, which in plating baths is greatly influenced by adsorbed species at the electrode, such as organic brighteners, stress reducers and many other possible molecules present in the bath. The adsorption isotherm of these molecules is a function of the potential, and therefore by modulating the current, the nature of adsorbed species can continuously fluctuate. Depending on the adsorption and desorption rate constants, for given pulse parameters, adsorption or desorption of a species present in the electrolyte will influence surface diffusion in a different way than in DC, thus leading to different crystallization mechanisms and therefore to different physical properties of the deposit.

Another phenomenon that might occur in pulse plating during the off-period is recrystallization. Small grains are thermodynamically less stable than large ones because of high surface energy, and as in bubble coalescence, small grains tend to recrystallize. For this to occur, the surface should remain active during the off-time. Again depending on different adsorbed species, the surface may be inhibited, in which case no recrystallization will occur, and the fine grains obtained during the on-time of electrolysis are stabilized.

One can think of using pulse plating to produce amorphous metals or alloys, for instance, from transition metals alloyed to metalloids. In pulse plating with very high pulse current densities, the concentration of adatoms at the interface is so high during the pulse that there is no time for ordering them into a crystal lattice, resulting in similar effects, as in rapid quenching.

5. Kinetics

In pulse plating, the electrodeposition rate can be considerably increased during the on-time as compared to DC. The kinetics of competitive reactions will therefore play a relevant role on the relative rate of possible competitive reactions, as in alloy plating or in plating with hydrogen codeposition.

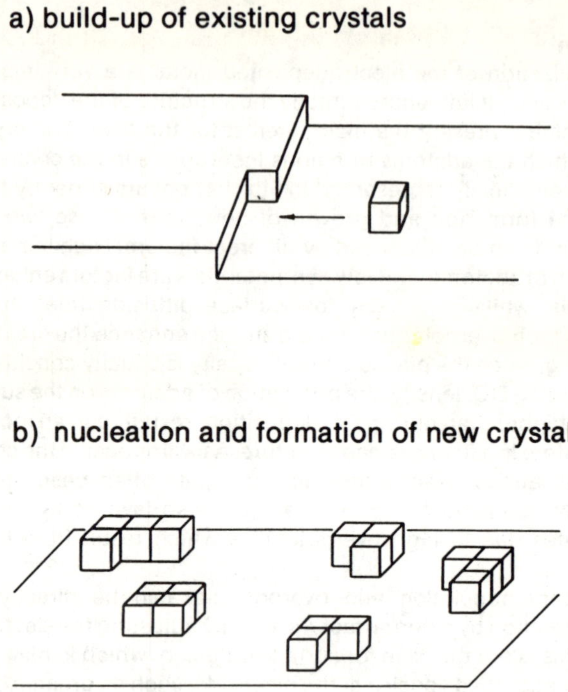

Fig. 1.4 Crystallization mechanisms.

Referring to alloy deposition (see Fig. 1.5), if the slope of the polarization curve (I = f(E)) increases faster for the deposition of metal I than for metal II, the current efficiency for metal I will usually increase with increasing pulse current density. Similarly, in metal deposition accompanied by hydrogen evolution, if the slope of the polarization curve for metal deposition increases faster than for hydrogen evolution, the current efficiency for metal deposition will usually increase with increasing pulse current density.

However, particularly in pulse electrolysis with short pulses, adsorption and desorption effects can lead to unexpected trends in current efficiencies. For example, with hydrogen, which tends to desorb during the off-time, a given coulombic charge is consumed at the beginning of a new pulse in building up a new adsorbed layer of hydrogen. If the charge needed for this purpose represents a substantial part of the total charge of the pulse, the current efficiency for metal deposition can show a significant fall, even if the slope of the polarization curves would suggest an increase in current efficiency. This effect is illustrated in the case of gold deposition from a typical acid bath containing cobalt as a hardening additive (3,4).

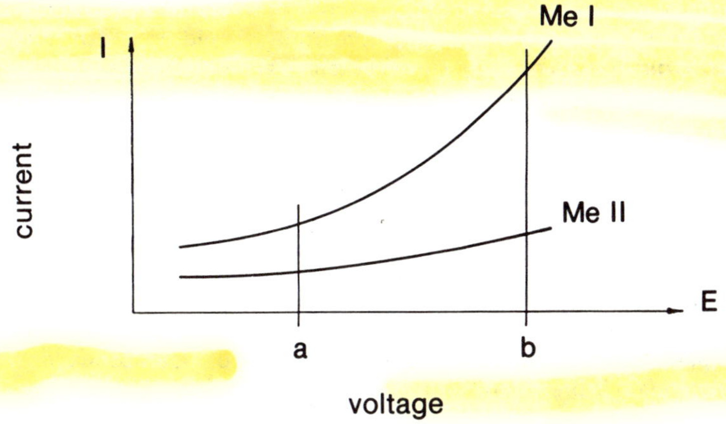

Fig. 1.5 Current-voltage curves for two competing reactions with different kinetics. An increase in voltage from a to b produces a change in current efficiency for reaction I (Me I) and reaction II (Me II).

Expected trends in current efficiency can also be vitiated by adsorption of metal species during the off-time. As an example, the deposition of rhenium can be mentioned (see Chapter 10). In rhenium deposition, rhenium is reduced from a heptavalent ion. Its reduction potential is much more negative than that of hydrogen ions, and thus the current efficiency for rhenium is usually very poor (less than 10 percent). Nevertheless, in pulse plating with very short pulses, the current efficiency increases up to 60 percent. The reason is that monovalent rhenium, the reduction potential of which is less negative than that of hydrogen, is adsorbed at the electrode during the off-time. If the on-time and pulse current density correspond approximatively to the coulombic charge needed for the reduction of a Re-I monolayer, the current efficiency for rhenium is substantially increased. Adsorption during the off-time in this case thus results in a trend in current efficiency that would be quite unexpected according to considerations of steady state kinetics.

6. Pulsed current vs pulsed voltage

Electrolysis can be controlled by regulation of either current or voltage. In current regulation mode, the reaction rate is kept constant and the potential varies as a function of time. In voltage regulation, the driving force for the reaction is kept constant and the reaction rate varies as a function of time. The advantages and disadvantages of these two modes of electrolysis are briefly discussed here for pulse plating. In current regulation, the average deposition rate can be very simply derived from the following equation:

$$j_m = j_p \frac{T_{on}}{(T_{on} + T_{off})}$$

while in voltage regulation the average deposition rate can be predicted only from computations of a speculative nature. The main advantage of voltage regulation is a better control of the current efficiency and of alloy composition. High overpotentials resulting from excessive concentration depletions are avoided. However, from a practical point of view, regulation of pulsed voltage is very difficult to achieve. To do it properly, a third electrode, a reference electrode, should be added to the system. Moreover, to get instantaneously a given potential at the electrode, the current should start theoretically from an infinite value, which is obviously not feasible because of apparatus limitations. On the other hand, at the end of a pulse, for instantaneous reestablishment of the starting potential (which might be the rest potential of the system), some metal should be redissolved; hence a constant voltage pulse requires a short inversion of current at the end of the pulse, which is not generally desirable. Furthermore, passivation may occur during the inversion. It should be pointed out that the above remarks apply to a regulated cathodic potential and not to the voltage between the negative and positive poles of the power supply. The latter is much easier to regulate, and sometimes, in particular for very high currents, the power supply engineer prefers it to the constant current regulation. For most applications, however, current regulation is preferred.

7. Pulse reverse

Among the different modes of current modulation, pulse reverse is a very early one and is very common. In early works, relatively long periods of forward and reverse cycles, longer than the usual on-time in pulse plating, were used. This kind of current is often called "periodic reverse." The term "pulse reverse" is not restricted to shorter cycles and applies to what is commonly understood under "periodic reverse" as well. The main purpose of inverting the current from cathodic to anodic during a short fraction of the total period is to remove metal preferentially from areas that tend to overplate during the cathodic part of the cycle. It is thus possible to considerably retard the development of dendrite formation or to improve the plating thickness distribution over complicated shapes. Changes in deposit structure, mainly grain size, can also be achieved because of forced nucleation at each new cathodic pulse. It is clear that adsorption-desorption as well as recrystallization phenomena will be quite different from those in pulse plating. It should be pointed out that pulse reverse is usually applied when the deposit is easily soluble in the electrolyte, otherwise passivation can occur. Sometimes, as in palladium deposition, the purpose of inverting the current is to remove codeposited hydrogen. In most cases the objective of the application of pulse reverse is to improve the plating thickness distribution.

As shown in Chapter 6 of this book, from theoretical considerations the current distribution in pulse plating is often poorer than in DC. At least, an improvement as compared to DC normally cannot be expected. The deterioration of current distribution in pulse plating is due to higher ohmic drops in the solution, leading to higher current density differences between peaks and recesses of the substrate profile. Another reason for deterioration of current distribution in pulse plating is the decrease in activation overpotential with increasing current density. Activation overpotential is known for equalizing the current distribution (5). Deterioration of current distribution with pulse plating occurs if the pulse current density is substantially increased as compared to the normal DC density. If the pulse current density is the same as the usual DC density, pulse plating could even improve the current distribution. Because of divergent opinions reported in the literature regarding this matter, it seems important to underline the importance of the pulse current density.

Pulse reverse can be of great help in equalizing the plating thickness distribution. Areas exposed to concentrations of current density are preferentially plated in the cathodic cycle, but for the same reason, metal is preferentially removed in the anodic cycle.

To take full advantage of pulse reverse, the anodic current density should be adjustable independently of the cathodic current density. The reason for this is to increase the dissolution rate of peaks by applying very high anodic current densities, but for short time durations. It is obvious that the balance of electrical charges over a total period should remain cathodic. The manufacture of power supplies offering this capability involves complications, but this remains, nevertheless, an important requirement.

8. Other current modulations

Square-wave pulse plating and pulse reverse plating represent most of the applications today, while in the early days of pulse plating, because of less developed electronics, modified sine-wave pulses were used. Modern trends show increasing use of more sophisticated square-wave current modulations such as the ones represented in Fig. 1.1.

In recent developments, improvements in physical properties of deposits have been observed by the use of a periodic reverse current mode with cycle times in the order of seconds, but with pulsing of current during both cathodic and anodic periods (pulsed pulse reverse). With optimization of electrical parameters, this approach offers a combination of the advantages of pulse plating with respect to modification of deposit structure and properties, with those of anodic deplating in improving the thickness distribution. The use of pulsed current allows a substantial increase in current density in both anodic and cathodic cycles. During the cathodic period this leads to grain refinement normally associated with pulse plating, which is further favored by the disturbance of growth steps by periodic inversion of the current. Thickness distribution is improved, not only by the normal anodic dissolution of metal from high spots or other exposed areas, but also because the preferential nature of this effect can be significantly enhanced by application of relatively short anodic pulses of high amplitude. Full advantage can be taken of this method only if all electrical parameters can be set independently, and in particular, if cathodic and anodic pulses can be independently regulated, e.g., with respect to pulse length and amplitude.

However, it should be kept clearly in mind that while in conventional pulse plating understanding has been gained based on theoretical models on the specific influence of the pulse parameters on-time, off-time and pulse current density, which enables pulse parameters to be set with due regard to the different physicochemical phenomena involved, the introduction of more independent electrical parameters (at least ten as compared to three in conventional pulse plating) renders the optimal choice of parameters for expected trends in deposit properties much more difficult.

9. Early results in pulse plating

The publications on pulse plating issued before 1979 are summarized here. The works published after this date are more specifically treated in the following chapters of this book as well as in the two Proceedings of AES Pulse Plating Symposia (6,7).

The use of modulated current in electrolysis is by no means new. The first application of periodically interrupted current dates back to the end of the last century, when it was used by Coehn (8) to produce coherent deposits of zinc. Baeyens (9) reviewed applications of various types of modulated current up to 1952, and Dini (10) has reviewed the main work carried out up to 1963 with reverse current, i.e., current comprising an anodic component in the interval between cathodic pulses. Studies on this subject have been covered by Cheh (11) and a review has also been given by Puippe et al. (12).

The principal advantage of pulsed current is that it can be used to improve deposit properties. Advantages have also been claimed in respect to saving of time and reduction in energy requirements, but the former cannot be relied upon, and the latter is certainly not achievable.

Half-rectified alternating current somewhat resembles pulsed current, with the difference that the descending portion of the current curve is rounded instead of vertical. By using this type of current, Romanov (13) obtained compact deposits of zinc, free from dendrites, although with continuous DC dendrite formation is inevitable. Gurovich and Krivtsov (14) also used this type of current in deposition of copper and zinc and found an increase in the number of grains, which they attributed to passivation of growth sites by adsorption of an inhibiting species during the off-time, thus forcing the formation of new nucleation centers for continued growth.

Decrease in crystallite size has often been observed (11,14-20), leading to a fine and more compact structure (21), and hence reducing the porosity of the deposit (16,22). Variations in pulsed current parameters also affect the dimensions of particles in the production of powder by electrolysis (23). Reduction in the internal stress of nickel deposits has also been recorded (24).

Not unnaturally, considerable attention has been directed to the effects of the pulsed current technique in gold plating, in view of the potential economic benefits that would accrue from improvements in properties. This aspect is of particular interest with respect to applications in the electronics industry, where annual global consumption of gold exceeds 100 tons (25). In this context Avila and Brown (17) have shown that it is possible by the use of pulsed current plating to increase the density and quality of gold coatings to an extent which permits a 50 percent reduction in thickness while maintaining the requisite conductivity of conductor patterns. Increase in ductility has been reported by Lendvay and Raub (26), while Reid (27) has shown that this method of electrolysis can eliminate nonmetallic inclusions in deposits from cobalt-hardened acid gold electrolytes (the so-called "polymer"), with substantial increases in density, ductility and conductivity of coatings. Reduced porosity has been observed by Mentone (28) and Rehrig (29), and other investigations concerning the effect of pulsed current parameters on the properties of gold coatings have been carried out by Angerer (3,30,31) and Bielinski (31).

Metals deposited by pulsed current exhibit smaller contents of adsorbed hydrogen than those produced with continuous current, as a result of desorption during the off-time (24,28,32). Since hardness is related to hydrogen content, the former is also affected. Leland (33), for example, deposited chromium with pulsed current, but with gradual reduction of the off-period during the course of electrolysis to finally approach continuous current conditions, and in this way obtained a coating in which both hydrogen content and hardness increased towards the outer layer.

Pulsed current has also been applied to alloy deposition (34,35,36), where it has been shown that the alloy composition approaches that of the electrolyte, providing a means of controlling the former, which is of particular importance in the deposition of colored alloys for decorative purposes.

Many workers have observed an increase in smoothness of deposits using pulsed current (13,14,37,38,39). Despic and Popov (40) found that surface roughness decreased with decreasing pulse length, which is attributable to mass transport phenomena. This effect has been investigated by other workers for copper (41-44), silver (41-42), and cadmium (45), the results obtained being explained in terms of a diffusional model developed by Ibl (46).

10. Concluding remarks

- It should be kept in mind that current "shaping" represents only one way of influencing a complicated system which contains many other dependent process parameters, such as electrolyte composition (concentration of electroactive ions, supporting electrolyte and additives), temperature, hydrodynamics and cell geometry. The interdependence of these process parameters with the pulse parameters should be at least qualitatively understood. One cannot expect that a given pulse shape will lead to the same deposit properties regardless of other process parameters. This fact could explain why sometimes contradictory observations are reported in the literature.
- Pulse plating, in allowing the use of simpler solutions and fewer additives, makes a useful contribution to the scientific basis of electroplating, and in this sense counters the "black art" reputation from which this general field still suffers to some extent.
- In order to set the electrical parameters, one has to remember the main limitations resulting from the charge and the discharge of the electrical double layer at the cathode interface and from mass transport effects. The on-time should be longer than the charging time of the electrical double layer and shorter than the transition time, the time at which the system starts working under limiting current conditions.
- The off-time between two pulses is not a "dead" time. Adsorption and desorption phenomena as well as recrystallization can occur.
- The main purposes of pulse plating are:
  - the improvement of deposit properties, namely, porosity, ductility, hardness, electrical conductivity, wear resistance and roughness;
  - deposition of alloys, the composition and structure of which are not obtainable with DC plating;
  - improvement of plating thickness distribution by periodic inversion of polarity;
  - increase in average deposition rate, although this can be achieved only to a rather limited extent.

References

1. H.Y. Cheh; *Journal Electrochemical Society,* **118**, (4) : 551 (1971)
2. N. Ibl; *Metalloberfläche,* **33** : 51 (1979) and *Surface Technology,* **10** : 81-104 (1980)
3. N. Ibl, J. Cl. Puippe and H. Angerer; *Surface Technology,* **6** : 287-300 (1978)
4. K. Hosokawa, H. Angerer, J. Cl. Puippe and N. Ibl; *Plating and Surface Finishing,* **67,** (10) : 52-55 (1980)
5. N. Ibl; *Comprehensive Treatise of Electrochemistry,* Vol. VI, Chap. 4; ed. by E. Yeager, J. O'M. Bockris, B.E. Conway, S. Sarangapani
6. *Proceedings of the AES First International Pulse Plating Symposium,* Boston, MA (April 19-20, 1979)
7. *Proceedings of the AES Second International Pulse Plating Symposium,* Rosemont, IL (Oct. 6-7, 1981)
8. A. Coehn; German Patent 75482 (1893)
9. P. Baeyens; *Transactions Institute of Metal Finishing,* **31** : 429 (1954)
10. J.W. Dini; *Metal Finishing,* **61**, (7) : 52 (1963)
11. C.C. Wan, H.Y, Cheh and H.B. Linford; *Plating,* **61**, (6) : 559 (1974)
12. J. Cl. Puippe, N. Ibl, H. Angerer and H.J. Schenk; *Oberfläche-Surface,* **20**, (4) : 77-85 (1979).
13. V.V. Romanov; *Zhurnal Prikladnoi Khimii,* **36**, (5) : 1050-1057 (1963)
14. R.I. Gurovich and A.K. Krivtsov; *Zhurnal Prikladnoi Khimii,* **41**, (6) : 1227 (1968)

15. A.P. Popkov; *Journal Applied Chemistry USSR,* **39** : 1632 (1966)
16. M. Viswanathan and Ch. J. Raub; *Galvanotechnik,* **66,** (4) : 277 (1975)
17. A.J. Avila and M.J. Brown; *Plating,* **57**, (11) : 1105 (1970)
18. A. Hickling and H.P. Rothbaum; *Transactions Institute of Metal Finishing,* **34** : 199 (1957)
19. A.M. Ozerov, N.P. Litvishko, I.N. Vavilina, P.M. Chetvertov, Y.E. Zhak; *Zhurnal Prikladnoi Khimii,* **40** : 1144 (1967)
20. M. Viswanathan and Ch. J. Raub; *Surface Technology,* **4** : 339 (1976)
21. W. Sullivan; *Plating,* **62,** (2) : 139 (1975)
22. K.I. Popov, D.N. Keca and B.I. Vuksanovic; *Journal Applied Electrochemistry,* **7** : 185 (1977)
23. K.I. Popov, M.D. Maksimovic, M.G. Pavlovic and G.R. Ostojic; *Journal Applied Electrochemistry,* **7** : 331 (1977)
24. W. Kleinekathofer and Ch. J. Raub; *Surface Technology,* **7** : 23 (1978)
25. D. Lloyd-Jacob; *Gold Bulletin,* **4**, (2) : 25 (1971)
26. J. Lendvay and Ch. J. Raub; *Metalloberfläche,* **29**, (4) : 165 (1975)
27. F.H. Reid; *Metalloberfläche,* **30**, (10) : 453 (1976)
28. F. Mentone; *Product Finishing,* (Cincinnatti), 78 (April, 1973)
29. D.L. Rehrig; *Plating,* **61**, (1) : 43 (1974)
30. H. Angerer; *Dissertation 6213,* ETH, Zurich (1978)
31. J. Bielinski, N. Ibl and H. Angerer; *Oberfläche-Surface,* **20**, (1) (1979)
32. C.L. Faust, G.R. Schaer and D.E. Sermones; *Plating,* **48** : 605 (1961)
33. M. James Leland; "Programmable Pulse Electroplating Process," U.S. Patent 3,886,053 (cl. 204-36; C 23b), (May 27, 1975), Appl. 411.962.01 (Nov., 1973)
34. H. Leidheiser Jr and A.R.P. Ghuman; *Journal Electrochemical Society,* **120** : 484 (1973)
35. J. Winkler; German Patent 576585, cl. 48a (May 13, 1933)
36. R. Girard; *Electrodeposition of Thin Magnetic Films,* French Atomic Energy Commission, Grenoble Center of Nuclear Studies, (June 10, 1965)
37. V.U. Polygraficky; British Patent 932,822 (1963)
38. N.P. Gnusin and Ya. Kovarski; *Elektrokhim,* **1** : 46 (1965)
39. G. Savic-Maglic and A.R. Despic; *Electrodeposition and Surface Treatment,* **1** : 429 (1972/73)
40. A.R. Despic and K.I. Popov; *Journal Applied Electrochemistry,* **1** : 275 (1971)
41. M. Braun; *Dissertation ETH 5015,* Zurich (1973)
42. N. Ibl and M. Braun; *Chemie Ingenieur Technik,* **45**, (4) : 182 (1973)
43. M. Braun and N. Ibl; *Proceedings of the 8th Congress of the International Union for Electrodeposition and Surface Finishing,* Basel (1972)
44. H.J. Schenk, *Dissertation 5566,* ETH, Zurich (1975)
45. J. Cl. Puippe; *Dissertation 6225,* ETH, Zurich (1978)
46. N. Ibl; *Proceedings of "Surface 66",* Basel : 48 (1966); *Galvanotechnik + Oberflächenschutz,* **7**, (11): 256 (1966)

Chapter 2

Electrochemical Kinetics in Pulse Plating

H.Y. Cheh

1. Introduction

The application of pulsed electrolysis to the electroplating of pure metals and alloys has attracted considerable attention since the early 1970s. There is a wealth of information on the mass transfer of pulsed electrolysis. Cheh (1,2) defined the concept of limiting current density and adopted a Nernst diffusion model to estimate the rate of mass transfer in pulsed and periodic reverse plating. Bedrossian and Cheh (3) conducted theoretical and experimental investigation on the reduction of concentration overpotential in pulsed electrolysis. Ibl (4) developed approximate models to describe the pulsating diffusion layer. Viswanathan et al. (5) rigorously analyzed mass transfer of pulsed electrolysis on a rotating disk electrode. More recently, Cheng and Cheh (6,7) reported their investigation on both theoretical modeling and experimental studies in complex plating systems which include processes with gas evolution and alloy deposition.

In contrast to the extensive literature for mass transfer studies, little is available on the kinetic aspects of pulse plating. Pulsed electrolysis affects the reaction kinetics in two ways. First, the abrupt changes in the applied current density at both the beginning and the end of a pulse cause a change in the state of the electrical double layer. The charge and discharge of the electrical double layer affect the faradic current which governs the rate of plating. Second, the change in the surface concentration of the metal ion influences both the kinetics and the structural aspects of the plating system.

2. Electrical double layer effect†

When a current is changed abruptly from one value to another, the electrode potential and the state of the double layer also change. The total applied current in a transient process is a sum of the nonfaradic and the faradic current,

$$j = j_C + j_F \qquad [1]$$

where j is the total current density, j_C is the nonfaradic current density, which is responsible for charging and discharging the electrical double layer, and j_F is a faradic current density, which represents the rate of the electrochemical reaction. For a transient process,

$$j_C = C \frac{d\eta_s}{dt} \qquad [2]$$

† See also Chapter 4.

where C is the electrical double layer capacity, η_s is the electrode surface potential, and t is time. In contrast to DC plating where the nonfaradic current is zero, this double layer current might be significant in pulsing. If the plating system does not involve side reactions, the portion of the faradic current that is lost at the onset of a pulse is recovered when the pulse is switched off, maintaining a 100 percent current efficiency for plating. Nevertheless, the kinetics of the process which depend on the magnitude of the faradic current can be altered. If the plating process involves more than one reaction, the faradic current is now a sum of the partial currents for the individual reactions. A time-varying total faradic current results in a transient electrode potential, thereby leading to a readjustment of the partial currents and affecting the current efficiency of the plating process. This point has been discussed qualitatively by Wan et al. (8). Assume that copper deposition proceeds with a two step reduction coupled to a disproportionation reaction,

$$Cu^{++} + e^- \rightarrow Cu^+ \quad [3]$$

$$Cu^+ + e^- \rightarrow Cu \quad [4]$$

and

$$2\,Cu^+ \rightleftharpoons Cu + Cu^{++} \quad [5]$$

Wan et al. (8) state,

"Let j_1 be the current density carried by reaction [3] and j_2 be the current density carried out by reaction [4]. The total applied current density j is then the sum of j_1 and j_2. The magnitude of j_1 and j_2 are in turn controlled by the concentration of the reacting species and the electrode kinetics.

"Consider a pulse with the following hypothetical characteristics. The pulse time is 1 sec and the applied current density is 2 mA/cm^2. During the first 0.1 sec, the rate of reaction [3] is faster than reaction [4] and we assume that the average value of j_1 is 1.5 mA/cm^2 and that of j_2 is 0.5 mA/cm^2. The difference between j_1 and j_2 diminishes as the pulse progresses. This is because of the decrease of Cu^{++} concentration and the increase of Cu^+ concentration at the interface, which result in a change of the rates of the two reactions. Consequently, let us assume the average j_1 is 1.2 mA/cm^2 and j_2 is 0.8 mA/cm^2 during the second 0.1 sec and both j_1 and j_2 are 1.0 mA/cm^2 for the remaining 0.8 sec. For a 1 cm^2 electrode area, the total amount of charge for copper deposition assuming a 100 percent current efficiency during the one second is (2 mA/cm^2 × 1 cm^2 × 1 sec)/2 which is 1×10^{-3} coulombs, whereas according to the present model, the total amount of Cu deposited is equivalent to [(0.5×0.1) + (0.8×0.1) + (1.0×0.8)] mA-sec which is 0.93×10^{-3} coulombs. Consequently, the current efficiency for copper deposition is 93 percent.

"This model is capable of explaining the experimental observations on the dependence of current efficiency on pulsed condition (9,10). For instance, if the pulse time is reduced to 0.5 sec, the current efficiency is then [(0.5×0.1) + (0.8 × 0.1) + (1.0 × 0.3)] [2 × (0.5/2)] × 100% or 86%. Also, if the relaxation time is lengthened, there is a general increase of Cu^{++} concentration and a decrease of Cu^+ concentration at the interface at the onset of each cycle. This results in a further increase of the difference between the rates of the two reactions, causing a further reduction of the current efficiency."

Experimental data on the current efficiency of copper deposition from a copper sulfate system has been presented by Cheh et al. (11). A theoretical analysis on this phenomenon has been presented by Wan et al. (12). The results show that the current efficiency drops continuously as the difference between the rate of the two reactions increases.

3. Surface concentration effect

Reaction rate and deposit structure depend on the surface concentration of the reacting species. For single reaction systems, a variation of surface concentration in pulse plating leads to a time-dependent overpotential of the deposition process. For complex reaction systems involving two consecutive reaction steps, Cheh et al. (11,12) have shown that pulsed electrolysis favors the first step in the reaction sequence, resulting in a loss of current efficiency. A quantitative understanding of the effect of pulsing on the kinetics of plating requires a detailed knowledge of the dependence of reaction rates on surface concentration, which in turn is a function of mass transport. For single electrochemical reactions, Cheh (1) derived theoretical expressions for calculating the surface concentration of the reacting species by using a Nernst diffusion model. More recently, Cheng and Cheh (6,7) obtained results on surface concentration calculations for multiple reaction systems by using numerical techniques. In principle, the value of the surface concentration can then be used to derive kinetic information for plating systems under a variety of conditions. Nevertheless, a combining of mass transfer, kinetics and pulsing characteristics leads to a formidable theoretical and experimental problem. In view of the immensely important role of kinetics in any deposition process, we must conclude that new research in this area is of the highest priority.

References

1. H.Y. Cheh; *Journal Electrochemical Society,* **118** : 551 (1971)
2. H.Y. Cheh; *Journal Electrochemical Society,* **118** : 1132 (1971)
3. A.A. Bedrossian and H.Y. Cheh; *Journal Electrochemical Society,* **128** : 2536 (1981)
4. N. Ibl, *Surface Technology,* **10** : 81 (1980)
5. K. Viswanathan, M.A. Farrell-Epstein and H.Y. Cheh; *Journal Electrochemical Society,* **125** : 1772 (1978)
6. T. Cheng and H.Y. Cheh; Paper presented at the Montreal, Canada, meeting of the Electrochemical Society, Abstract 414, (May 1982)
7. T. Cheng and H.Y. Cheh; Paper presented at the Washington, DC, meeting of the Electrochemical Society, Abstract 248, (Oct., 1983)
8. C.C. Wan, H.Y. Cheh and H.B. Linford; *Journal Chinese Institute of Chemical Engineers,* **9** : 69 (1978)
9. P.A. Danna; "Overpotential Charateristics and Kinetic Parameters for the Simultaneous Deposition of Nickel and Hydrogen on Clean and Soiled Nickel Electrodes," doctoral dissertation, Columbia University, New York, (1967)
10. H.S.-C. Chang; "Galvanostatic Studies of Current Efficiencies on Clean Copper Surfaces," MS thesis, Columbia University, New York (1972)
11. H.Y. Cheh, H.B. Linford and C.C. Wan; *Plating and Surface Finishing,* **64,** (5) : 66 (1977)
12. C.C. Wan, H.B. Linford and H.Y. Cheh; *Journal Applied Electrochemistry,* **9** : 29 (1979)

Chapter 3

Influence of Pulse Plating on Crystallization

J.Cl. Puippe

1. Introduction

In an electrodeposition process, ions are transferred towards the cathode by migration, convection and diffusion; they cross the electrified interface where the charge-transfer reaction occurs, and are finally incorporated into the crystal lattice of the cathode, the final step being crystallization. This occurs by building up existing crystals or growing new ones. These two processes are competitive, depending on chemical and electrochemical parameters, and are related to the grain size and the piling up of grains, i.e., the morphology and porosity of deposits. The influence of pulse parameters on these two factors can be qualitatively understood with the help of the simple model "building-up of existing crystals—formation of fresh nuclei;" trends in morphology and porosity are predictable, but other crystallization features such as crystal orientation, concentration of defects, incorporation of hydrogen or bath additives, phase formation (especially in alloy plating) and deposition of amorphous metals are less clearly correlated with the pulse plating parameters.

2. Crystal growth vs. nucleation in pulse plating

In the electrodeposition process after the charge-transfer reaction, adatoms are formed. There is very little probability that they meet exactly at a growing step and therefore that they are immediately incorporated in the crystal lattice. Adatoms most probably land far from growing steps and have to move on the surface to reach growing steps. The adatom movement occurs by surface diffusion. Let us assume (1) a constant distance $2x_0$ between two growing steps and a constant overpotential applied to the cathode. At the growing steps there is an adatom concentration of c°_{ad} and between the growing steps there is a higher adatom concentration c_{ad} reaching a maximum in the middle of the two growing steps at x_0. Adatom concentration profiles are schematically represented in Fig. 3.1. For high surface diffusion rates the adatom concentration profile between two growing steps is relatively flat, but for low surface diffusion rates it is much more curved, showing c_{ad} values much higher than the c°_{ad} one. The adatom concentration ratio c°_{ad}/c_{ad} is related to the crystallization overpotential η_k by the following relationship:

$$\eta_k = \frac{RT}{ZF} \ln \frac{c^{\circ}_{ad}}{c_{ad}} \qquad [1]$$

From eqn. [1], the higher c_{ad}, the more negative η_k will be. From these considerations it is clear that high surface diffusion rates are accompanied by low crystallization

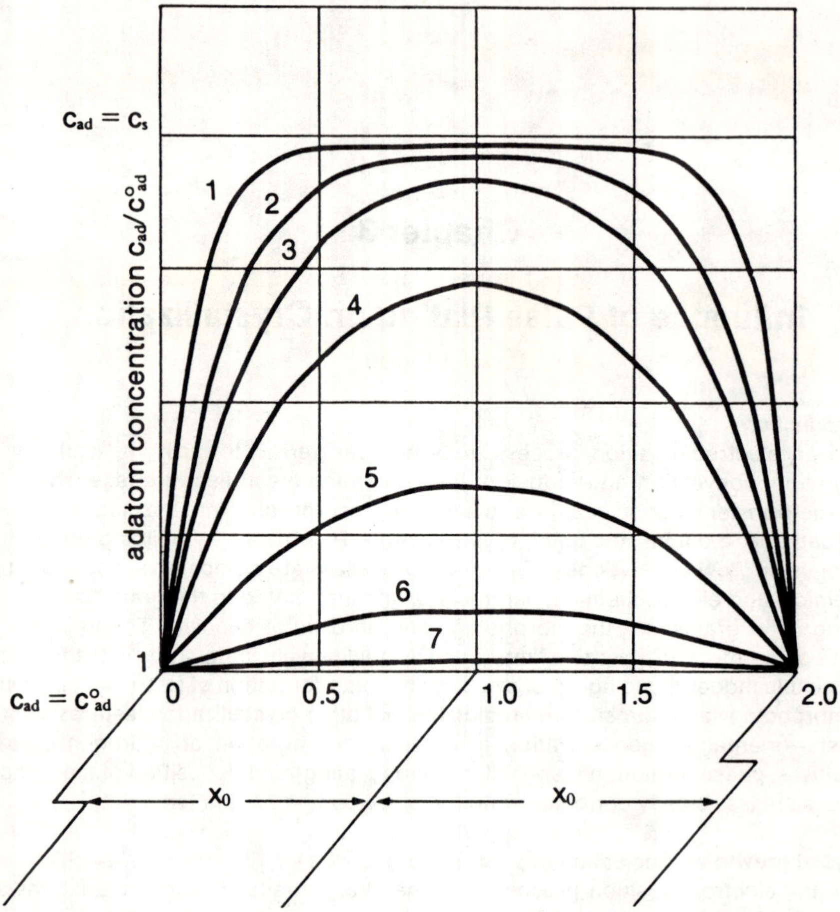

Fig. 3.1—Adatom concentration profile.
From 1 to 7: increase in surface diffusion rates
$2x_0$ distance between two growth steps;
C^{o}_{ad} adatom concentration at a growing step;
C_{ad} adatom concentration between growing steps;
C_s saturation concentration of adatoms between growing steps.

overpotentials, and vice versa. For a given surface diffusion rate and from eqn. [1], very high negative values of η_k can be expected. In practice, however, this will not occur because a competitive crystallization process, nucleation, is enhanced by increasing overpotentials. The nucleation rate (v) is given by the following equation from Ref. (2):

$$v = k_1 \exp(-k_2 / |\eta|) \qquad s^{-1}\ cm^{-2} \qquad [2]$$

where k_1 is a proportionality constant and k_2 is related to the amount of energy needed for the two-dimensional nucleation. According to eqn. [2], the nucleation rate

TABLE 3.1

Experimental Conditions for Metal Deposition

Bath	Composition		Temperature ° C	Hydrodynamic Conditions
A	$CdSO_4$	0.75 M	20	flow channel
	H_2SO_4	1.00 M		
B	Cd^{+2}	0.23 M		agitated with
	NaCN	170 g/l	20	stirrer
	NaOH	18 g/l		
	+ organic additives			
C	$[Pd(NH_3)_4]Cl_2$	0.1 M	20	rotating disk
	NH_4Cl	1.0 M		(15 rpm)
D	$CuSO_4$	0.75 M	20	flow channel
	H_2SO_4	1.00 M		
E	$KAu(CN)_2$	~ 15 g/l		
	citric acid	~ 80 g/l	27	flow channel
	base metal (Co)	0.1 g/l		
	pH (adjusted with KOH)	3.7		

increases exponentially with increasing overpotential. Nucleation decreases the average distance between neighboring growth steps and therefore reduces the surface diffusion distances and decreases the value of the maximum of the adatom concentration profile. Thus nucleation can counteract an excessive increase in crystallization overpotential resulting from high adatom concentrations. In fact, in any system a balance will be established between the two crystallization modes, building up of existing crystals or formation of fresh nuclei, such that the crystallization overpotential is as low as possible. The two competitive steps of crystallization, building up of existing crystals and formation of fresh nuclei, are schematically illustrated in Fig. 1.4.

In addition to surface diffusion rate and to applied overpotential, other parameters, such as the exchange current density between adatoms and cations, can also influence the adatom concentration profile. In practice, for a given system (cathode-electrolyte), the applied overpotential is the only parameter that can be easily varied externally. In pulse plating, full advantage of this possibility is taken by varying the amplitude of the pulsed voltage or pulsed current. In the simple case of a cathode-electrolyte system, where surface diffusion rate and exchange current density between adatoms and cations are constant, the influence of pulse plating on the nucleation rate can be qualitatively understood. Increased pulsed voltage or pulsed current amplitude, as compared to usual DC amplitudes, leads to increased adatom concentrations, corresponding to equation [1] to higher overpotentials and therefore, according to equation [2], to higher nucleation rates.

To sum up, pulse plating, when the pulse amplitude is increased as compared to DC amplitude, favors the formation of new crystal nuclei rather than the building up of existing crystals. In pulse plating, a decrease in the grain size of deposits can therefore be expected.

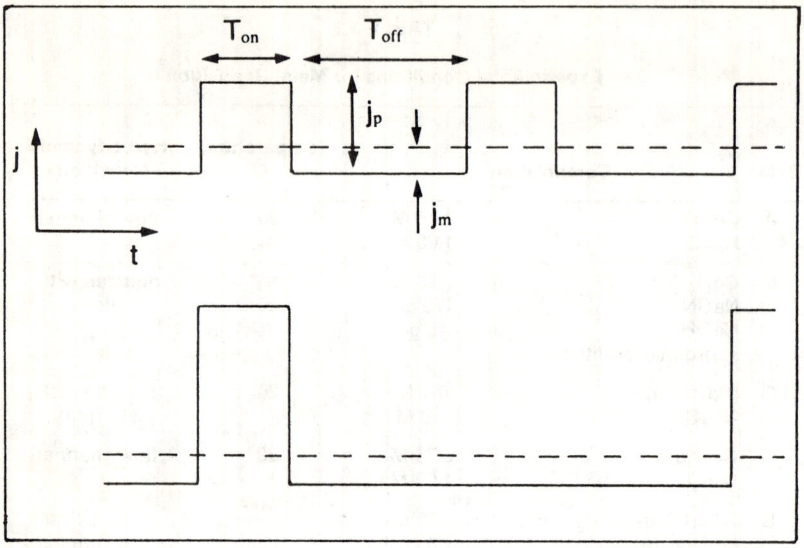

Fig. 3.2—Scheme for investigating changes in PC density with constant pulse time and constant average current density.

3. Influence of pulse parameters on the morphology of deposits

The results presented in this section refer to deposits made well below the diffusion limited current, hence the effects observed cannot be attributed to mass transport effects but to crystallization-related phenomena.

3.1 *Effect of pulsed current density*

A decrease in grain size by application of a pulsed current has often been reported in the literature (see Ref. 11, 14-20 of Chapter 1). This phenomenon is worthy of more attention. The experimental results reported in this chapter refer to the experimental conditions summarized in Table 3.1.

There are several ways of examining the effect of pulsed current density on the grain size, of which two methods are:

a) increase in pulsed current density at constant pulse length and constant average current density,
b) increase in pulsed current density at constant pulse charge (number of coulombs per pulse is kept constant) and at constant average current density.

3.1.1 An increase in pulsed current density at constant pulse length and constant average current density is illustrated in Fig. 3.2. Results are shown for copper plating (Fig. 3.3), cadmium plating (Fig. 3.4) and palladium plating (Fig. 3.5). A decrease in grain size with increasing pulsed current density is systematically observed. One disadvantage of this method of investigation is that an increase in pulsed current density implies an increase in the off-time. Since the off-time may also influence crystallization (see section 3.2 of this chapter), it is *a priori* not

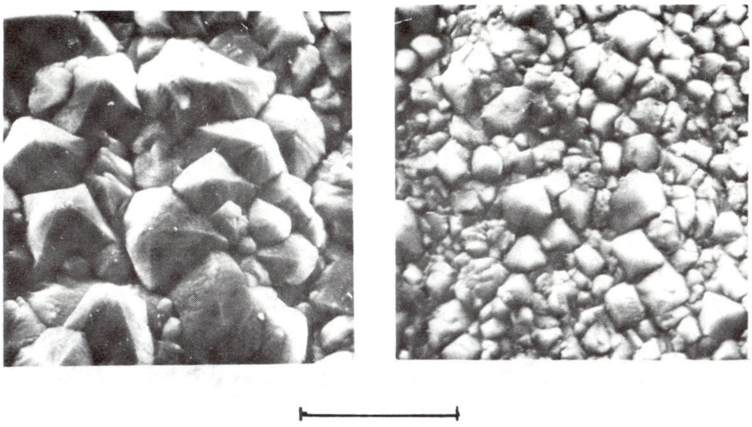

Fig. 3.3—Influence of PC density on the morphology of 10 μm copper deposits from bath D, Table 3.1, with a pulse time of 30 μs and an average current density of 37.5 A/dm^2; left, PC density: 1000 A/dm^2; right, PC density: 4000 A/dm^2.

clear whether the grain refinement is a result of changes in pulsed current density or of changes in off-times. To answer this question, another mode of investigation has been devised, as described in the following section.

3.1.2 An increase in pulsed current density while keeping both the average current density and the number of coulombs per pulse constant is illustrated in Fig. 3.6. This method of investigation has the advantage that the off-time is only very slightly changed by increasing the pulsed current density, and ambiguity in interpretation of the results due to the influence of the off-time can therefore be avoided. Physically, this mode of investigation corresponds to the situation where constant numbers of atoms are supplied to the surface with different energies, resulting in differences in crystal growth. The results for cadmium are shown in Fig. 3.7. As expected from the model described in section 2, an increase in pulsed current density or in pulsed voltage results in a grain refinement.

In summary, the method of investigation described under 3.1.2 shows clearly the specific influence of the pulsed current density on the grain size. A decrease in grain size with increasing pulsed current density occurs in agreement with the above model. Similar results were obtained by the method of investigation described in 3.1.1 where, however, an ambiguity of interpretation remained because of the unknown effect of the off-time, which is discussed in more detail in the following section.

3.2 *Effect of off-time*

The interval between two pulses, the off-time, has sometimes been called the "dead time." Evidence has shown, however, that the off-time can in fact

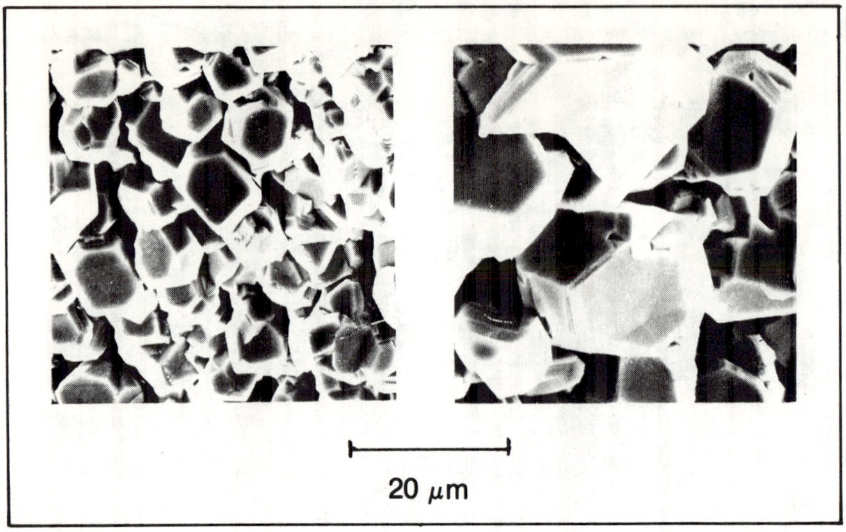

Fig. 3.4—Influence of PC density on the morphology of 10 μm cadmium deposits from bath A, Table 3.1, with a pulse time of 0.7 μs and an average current density of 37.5 A/dm^2; left, PC density: 1000 A/dm^2; right, PC density: 5000 A/dm^2.

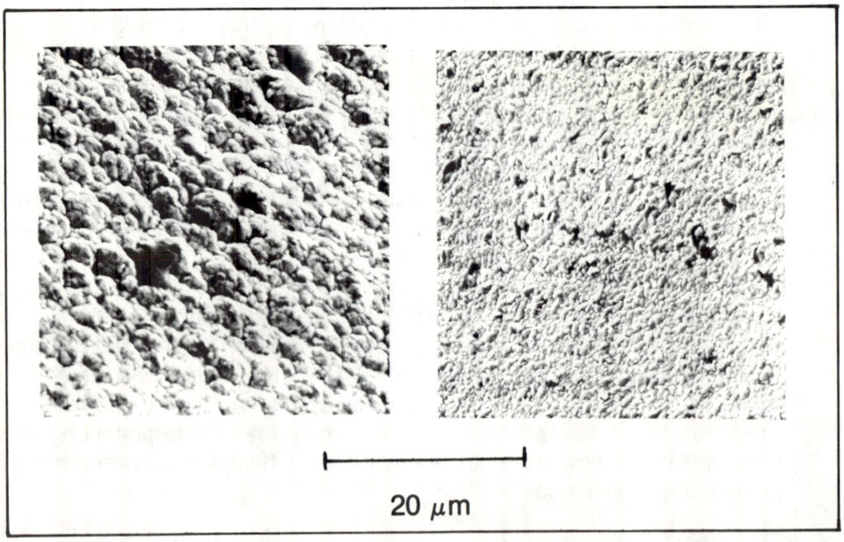

Fig. 3.5—Scanning electron micrographs of 10 μm palladium deposits from bath C, Table 3.1: left, DC density of 1 A/dm^2; right, PC density of 5000 A/dm^2, average current density of 1 A/dm^2, pulse time of 1 ms.

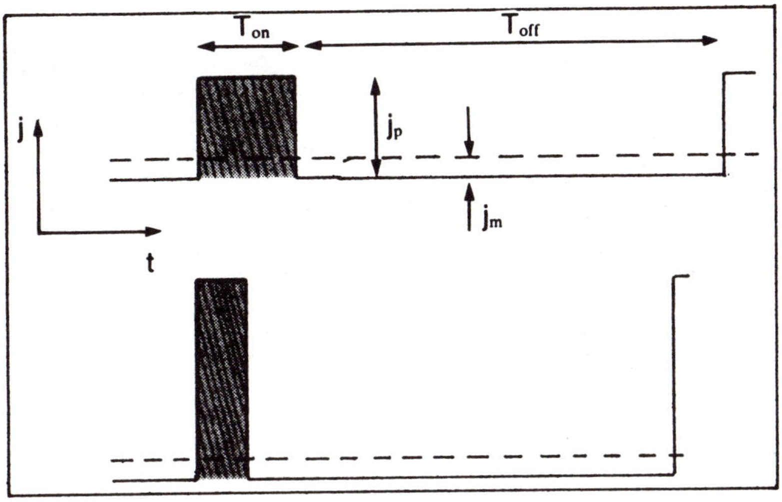

Fig. 3.6—Scheme for investigating changes in PC density with constant (j_p T_{on}) and T_{off}. (If T_{off} is much greater than T_{on}, j_m remains nearly constant at increasing j_p.) The shaded areas are equal.

sometimes be a very active period with respect to crystallization. The designation "off-time" is therefore more appropriate. Scanning electron micrographs of deposits of cadmium, copper and gold produced with constant pulsed current densities and constant pulse lengths, but with different off-times, are shown in Figs. 3.8, 3.9, and 3.10 respectively. An increase in off-time results in grain refinement for cadmium, but in grain growth for copper and for gold. The most plausible explanation is that for copper and gold a *recrystallization* occurs during the off-time. The bigger grains are thermodynamically more stable and, if enough time is allowed to the system, the most stable state will be reached. For cadmium, the opposite trend in grain sizes is observed, most probably due to blocking of growth centers of the cathode during the off-time as a result of adsorption of inhibiting species. This forces the system to create new nuclei at each new pulse. In cadmium deposition, with electrolyte B described in Table 3.1, the inhibitors are probably sulfate ions. Indeed, the equilibrium potential of cadmium in the acid sulfate solution is 0.5 V more positive than the zero charge potential of this metal (3), hence electrodeposition can occur readily at potentials allowing the adsorption of anionic particles. *A fortiori* adsorption of sulfate ions will occur during the off-time when the current is switched off. The phenomenon of inhibited recrystallization of cadmium has been further studied by investigating the early stages of electrodeposition. Scanning electron micrographs of cadmum and copper deposits with an equivalent thickness of 1 μm are shown in Fig. 3.11. The cadmium deposit grew on some isolated centers while copper grains spread over the whole of the substrate surface. It is clear from this observation that crystallization of cadmium is strongly inhibited as compared to copper.

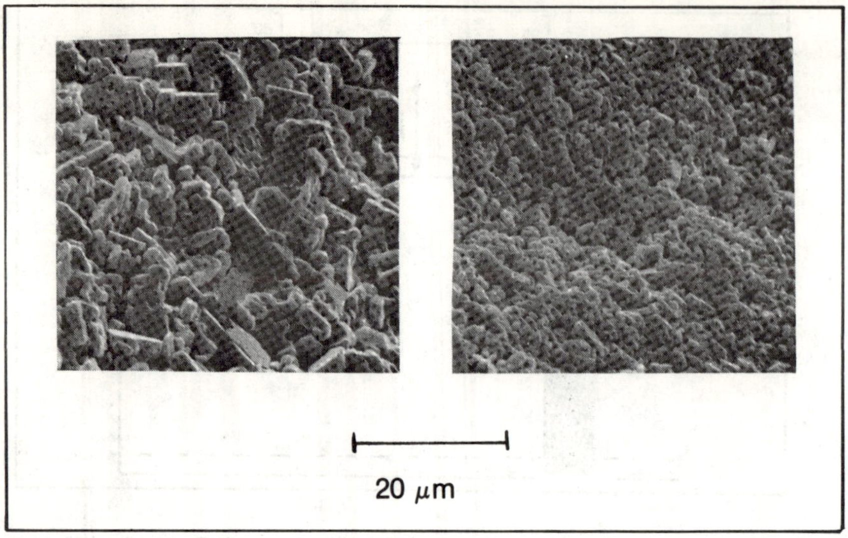

Fig. 3.7—Influence of PC density on the morphology of 10 μm cadmium deposits from bath A, Table 3.1, at constant (j_p T_{on}) and T_{off} with an off-time of 66 ms and an average current density of about 1.5 A/dm^2: left, pulse time of 1,000 μs and PC density of 100 A/dm^2; right, pulse time of 100 μs and PC density of 1000 A/dm^2.

3.3 *Effect of on-time*

The influence of the pulse duration T_{on} on the morphology of cadmium deposits, with constant pulsed current density and off-time, is illustrated in Fig. 3.12, showing that an increase in on-time results in an increase in grain size of cadmium crystals. A possible explanation for this observation is related to inhibition during the off-time. The crystal growth is not much disturbed during the on-time by anionic adsorbing species, but at each new pulse because of inhibition of growth centers during the off-time new nuclei are built. Therefore, the higher the ratio T_{on}/T_{off} is, the less disturbed the crystallization will be. By increasing the on-time, the crystals can grow to a larger extent without being blocked by inhibiting species, resulting in larger grains. This interpretation is restricted to cadmium deposition with bath B of Table 3.1 and cannot be generalized to other metals.

4. Influence of pulse plating on the crystal structure of deposits

While considerable attention has been paid to the morphology of deposits (topography and grain size of deposits as apparent under SEM inspection at 1000 X magnification, for example), much less information is available concerning the crystal structure of deposits on a much smaller dimensional scale. Some of the studies reported in this context will be discussed here.

4.1 *TEM studies (4)*

Studies of the effect of pulsed current plating on the structure of gold-cobalt deposits (bath E, Table 3.1) have been carried out by transmission electron microscopy on coatings after thinning by anodic dissolution (4). The conditions used were:

a) direct current, 1 A/dm^2
b) pulsed current (PC 10/20), i.e., 10 ms on, 20 ms off; average current density 1 A/dm^2
c) pulsed current (PC 10/100), i.e., 10 ms on, 100 ms off; average current density 1 A/dm^2

Typical structures of the three types of deposits are shown in Figs. 3.13, 3.14 and 3.15.

In the case of the PC coating (Fig. 3.13), pores or inclusions† in the metallic matrix appear clearly, with diameters ranging from about 5 to 50 Å, and with a density of about 10^{19} per cm^3. Some typical growth faults (twins) also are visible, but dislocations are absent. These features are responsible for the low ductility of such a coating. Dislocations cannot readily be formed in the small grains, and, even if some are present, their movement is impeded by the numerous inclusions.

The structure of a deposit produced with PC 10/100 is completely different, with a marked increase in grain size, and complete absence of pores or inclusions. Many growth faults remain, and dislocations within the crystallites are clearly visible (Fig. 3.15). The structure resembles that of a pure wrought metal and is characteristic of a highly ductile metal.

In the intermediate case (PC 10/20), grain size is greater than that for the DC deposit, but remains relatively small (about 500 Å). There are extended growth faults in each crystallite (Fig. 3.14) and less distortion than that observed in the DC deposit. Inclusions are plainly visible, but with about half the density of those in the DC deposit. This structure may be associated with a transitional stage between the brittle DC coating and the ductile PC deposit of Fig. 3.15, and raises the interesting possibility of tailoring the properties of deposits to achieve optimum performance for a specific application by suitable choice of the PC cycle.

Analysis of the pore-size frequency distribution gave a weighted (volume) average diameter of about 23 Å for both DC and PC 10/20 deposits. In conjunction with the observed pore density, this corresponds to a relative pore (inclusion) volume of about 10 percent, which compares well with the observed gravimetric density defect of about 13 percent (Table 3.2).

4.2 *X-ray diffraction analysis*

4.2.1 Silver deposits (6)

Silver deposited by direct current from an ammoniacal bath (bath A of Table 3.3) shows a preferred orientation of the [111] plane for the low current density range and a progressive change to [110] and [311] planes with increasing current density. In pulse plating, no significant influence resulting from the variation of the pulsed current density was observed when the average current density was kept constant at 1.5 A/dm^2. An increase in average current density

† The occluded material, often referred to as "polymer," following the original observations of Munier (29), has, in fact, been identified as a cobalt complex (4), most probably potassium hexacyanocobaltate III (5).

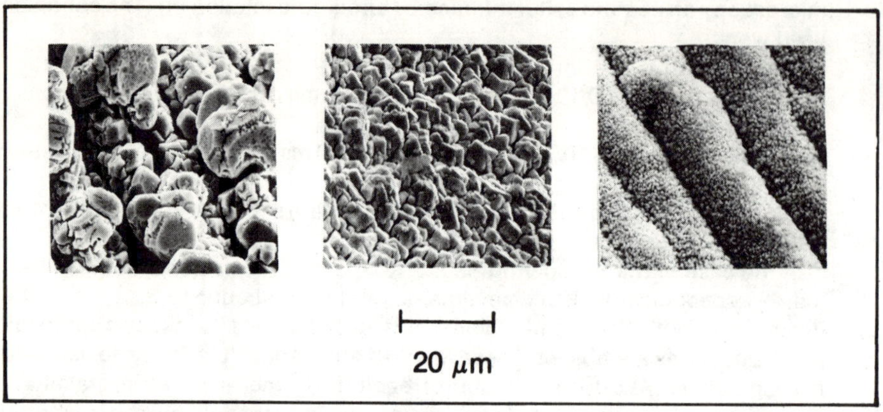

Fig. 3.8—Influence of off-time (T_{off}) on the morphology of 10 μm cadmium deposits from bath A, Table 3.1, with a pulse time of 30 μs and a PC density of 5000 A/dm^2; from left to right: T_{off} = 4ms; T_{off} = 20 ms; T_{off} = 100 ms.

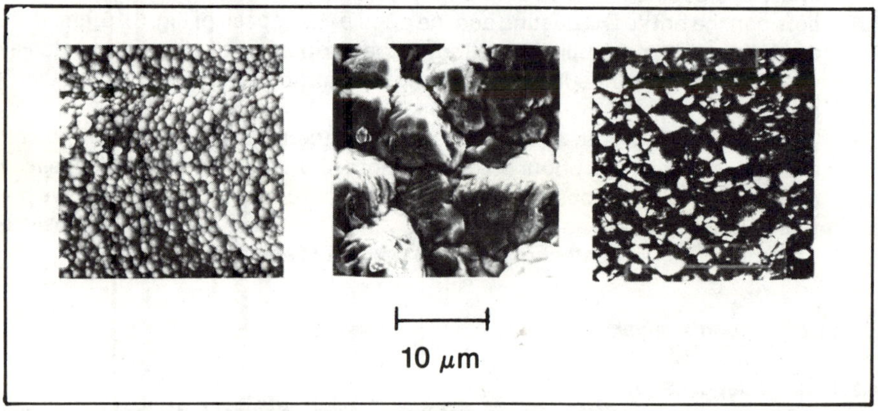

Fig. 3.9—Influence of off-time (T_{off}) on the morphology of 10 μm copper deposits from bath D, Table 3.1, with a pulse time of 30 μs and a PC density of 4000 A/dm^2; from left to right: T_{off} = 3.2 ms; T_{off} = 80 ms; T_{off} =400 ms.

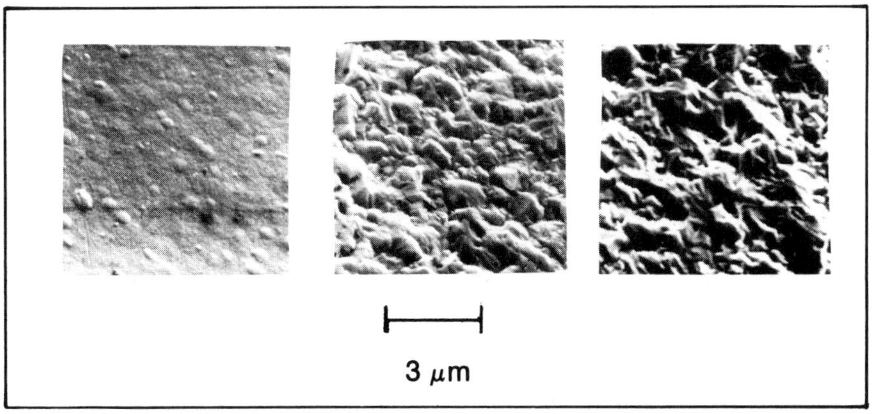

Fig. 3.10—Influence of off-time (T_{off}) on the morphology of 2 μm gold deposits from bath E, Table 3.1, with a pulse time of 1 ms and a PC density of 10 A/dm^2; from left to right: T_{off} = 9 ms; T_{off} = 49 ms; T_{off} = 99 ms.

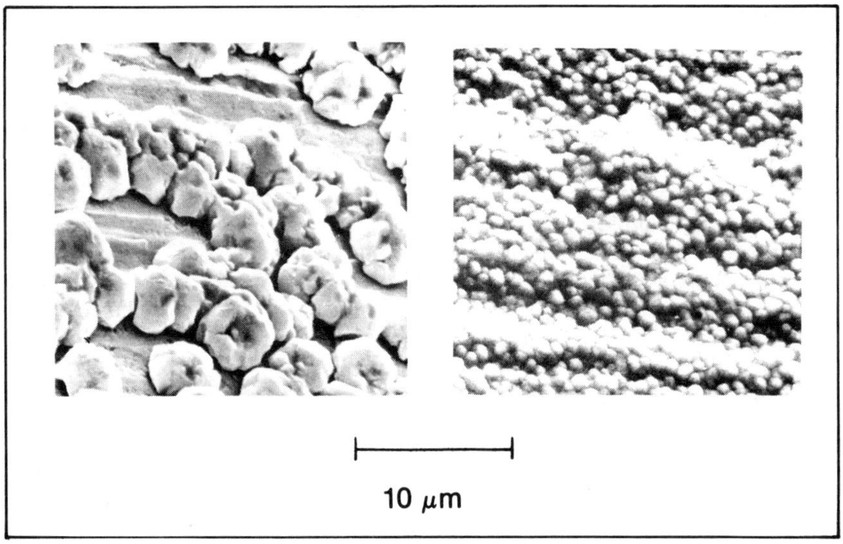

Fig. 3.11—Initial stages of cadmium and copper deposition (1 μm) with PC density of 1000 A dm^2 and a pulse time of 30 μs; left: cadmium, bath A, Table 3.1, average current density of 1.5 A/dm^2; right: copper, bath D, Table 3.1, average current density of 0.3 A/dm^2.

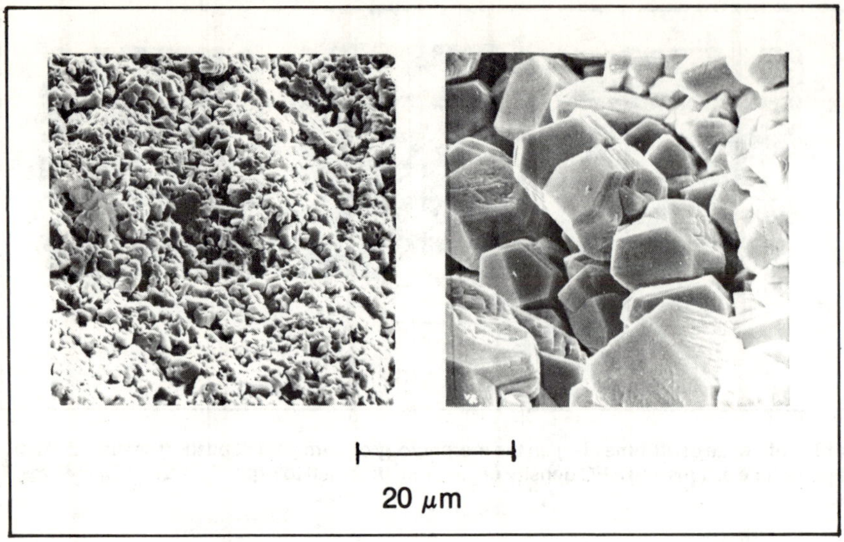

Fig. 3.12—Influence of pulse time (T_{on}) on the morphology of 10 μm cadmium deposit from bath A, Table 3.1, at a PC density of 400 A/dm^2 and an off-time of 9 ms; left: $T_{on} = 30\ \mu s$; right: T_{on} 1,000 μs.

up to 2.5 A/dm^2 resulted in a change in preferential orientation of the [111] plane to [110] and [311]. In fact, an increase in average current density in pulse plating leads to the same trend as an increase in DC current density. An increase of the on-time on the preferred orientation of deposits was, however, found to be significant; an increase from 1 to 10 ms resulting in preferred orientation of the [110] and [311] planes which was much more pronounced than could be expected from the corresponding increase in average current density. The change in the preferred orientation of the crystallization planes was explained in terms of increase in polarization overpotentials.

In DC plating, silver deposits from a cyanide bath (bath B of Table 3.3) show a decrease in the preferred orientation of the [111] plane as well as of [110] and [311] planes with increasing current density. In pulse plating the preferred orientation of the [111] plane increased with increase of the pulsed current density, for on-times of 1 and 10 ms.

4.2.2 Palladium-nickel alloy deposits (7)

The crystallization plane orientation has been investigated by X-ray diffraction for deposits obtained from the bath composition given in Table 3.4. The influence of the average current density on the orientation index of Pd-Ni deposits is shown in Fig. 3.16. For small j_m, the preferred orientation is [111]. An increase in j_m resulted in a decrease in preferred orientation [111], an increase to a maximum and further decrease in [100] plane and an increase in [110] plane preferred orientation. The trends observed were explained in terms of polarization overpotentials. The energy required for the formation of secondary nuclei increased for [111] to [100] and to [110] planes. This sequence corresponds exactly to that observed above with increasing average current density in pulse plating of Pd-Ni alloys.

TABLE 3.2

Effect of Pulsed Current on Composition and Structural Properties of Deposits

Deposit	Cobalt Content %	Density g/cm^3	Grain Size Å	Observed Structure Defects	Pore (inclusion) density, number/cm^3*
DC	0.18-0.20	16.6-17.1	250	Pores, few stacking faults and twins	10^{19}**
PC 10/20	0.23	17.3-17.4	550	Pores, stacking faults and twins	0.5 x 10^{19}**
PC 10/100	0.25	19.2	1050	Dislocations, many stacking faults and twins	no pores

* 10 to 12 Å was the most frequent size in the distribution curve corrected for diffraction effects; average volume pore size was about 21 Å.

** Maximum distribution frequency for pores of 10 to 12 Å in diameter.

Fig. 3.13—Transmission electron microscopy (TEM) of DC plated coating. Structure has small grains, numerous inclusions, twins (seen as bands) and absence of dislocations.

4.2.3 Cu-Zn-Sn alloy deposits (8)

X-ray diffraction studies have also been made on electrodeposited ternary alloy (Cu 55%, Zn 20%, Sn 25%) coatings produced by DC and pulse plating. The corresponding d values for as-plated deposits are given in Table 3.5, d being the distance between two crystallographic planes of same indices. The diffraction lines of the DC plated alloy correspond to a single-phase orthorhombic structure. The pulse plated alloy phases are quite different. While they were not identified, this example illustrates the potential usefulness of pulsed current plating, where higher free energy is available for the deposition reaction, as a means of influencing the structure and properties of alloy deposits.

4.3 *Nuclear Magnetic Resonance (NMR) studies (9)*

NMR studies of nickel-phosphorus alloys, both electroless and electrodeposited, have been reported in the literature (9) and a review of this technique as applied to alloys has recently been presented by Bennett (10). The method has been used to examine the local neighborhood of phosphorus atoms by

Fig. 3.14—Transmission electron microscopy of PC 10/20 deposit. The inclusion density of this structure is lower than for DC; grains are larger and twins (t) are present.

observation of the Knight shift in alloys produced by pulsed and direct current deposition (11, 11a), using the bath composition of Table 3.6, and by electroless deposition. Quenching from the melt was also used. A wide range of pulse plating conditions, varying the on- and off-times from 0.2 to 100 ms, was investigated. The specimens were prepared by plating copper or brass foils to a thickness of about 50 μm and then stripping off the copper or brass. The resultant nickel alloy was then ground to a coarse powder using a mortar and pestle.

Phosphorus Knight shifts were measured and the results are shown in Fig. 3.17. It is clear that two different Knight shift (K) values are obtained in the range of 14 to 25 at.% P; the DC plated samples give a K value equal to about 0.15 percent, and all other preparation techniques give higher values. Hence at least two unique metastable amorphous phases exist at compositions between 14 and 25 at.% P. One can be produced by autocatalytic or by pulse plating techniques and the other one by DC electrolysis. From this study it is evident that pulse plating can be used as a means of producing structures not obtainable by DC electrolysis.

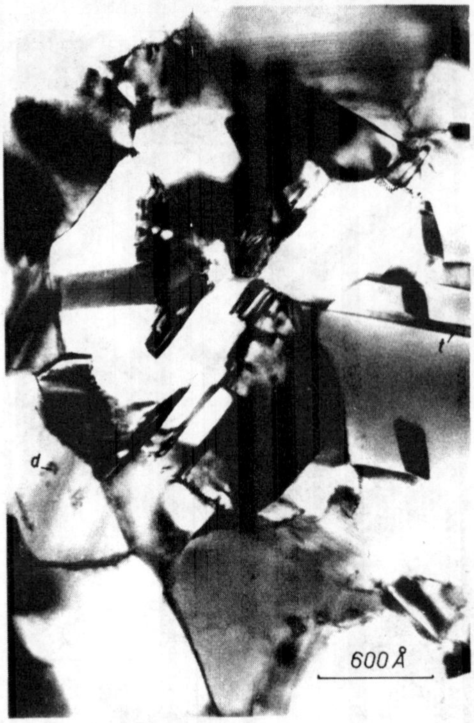

Fig. 3.15—Transmission electron microscopy of PC 10/100 deposit. Pores or inclusions are absent; in larger grains appear twins (t) and dislocations (d).

Table 3.3

Silver Electrolytes

A		B	
$AgNO_3$	0.3 M	AgCN	0.3 M
NH_4OH	0.75 M	KCN	0.9 M
NH_4NO_3	1 M	K_2CO_3	0.3 M
Temperature	30° C	Temperature	30° C
pH	9.5	pH	11

TABLE 3.4

Pd-Ni Bath Composition

$Pd(NH_3)_2Cl_2$	0.1 M
$NiSO_4 \cdot 6\ H_2O$	0.2 M
$(NH_4)_2SO_4$	0.38 M
pH (adjusted with NH_3)	7.5

TABLE 3.5

Experimental d Values in Å (For 2 θ between 25° and 60°) for Pulse Plated and DC Plated Cu-Zn-Sn Alloys

Pulse Plating	DC Plating
2.33 w	3.23 vw
2.16 w	3.01 w
2.10 st	2.94 st
1.95 w	2.71 w
	2.11 vw
	1.65 w

st = strong; w = weak; vw = very weak

TABLE 3.6

Ni-P Alloy, Bath Composition

$NiSO_4 \cdot 6\ H_2O$	150 g/l
$NiCl_2 \cdot 6\ H_2O$	45 g/l
H_3PO_4	50 g/l
H_3PO_3	40 g/l
Temperature	78° C
pH	0.8

TABLE 3.7

Electrolyte Composition for Pd Deposition

$Pd(NH_3)_2(NO_2)_2$	0.33 M
NH_4NO_3	1.25 M
$NaNO_2$	0.15 M

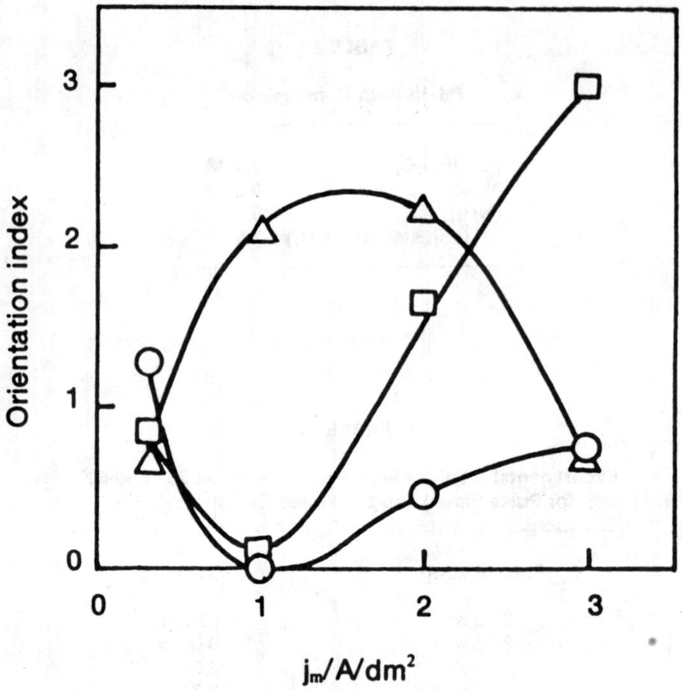

Fig. 3.16—Effect of average current density j_m on the orientation index of pulse plated Pd-Ni alloys. $T_{on} = 0.1$ ms; $T_{off} = 9.9$ ms.

5. Topics related to crystallization

5.1 *Porosity of deposits*

Porosity may be defined in a number of ways (12). Bulk porosity is defined as the volume fractions of the deposit occupied by voids and discontinuities of all types. Transverse porosity is the fraction of the total area at which the basis metal substrate is exposed. Transverse pores form a continuous path from the substrate to the surface. The transverse porosity is quantified by the number of pores per cm^2. In this section only the latter will be discussed; the bulk porosity will be considered in the next section dealing with inclusions.

The transverse porosity of electrodeposits depends on a number of parameters, including substrate roughness (12,13,14), surface defects of the substrate (15), bath parameters (16), hydrodynamics (17), thickness of deposit (12,17,18) and current density (17). From the point of view of the mechanism of formation of transverse porosity, superficial density of nuclei on the substrate and the ability for intergranular growth with increasing thickness are determining factors. Since pulse plating enhances the nucleation rate, both improved surface

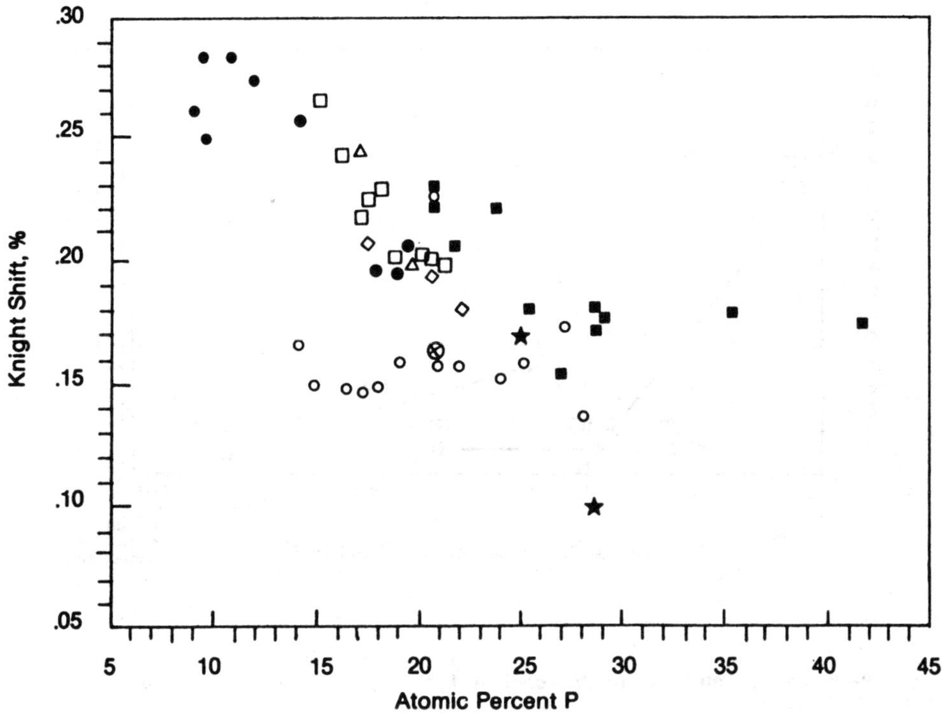

Fig. 3.17—Room temperature Knight shift data for Ni-P amorphous alloys and for crystalline Ni_3P and Ni_5P_2. ★, crystalline; ■, pulse plated (2 ms on, 0.2 ms off); ⊗, pulse plated (100 ms on, 0.1 ms off); △, splat quenched; ○, DC plated; □, electroless; ●, electroless; ◇, DC plated. Typical standard deviations of the Knight shift were K ± 0.01%.

coverage with nuclei and denser build-up of grains are to be expected. Lower porosity in pulse plating as compared to DC has been reported in the literature (see Ref. 16, 21, 22, 28 and 29 of Chapter 1 as well as Ref. 19 and 20 of this chapter). The study of Rehrig (Ref. 29 of Chapter 1) makes direct comparisons between different deposition modes, including electron gun evaporation, asymmetric AC plating, DC plating and pulse plating. The results are reported in Fig. 9.1. The decreased porosity in pulse plating is interpreted in terms of enhanced intergranular growth due to higher nucleation rate. This example illustrates well the possibilities of pulse plating for decreasing the porosity of gold deposits.

5.2 *Inclusions in deposits*

Inclusions can be considered as bulk porosity. They will greatly influence the density of the deposit. Voids in the deposit may be filled with hydrogen, or residual liquids, or other materials from various processing baths. Gold deposited from baths containing gold as cyanide complex may contain a significant quantity of codeposited polymeric material. This particular inclusion type has already been discussed in section 4.1 of this chapter. The influence of pulse plating on "polymer" inclusions has been studied over a wide range of

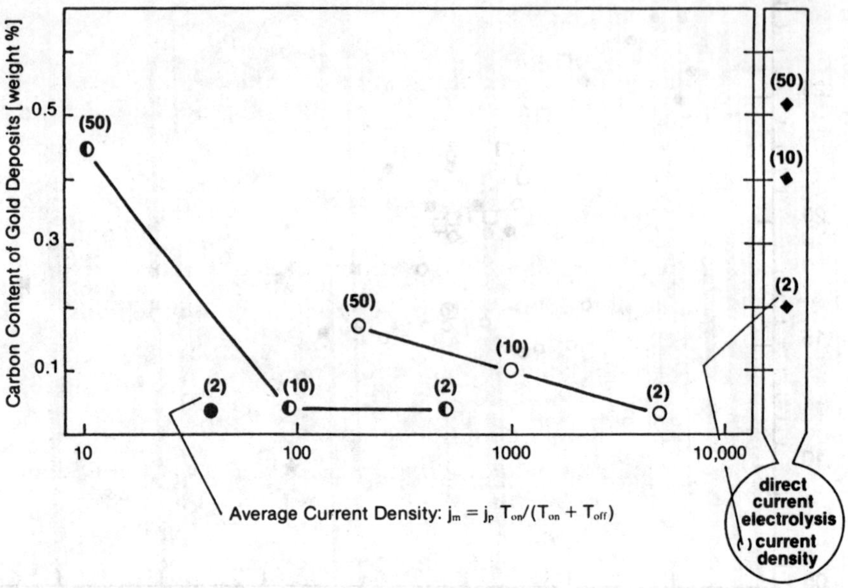

Fig. 3.18—Influence of off-time on carbon content of gold deposits for T_{on} = 10 ms. j_p: ● 1 A/dm²; ◐ 10 A/dm²; O 100 A/dm².

pulsed current parameters (21,22). The polymer content has been measured by carbon content analysis. As shown in Fig. 3.18, pulse plating can greatly decrease carbon content in gold deposits obtained from bath E of Table 3.1. These results are in agreement with TEM investigations (4). Inclusions can be of a molecular dimension, and may consist in organic molecules of brighteners or other addition agents in the plating bath.

Particular attention has been directed to the influence of pulse plating on gas content in electrodeposits. Hydrogen, nitrogen and oxygen contents in gold-cobalt and in gold-nickel deposits have been measured as a function of pulsed current parameters and compared with gas contents in deposits made with DC (23). Gas contents were decreased by a factor up to 5, but generally of 3, by using pulse plating as compared to DC. The influence of pulse plating on hydrogen content has been investigated for nickel deposits from different plating baths (24). It was demonstrated that the hydrogen content in nickel deposits can be significantly decreased by using pulsed current. According to Snavely (25), hexagonal chromium hydride is electrodeposited at the cathode from chromic acid electrolytes. This hydride "alloy" is unstable and decomposes to metal during or after deposition with a change in structure from hexagonal to body-centered cubic, resulting in crack formation. In line with this hypothesis, it might be expected that the use of pulsed current, providing an off-time to allow for the decomposition of the layer of CrH_x deposited during the cathodic period, would result in reduced cracking, and in fact crack-free coatings have been produced by this approach (26). Other studies related to hydrogen content in pulse plating are referred to in Chapter 1 (Ref. 28 and 33).

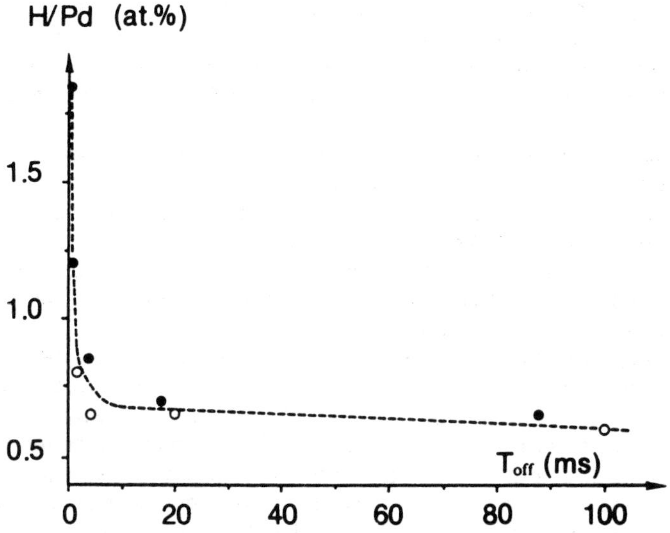

Fig. 3.19—Hydrogen content in Pd pulse plated deposits as a function of off-time. $j_m = 15$ A/dm^2; ● $j_p = 100$ A/dm^2; O $j_p = 500$ A/dm^2; pH = 9.

Since palladium is well-known for its ability to absorb large amounts of hydrogen, special attention has been focused on the possibility of reducing the hydrogen content in Pd deposits by pulse plating methods (27). A wide range of pulse parameters has been investigated for deposits from a solution described in Table 3.7. The hydrogen content was determined electrochemically by the method described in Ref. 28. Figure 9.36 shows the hydrogen content as a function of the on-time for different j_p and T_{off}/T_{on} ratios and compares the pulse plating values with DC. A decrease in hydrogen content is observed for increasing on-times. The lowest values of hydrogen content are obtained for higher T_{off}/T_{on} ratios which also correspond to the higher j_p values. When compared with DC, pulse plating results in a much lower hydrogen content. In Fig. 3.19 the hydrogen content is plotted as a function of the off-time, a sharp decrease in hydrogen content is observed with increasing off-time. Almost the same curves are obtained for different j_p values, indicating that the off-time is the most important parameter in decreasing the hydrogen content. The reduced hydrogen content in pulse plating can therefore be explained by desorption during the off-time. With pulses leading to plating thicknesses of tens of Å units, the desorption time can be derived from Fig. 3.19, and estimated to be 5 μsec.

6. Concluding remarks

Attempts have been made to identify specific trends associated with the different pulse plating parameters. For T_{off} and T_{on} this was not possible because other parameters related to the particular electrochemical systems, such as inhibition, can produce opposite trends from one system to another in terms of crystallization mechanisms. For j_p, however, it seems that its effect on grain refinement is systematic and does not depend upon the electrolyte. This trend can be qualitatively understood in terms of the presented model "building-up of existing crystals and formation of new nuclei".

When the grain size is referred to, it is important to distinguish between the grains as apparent on the SEM-pictures at 1000 X magnification, and the grains as apparent on TEM-pictures at 10^6 X magnification, for example. The SEM show polycrystalline grains while the TEM can show monocrystals.

From the experimental results reviewed in this chapter, it is clear that pulse plating can be used as a means of producing unique structures, i.e., coatings with properties not obtainable by conventional DC plating.

This chapter points out a number of different techniques used for structure investigations, including scanning electron microscopy, transmission electron microscopy, Rõntgen diffraction, and nuclear magnetic resonance as well as electrochemical methods for porosity and hydrogen content measurements.

Because the structure of deposits can be significantly influenced by pulse plating parameters, examination of structure can be a useful diagnostic tool for pulse plating studies. Due to the potential benefits of pulse plating in terms of deposit properties, industrial use of pulse plating should be extended, but lack of diagnostics often inhibits decisions.

References

1. H. Hamann and H. Vielstich; *Elektrochemie II*, Verlag Chemie, Weinheim (1981)
2. B.R. Kaischew, E. Budewski and J. Malinowski; *Zeitschrift für Physikoliscje Chemie* (Leipzig), **204** : 348 (1955)
3. L. Domnikov; *Metal Finishing*, **61**, (6) : 96 (1963)
4. W. Flühmann, F.H. Reid, P.A. Mäusli and S.G. Steinemann; *Plating and Surface Finishing*, 67(6): 62-65 (1980)
5. Y. Okinaka et al.; *Journal Electrochemical Society*, **125** : 1745 (1978)
6. T. Hayashi, H. Kitanishi and Y. Fukumoto; *Proceedings of the AES Second International Pulse Plating Symposium*, Rosemont, IL, (October 6-7, 1981)
7. Y. Fukumoto, Y. Kawashima and T. Hayashi; *Proceedings of the AES 71st Annual Technical Conference*, New York, (July 16-19, 1984)
8. J.Cl. Puippe; *Proceedings of Interfinish 84*, Jerusalem (October 21-26, 1984)
9. L.H. Bennett, H.E. Schone and P. Gustafson; *Physical Review*, Series 3, B-Condensed Matter, **18**, (5) : 2027 (1978)
10. L.H. Bennett; *Proceedings Nuclear and Electron Spectroscopy Applications to Materials Science*, Materials Research Society (1980)
11. D.S. Lashmore, L.H. Bennett, H.E. Schone, P. Gustafson and R.E. Watson; *Physical Review Letters*, **48**, (25) : 1760-1763 (1982)

11a. D.S. Lashmore and J. Weinroth; *Proceedings of the AES Second International Pulse Plating Symposium*, Rosemont, IL, (October 6-7, 1981)

12. F.H. Reid and W. Goldie; *Gold Plating Technology*, Electrochemical Publications Ltd., Glasgow (1974)
13. S.M. Garte; *Plating and Surface Finishing*, **53** : 1335 (1966)
14. S.M. Garte; *Plating and Surface Finishing*, **55** : 946 (1968)
15. F.I. Nobel, B.D. Ostrow and D.W. Thompson; *Plating and Surface Finishing*, **52** : 1001 (1965)
16. J.M. Leeds and M. Clarke; *Transactions Institute of Metal Finishing*, **46**, (1) (1968)

17. J. Bielinski, N. Ibl and H. Angerer; *Oberfläche-Surface,* **20**, (1) (1979)
18. R. Sard, Y. Okinaka and J.R. Rushton; *Plating and Surface Finishing,* **58**, (9) : 893-900 (1971)
19. F.H. Leaman and B.A. Kownacki; *Proceedings of the First AES International Pulse Plating Symposium,* Boston, MA (April 19-20, 1979)
20. Y.Y. Wang, C.S. Tung and C.C. Wan; *Metal Finishing,* **67**, (9) (1980)
21. H.F. Angerer; *Dissertation 6213,* ETH Zurich (1978)
22. N. Ibl, J.Cl. Puippe and H.F. Angerer; *Surface Technology,* **6** : 287-300 (1978)
23. Ch. J. Raub and A. Knödler; *Gold Bulletin,* **10**, (2) : 38-44 (1977)
24. W. Kleinekathofer and Ch.J. Raub; *Surface Technology,* **7** : 23-34 (1978)
25. C.A. Snavely; *Transactions Electrochemical Society,* **92** : 537 (1947)
26. C.L. Faust, G.R. Schaer and D.E. Semones; *Plating and Surface Finishing,* **48**, (6) : 605-612 (1961)
27. J.M. Locarnini; *Dissertation 7163,* ETH Zurich (1982)
28. H.D. Hedrich and Ch.J. Raub; *Metalloberfläche,* **31** : 11 (1977)
29. G. Munier; *Plating and Surface Finishing,* **56** : 1151 (1969)

Chapter 4

Influence of Charge and Discharge of Electrical Double Layer in Pulse Plating

J.Cl. Puippe

1. Introduction

The influence of capacity effects of the electrical double layer at the metal/solution interface on pulsed current has been discussed qualitatively in Chapter 1. The charging and discharging times of the double layer are a function of a number of electrical parameters, including the exchange current density, capacitance, and pulse current density; the purpose of the present chapter is to discuss capacitance effects and their role in pulse plating in quantitative terms (see also Ref. 1). Only galvanostatic pulses with a period of zero current will be considered, as illustrated in Fig. 4.1.

Any current density for metal deposition, j_F, corresponds to a certain overpotential. Hence, at the beginning of each pulse the cathode potential must be raised to this value before metal deposition occurs. This process takes time because the electrical double layer at the interface represents a capacitor with plate separation of molecular dimensions and therefore with very high capacity. At the beginning of the pulse, then, the total current supplied by the generator, j_p, consists of two parts, i.e., a capacitive current, j_C, which charges the double layer, and a faradic current, j_F, which corresponds to the rate of metal deposition:

$$j_p = j_C + j_F = j_t \qquad [1]$$

The charging time is the time required for the cathode potential to reach the value corresponding to the pulse current, i.e., until j_F becomes equal to the total current supplied by the generator. Figure 4.2 illustrates various current profiles associated with different ratios of charging time to total pulse time. If charging time is negligible as compared to pulse time, ideal pulses are obtained as in Fig. 4.2a. In the practical case, however, where charging time is often a significant fraction of the pulse time, the pulse shape is slightly distorted as in Fig. 4.2b. If the charging time is longer than the pulse length, the potential corresponding to the pulse current density is never reached, and j_F remains smaller than j_p during the whole of the pulse (Fig. 4.2c, 4.2d). Similar effects occur at the end of a pulse. The double layer must be discharged and some time is required before the potential falls to the value corresponding to zero current (Fig. 4.2b). If this time is longer than the off-time, the interval between pulses, the double layer is not completely discharged and j_F never decreases to zero (Fig. 4.2d).

To sum up, if the charge and discharge times are negligibly small in relation to the pulse and off-time respectively, j_F is virtually equal to the current supplied by the generator (Fig. 4.2a). Otherwise the case is as shown in Fig. 4.2b where, although the current fed to the cathode is a rectangular pulse, that depositing metal is more or less

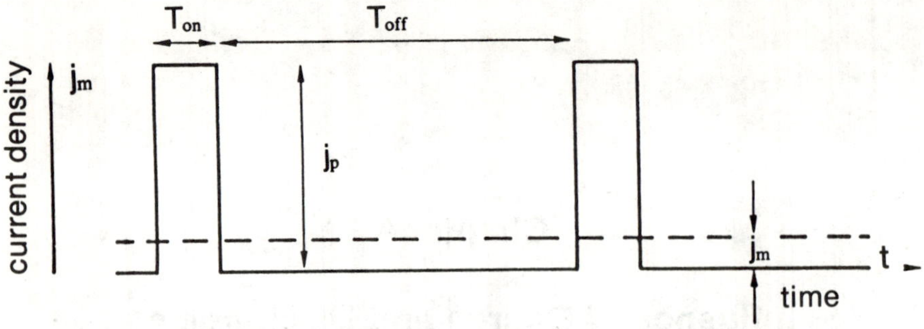

Fig. 4.1—Time dependence of applied current in pulse electrolysis $j_m = j_p T_{on}/(T_{on} + T_{off})$.

damped by the capacitance effect. If the latter is strong, j_F only oscillates around an average value and never falls to zero (Fig. 4.2d).

In the following sections, methods for calculation of charge and discharge times are presented, and the damping effect is discussed quantitatively by introducing the concept of "degree of flattening," with particular reference to conditions under which the effect can be expected to be significant in plating.

2. Calculation of charge and discharge time

2.1 *Principle of computation*

The principle of the method described in the following paragraph was established by the late Professor N. Ibl. As a matter of interest, the manuscript of the draft for computation, written at the coffee table, is reproduced in Fig. 4.3.

It is assumed that the overpotential associated with metal deposition is due solely to charge transfer and is linked with the faradic current by the relation:

$$j_F = j_0[\exp(\alpha nF\eta_a /RT)-\exp(-(1-\alpha)nF\eta_a /RT)] \quad [2]$$

This implies that there is no concentration overpotential, i.e., that the concentration of metallic ions at the metal/solution interface remains virtually the same as in the bulk of the solution during the pulse. The conditions under which this assumption is fulfilled can be calculated along the lines of Ref. 1a dealing with mass transport phenomena (see also Chapter 5).

In visualizing the system, the electrode/solution interface with its double layer may be represented by an equivalent circuit consisting of a capacitor C and resistor R_a (Fig. 4.4), where C is the capacity of the double layer and R_a is the resistance to charge transfer at the interface (given by $d\eta_a /dj_F$). According to this model the current supplied divides into two parts, one flowing through the capacitor and one through the resistor. It is to be noted, however, from equation [2], that $d\eta_a /dj_F$, and hence R_a, depends on j_F.

The capacitive current is given by:

$$j_C = dQ/dt = C\, d\eta_a/dt \quad [3]$$

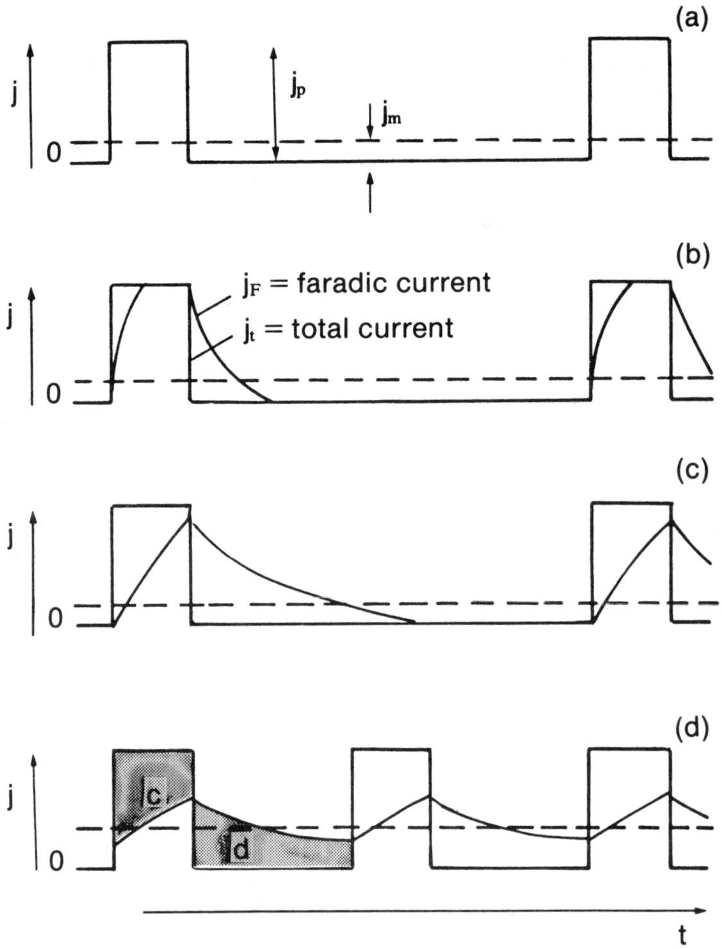

Fig. 4.2—Damping of faradic current.
(a) $t_c \leqslant T_{on}$; no damping (ideal situation);
(b) $t_c < T_{on}$; small damping;
(c) $t_c > T_{on}$;
(d) $t_c \geqslant T_{on}$ and $t_d \geqslant T_{off}$;
(c) and (d) strong damping.

The total current is the sum of the faradic and capacitive components:

$$j_t = j_F + j_C = j_p \quad [1a]$$

Since a galvanostatic pulse is considered,

$$dj_t = 0$$

$$dj_F = -dj_C \quad [4]$$

If sufficiently small increments of potential are considered, the variation of the faradic current with potential can be regarded as linear during small intervals involved. Thus, in the nth interval

$$j_F = \frac{1}{A_n} \eta_a + B \quad [5]$$

where A_n is a proportionality factor between current and potential in the nth interval, and B is a constant.

$$A_n = \frac{\eta_{an} - \eta_{a(n-1)}}{j_{Fn} - j_{F(n-1)}} \quad [6]$$

$$d\eta_a = A_n \, dj_F \quad [7]$$

From equations [3], [4] and [7],

$$j_C = -CA_n(dj_C/dt) \quad [8]$$

$$t_n = C \frac{\eta_{an} - \eta_{a(n-1)}}{j_{Fn} - j_{F(n-1)}} \ln \frac{j_{C(n-1)}}{j_{Cn}} \quad [9]$$

Equation [9] gives the time required to change the potential for the nth interval.

2.1.1 Charging of the double layer

We have

$$j_t = j_p \quad [10]$$

and

$$j_F = j_p - j_C \quad [11]$$

Defining the charging time, t_c, as the time required for the faradic current to reach 99 percent of j_p, t_c has been calculated by the above method, using a computer to sum the times t_n (Eqn. [9]), until j_F reaches the prescribed value. In reality, the faradic current, j_F, tends asymptotically to j_t, and the charging and discharging times up to 100 percent are theoretically infinite. It is, therefore, necessary to define arbitrary charging and discharging times for the double layer. The computation was based on potential increments of 1 mV, for each of which dn_a/d_j was calculated from equation [2]. At the end of the charge the increment of potential was reduced until the end value of j_F lay between 0.99 j_p and 0.991 j_p.

The charging time may also be calculated in more simple ways, for example,

(a) by assuming that the whole of the current is capacitive until the double layer is charged. Denoting this charging time as t_c^*, then

$$t_c^* = \frac{C\eta_{a\infty}}{j_p} \quad [12]$$

(b) by taking the charging time t_c^{**} as the time needed to increase the potential η_a to 98.2 percent of $\eta_{a\infty}$. For a resistance R_a and capacity C in parallel (2),

$$t_c^{**} = 4 \, R_a C \quad [13]$$

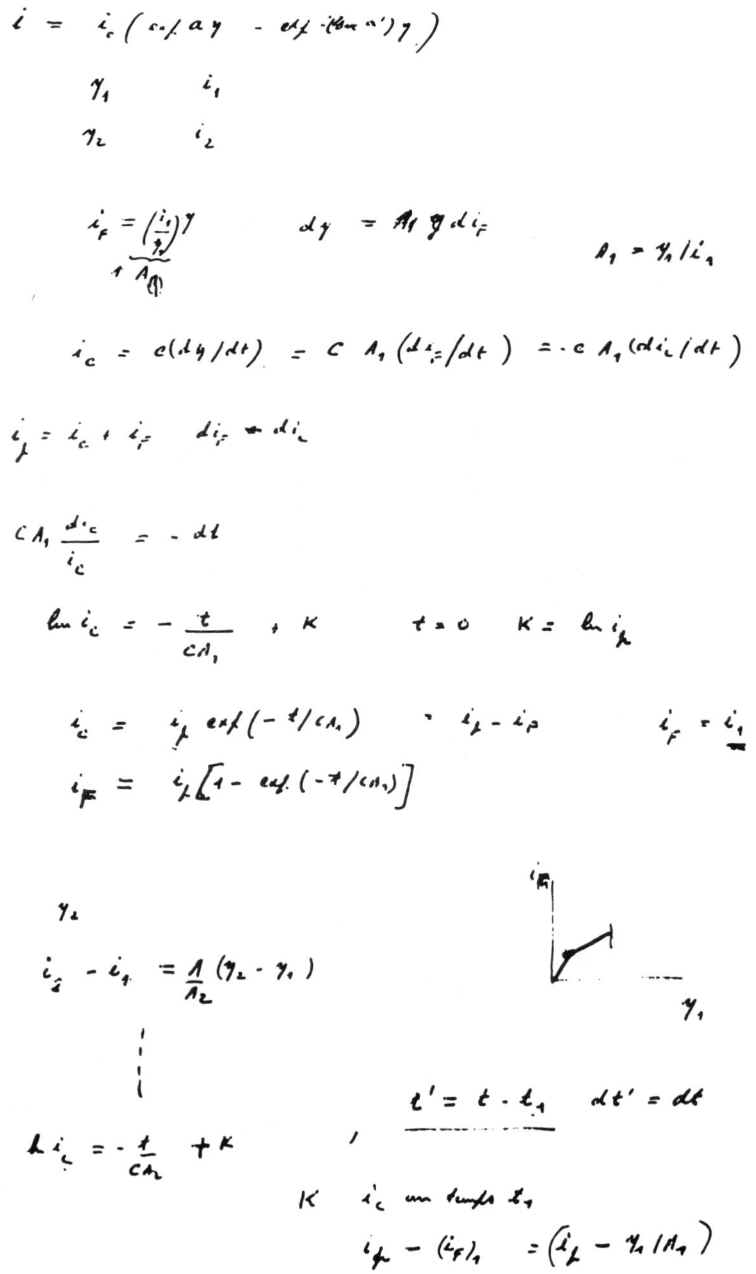

$$i = i_c\,(\exp a\eta - \exp -(\ldots)\eta)$$

$\eta_1 \quad i_1$

$\eta_2 \quad i_2$

$$i_F = \left(\frac{i_1}{\eta_1}\right)\eta \qquad d\eta = A_1\, d i_F \qquad A_1 = \eta_1 / i_1$$

$$i_c = C(d\eta/dt) = C A_1 (d i_F/dt) = -C A_1 (d i_c/dt)$$

$$i_f = i_c + i_F \qquad d i_F = -d i_c$$

$$C A_1 \frac{d i_c}{i_c} = -dt$$

$$\ln i_c = -\frac{t}{C A_1} + K \qquad t = 0 \quad K = \ln i_f$$

$$i_c = i_f \exp(-t/CA_1) \qquad i_f - i_F \qquad i_F = i_1$$

$$i_F = i_f\left[1 - \exp(-t/CA_1)\right]$$

η_2

$$i_2 - i_1 = \frac{1}{A_2}(\eta_2 - \eta_1)$$

$$t' = t - t_1 \qquad dt' = dt$$

$$\ln i_c = -\frac{t}{C A_2} + K$$

$K \quad i_c$ au temps t_1

$$i_f - (i_F)_1 = (i_f - \eta_1/A_1)$$

Fig. 4.3—Manuscript written by Prof. Ibl at the coffee table of his Institute, showing the principle of computation of the charge and discharge time of the electrical double layer at a metal/electrolyte interface.

where R_a is given by

$$R_a = d\eta_a/dj_F \qquad [14]$$

In the Tafel region of the polarization curve,

$$j_F = j_0 e^{(\alpha n F/RT)\eta_a} \qquad [15]$$

$$\frac{dj_F}{d\eta_a} = \frac{\alpha n F}{RT} j_0 e^{(\alpha n F/RT)\eta_a} \qquad [16]$$

$$\frac{d\eta_a}{dj_F} = \frac{RT}{\alpha n F j_p} = R_a \qquad [17]$$

where η_a is put equal to $\eta_{a\infty}$, corresponding to $j_F = j_p$.

This calculation does not take into account the change in resistance R_a during the charge. Since the activation resistance decreases with increasing j_F, the computed resistance R_a for large values of j_F is much lower than the average resistance which applies in reality and hence the calculated value of t_c^{**} is too small. In general the values computed from equations [13] and [17] are more accurate for small values of j_F.

2.1.2 Discharge of double layer during the off-time

In this case,

$$j_t = 0 \qquad [18]$$

and

$$j_F = -j_C \qquad [19]$$

At the end of a pulse (when the circuit of Fig. 4.4 is in position 'a') the accumulated charge in the capacitor flows through resistance R_a. Defining the discharge time, t_d, as the time required for decrease of j_F from 0.999 j_p to 0.01 j_p, the computation for t_d is similar to that described above for t_c. The increments of potential are again 1 mV and the increment at the end of the discharge is decreased in order that the final j_F value should lie between 0.011 j_p and 0.010 j_p. However, in computing t_d, equation [1a] is replaced by equation [19] and the increments of potential are subtracted instead of added.

2.2 *Results*

In Fig. 4.5, t_c, t_c^*, and t_c^{**} are plotted as functions of j_p for $\alpha = 0.5$, $n = 2$, $C = 50$ $\mu F/cm^2$ and $j_0 = 0.5$ A/dm^2. These latter are typical values, which would apply, for instance, to a concentrated solution of copper sulfate in sulfuric acid (3). It is to be noted that the charging time t_c^* is always less than t_c, due to the assumption used in the computation of the former that the total current charges the double layer, whereas in reality only a part of the current is capacitive during the charge. Nevertheless, for the conditions considered in the computation of Fig. 4.5 the

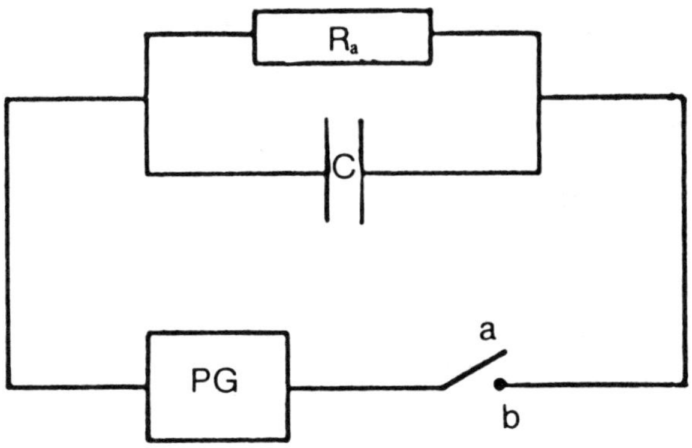

Fig. 4.4—Equivalent circuit of an electrode.
Position a: off-time, external circuit open;
Position b: on-time, external circuit closed;
PG, pulse generator;
C, capacitor;
R_a, resistance.

simple calculation gives relatively good results, with $t_c{}^*$ and t_c differing only by a factor of 2. This factor increases markedly for small j_p values, e.g. for $j_p/j_0 < 100$, hence the approximation for $t_c{}^*$ is much better for high j_p than for small values. For $t_c{}^{**}$ the opposite is the case, as already noted, because the hypothesis of constant resistance involves larger errors at higher j_p values. It is interesting to note, however, that the simple calculations for $t_c{}^*$ and $t_c{}^{**}$ give a good approximation for practical purposes, for example, where only the order of magnitude of the charging time is of interest.

Figure 4.5 is only useful for a given system with assigned values of the relevant parameters. In Fig. 4.6, on the other hand, t_c and t_d are included in a dimensionless number and can be read for large ranges of j_p, α, n, j_0 and C. For $j_p/j_0 > 100$ the log-log plot of Fig. 4.6 shows straight lines of slope -1, indicating that charging and discharging time are independent of j_0. This fact is of great importance, because the exchange current density is often not known with any accuracy. For metal deposition on polycrystalline substrates, the range of possible exchange current densities is very large (a few nA/cm^2 up to a few tens or a few hundreds of mA/cm^2) (4-6). With additives, j_0 of the system $Cu/CuSO_4/H_2SO_4$ can reach several A/cm^2. Since, however, in pulse plating instantaneous current densities are commonly much higher than in DC plating, the linear domain of the curves of Fig. 4.6 is often reached, when values of t_c and t_d can be read without precise knowledge concerning j_0.

When the pulse current density is close to the exchange current density ($j_p/j_0 <$ 100) the dimensionless terms on the ordinate of Fig. 4.6 are no longer simply inversely proportional to j_p/j_0, and t_c and t_d thus depend on j_0, which in these conditions must be known with reasonable accuracy. When j_p tends to j_0, t_c and t_d tend to a common constant value. For $j_p/j_0 > 100$, the discharging time is always about seven times longer than the charging time of the double layer.

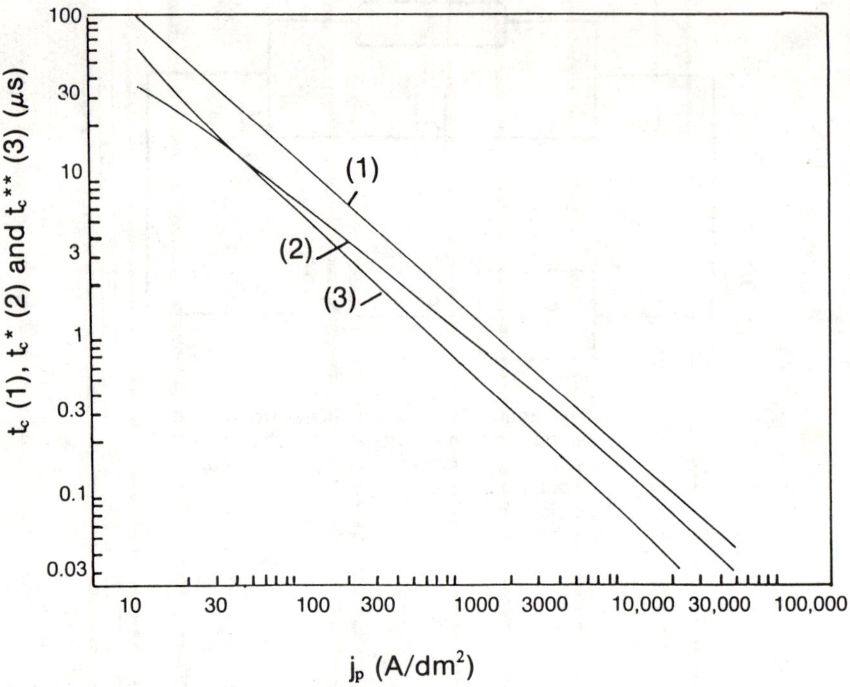

Fig. 4.5— Charging time of the double layer as a function of pulsed current density.
(1) t_c (charging until $j_F = 0.99\ j_p$) (Eqn. [9]);
(2) t_c^* ($j_c = j_t$ and $j_F = 0$ during the charge) (Equation [12]);
(3) t_c^{**} (charging until $\eta = 98.2\%\ \eta_{a\infty}$ and constant R_a (Equation [13]);
$j = 0.5$ A/dm^2, $C = 50\ \mu$F/cm^2, $\alpha n = 1$, temperature = 298 K.

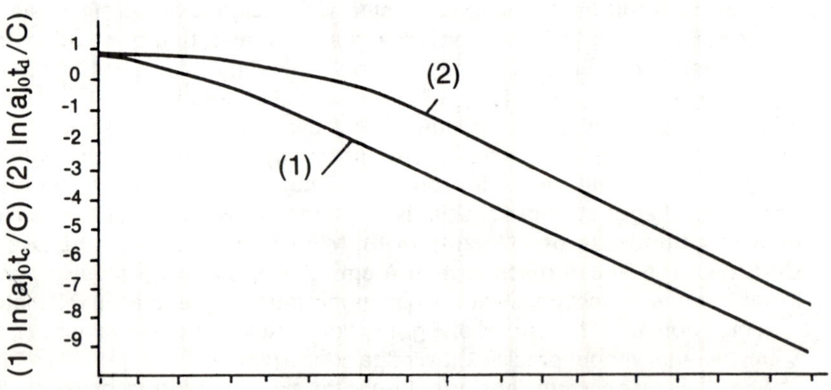

Fig. 4.6—Dimensionless representation of t_c and t_d as a function of the ratio j_p/j_0.
$a = \alpha nF/RT$.

The values t_c and t_d are always proportional to C and to the temperature, and inversely proportional to α and n. The proportionality of t_c and t_d with C is rigorous, with α, n and temperature it is not, but can be accepted with good approximation. The value n is usually known, and a value of 0.5 can reasonably be assigned to α. While C has often been measured on mercury drops, there is relatively limited information referring to solid electrodes. Factors affecting capacitance have been discussed by McMullen and Hackerman (7). Among these, the potential of the electrode is important, C showing different values on either side of the potential of zero charge (pzc). Thus for the system Pt/Na_2SO_4, 1 N, C has values of 22 $\mu F/cm^2$ and 7 $\mu F/cm^2$ at potentials negative and positive to the pzc respectively. The surface state is also important. Aluminum, for example, is easily covered by oxide layers, the capacity of which is very low, and C can vary by a factor of 10 depending on whether or not the electrode has been cathodically pretreated. Capacitance is sensitive to covering of the surface by organic additives and can therefore be influenced by changes in the nature and concentration of such materials which may result from aging of an industrial bath. For the system Cu/Na_2SO_4, 1 N, the curve C = f(E) goes through a sharp minimum at the pzc (21 $\mu F/cm^2$) and rapidly rises to 50 $\mu F/cm^2$ on either side of this. These values are based on the projected surface area, while the true area can be 2 or 3 times larger than this if roughness is taken into account (8). Other measured values of C include 80 $\mu F/cm^2$ for a rough nickel surface (in Ni/H_2SO_4, 1 N) (9), and values ranging from 45 $\mu F/cm^2$ (7) to 100 $\mu F/cm^2$ (10) for a silver surface (in Ag/Na_2SO_4, 1 N). C values measured in connection with the use of equipment for surface area measurement based on capacity effects have been reported as follows (11): Nickel PC 3 (Watts), 25-30 $\mu F/cm^2$; acid copper "CE", 16-17 $\mu F/cm^2$; copper "Ultimal", 39-42 $\mu F/cm^2$; "Aurobond TN", 36 $\mu F/cm^2$; "Kataclad 2N18", 20-40 $\mu F/cm^2$.

It is apparent from the relatively large range of values quoted (10-100 $\mu F/cm^2$) that the assignment of a value for C provides the largest degree of uncertainty in the estimation of t_c and t_d from Fig. 4.6. Although the range of possible values of j_0 is much larger than that of C, the imprecision in the calculation of t_c and t_d is more affected by C than by j_0 because when $j_p \geqslant j_0$, t_c and t_d are no longer dependent on j_0. Several methods for measurement of C are given in Ref. 7, but in the absence of a specific determination a typical value of 40-50 $\mu F/cm^2$ may be assumed. In the linear domain of the curve of Fig. 4.6, t_c and t_d can then be estimated as follows:

$$t_c = 17/j_p \qquad [20]$$

$$t_d = 120/j_p \qquad [21]$$

where t_c and t_d are in microseconds and j_p is in A/cm^2, C being taken as 50 $\mu F/cm^2$.

Equations [20] and [21] allow the estimation of t_c and t_d in the absence of any other knowledge concerning the system involved, and provide a rapid method for the selection of reasonable values of T_{on} and T_{off} in relation to the applied pulsed current j_p. It might be surprising that t_d decreases with increasing j_p. This is related to the choice of the integration limits: $j_F = 0.01\ j_p$. The tail of the curve $j_F = f(t)$ for the discharge is very flat; therefore small variations in the lower limit for j_F lead to important changes in discharging times. A precise knowledge of C and j_0 is necessary for accurate calculation of t_c and t_d, but in most technical

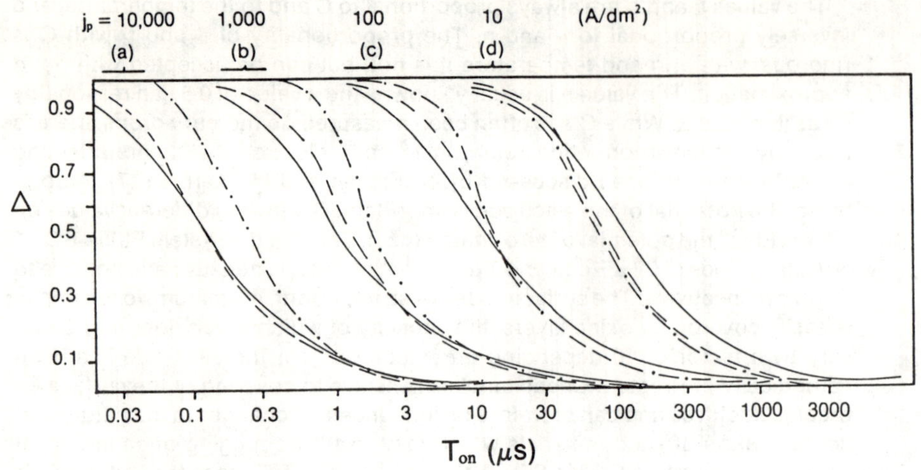

Fig. 4.7—Degree of flattening Δ of the faradic current as a function of j_p, T_{off}/T_{on} and T_{on}.
——— $T_{off}/T_{on} = 1$;
– – – – $T_{off}/T_{on} = 10$;
–·–·– $T_{off}/T_{on} = 1000$.
$j_0 = 0.5$ A/dm², $C = 50\ \mu F/cm^2$, $\alpha n = 1$, temperature = 298 K.

applications only the order of magnitude of these quantities is required as an indication that the pulse is not appreciably disturbed.

3. Degree of flattening of the faradic current

In the foregoing calculation of t_c, t_c^*, t_c^{**}, and t_d, the actual length of the pulse and off-time are ignored, i.e., it is assumed that T_{on} and T_{off} are sufficiently large to allow j_F to reach the value of j_t (Fig. 4.2b). However, this is not always the case (Fig. 4.2c and 4.2d). In Fig. 4.2d the overpotentials corresponding to $j_F = j_t$ and $j_F = 0$ are never reached and the upper and lower limits in the integration of equation [9] are no longer given by j_p alone, but depend on T_{on} and T_{off}, which complicates the problem mathematically.

In the cases of Fig. 4.2c and 4.2d the current for metal deposition is strongly damped and, in particular, in the case of Fig. 4.2d, conditions approach those of DC plating, the advantage of pulsed current virtually being lost. The degree of approach to DC (or departure from a rectangular pulse) may be described by introducing the concept of "degree of flattening" (Δ), defined as:

$$\Delta = \frac{\int_{T_{on}}^{T_{off}} j_F \, dt}{j_m T_{off}} = \frac{j_p T_{on} - \int_0^{T_{on}} j_F \, dt}{j_m T_{off}} \qquad [22]$$

where the numerator represents the amount of metal deposited during the off-time, which is zero when the pulse is not disturbed, and the denominator represents the

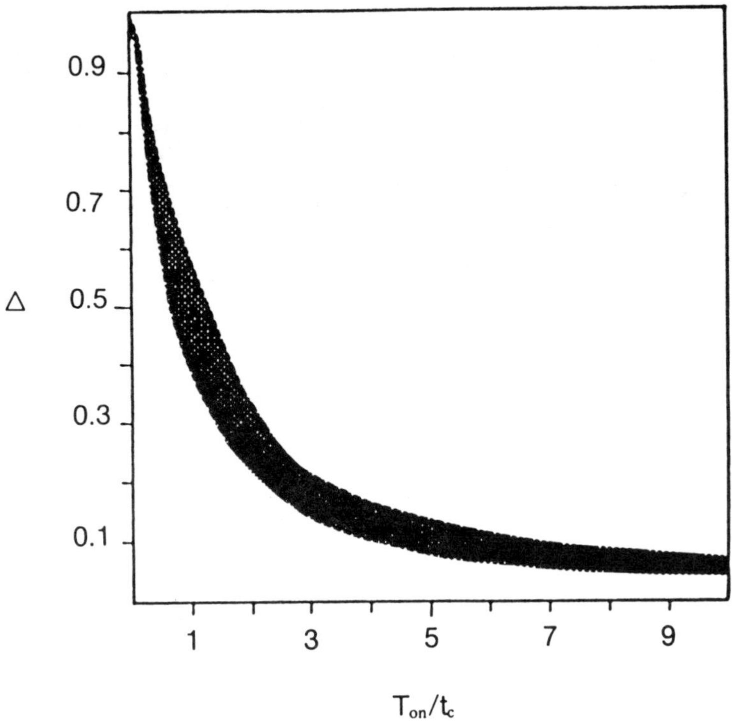

Fig. 4.8—Degree of flattening Δ of the faradic current as a function of T_{on}/t_c (approximation).

amount which would be deposited if a current equal to the average value j_m were to flow during the interval between pulses. In the limiting case, where the pulsed current is completely flattened by capacitive effects, Δ of equation [22] would be equal to 1; where no flattening at all occurs, $\Delta = 0$.

The degree of flattening, Δ, has been computed using equations [1], [2] and [9]. By incrementing the potential, the sum of j_F dt is computed until the sum of dt equals T_{on} but is smaller than 1.001 T_{on}, or until j_F equals j_p, whether the pulsed current corresponds to Fig. 4.2d or to 4.2c respectively. An analogous computation may be made during the off-time, starting with the j_F reached at T_{on} and summing the j_F dt, increment by increment, until the sum of dt equals T_{off} but is smaller than 1.01 T_{off}. The increments of potential were initially 1 mV, but were adjusted towards the end in order to satisfy the above limiting conditions with respect to T_{on} and T_{off}. The potential at the beginning of the pulse was also adjusted in order that the surface area 'c' (Fig. 4.2d), representing the capacitive part of the pulse, should be equal to the area 'd', representing the faradic current during the discharge of the double layer.

In Fig. 4.7 the degree of flattening, Δ, is shown as a function of pulse length. The computation has been made for a given system, with $j_0 = 0.5$ A/dm^2, $RT/(\alpha nF) = 0.026$ V, and $C = 50\,\mu F/cm^2$; four groups of curves are shown, each group corresponding to a given j_p and each curve in a group corresponding to a given ratio T_{off}/T_{on}. When T_{on} decreases from 10 t_c to t_c, Δ increases more and more rapidly, up to a value between 0.4 and 0.6 at $T_{on} = t_c$. At $T_{on} = 0.1\ t_c$, Δ lies between 0.9 and 1, depending on T_{off}/T_{on}. In a

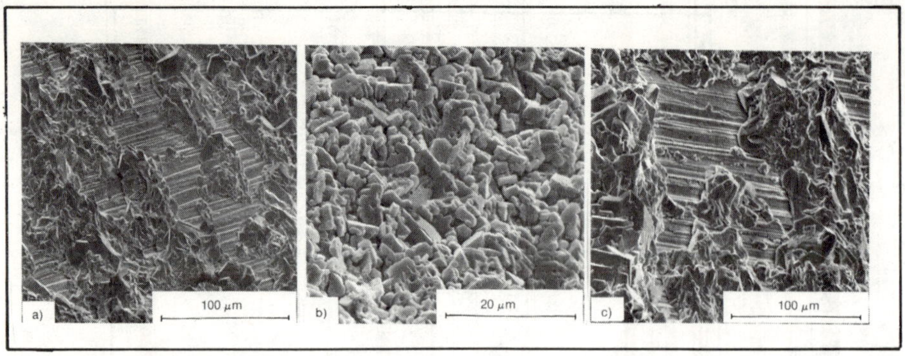

Fig. 4.9—Influence of the damping of the faradic current on the morphology of cadmium deposits. Bath compositions: 0.75 M $CdSO_4$ / 1 M H_2SO_4. Temperature: 20° C. Deposit thickness: 10 μm.

(a) Pulsed current:
$j_p = 100$ A/dm^2, $T_{on} = 0.7$ μs, $j_m = 1.5$ A/dm^2, $\Delta \cong 1$;

(b) Pulsed current:
$j_p = 100$ A/dm^2, $T_{on} = 1000$ μs, $j_m = 1.5$ A/dm^2, $\Delta \cong 0$;

(c) DC: $j = 1.5$ A/dm^2.

group of curves in which j_p is kept constant, the values of Δ do not depend significantly on the ratio T_{off}/T_{on} when $T_{on} > t_c$, which is precisely the range of interest when damping of the faradic current is to be avoided. For higher values of Δ, the influence of the T_{off}/T_{on} ratio is stronger.

For the groups a, b and c ($j_p = 100$, 1,000 and 100 A/dm^2), j_0 has no influence on Δ, which can then be read for different systems under the condition that $RT/(\alpha nF) = 0.026$ V and $C = 50$ μF/cm^2.

Fig. 4.8 shows Δ as a dimensionless function of T_{on}/t_c, a representation which has the advantage of being valid for different values, not only of j_0, but also of C and a (= $\alpha nF/RT$). However, t_c must be known, and may be computed by the method given in Section 2, or estimated from equation [20].

4. Effects of damping of faradic current

As shown by Ibl (1a), damping of the faradic current does not decrease the current efficiency for metal deposition because the charge accumulated during the pulse is recuperated during the off-time. This fact, which constitutes the fundamental hypothesis in the above computation of Δ, has been checked experimentally (Table 2 of Ref. 1) and has also been noted by Baraboshkin and Isaev (12).

The nucleation and relative speed of two simultaneous reactions are, however, strongly influenced by damping effects, since these two phenomena are related to the instantaneous overpotential of the electrode, and therefore to the shape of the faradic response. For $\Delta \cong 0$, $\eta_{a(t)}$ always equals η_{a_∞}, whereas for $\Delta \cong 1$, $\eta_{a(t)} = \eta_a$ corresponding to $j_F = j_m$. The influence of Δ on nucleation is illustrated in Fig. 4.9, showing scanning electron micrographs of the surfaces of cadmium deposits produced under three different conditions, i.e., (a) by pulsed current with Δ near 1, (b) by pulsed current with Δ near 0, and (c) by direct current. Deposit (c) shows the same morphology as (a), where the pulsed current is strongly flattened, whereas deposit (b), where there is no flattening ($\Delta = 0$), shows much finer grain size and good coverage of the substrate, due to the enhanced nucleation, as expected in pulsed current deposition (13).

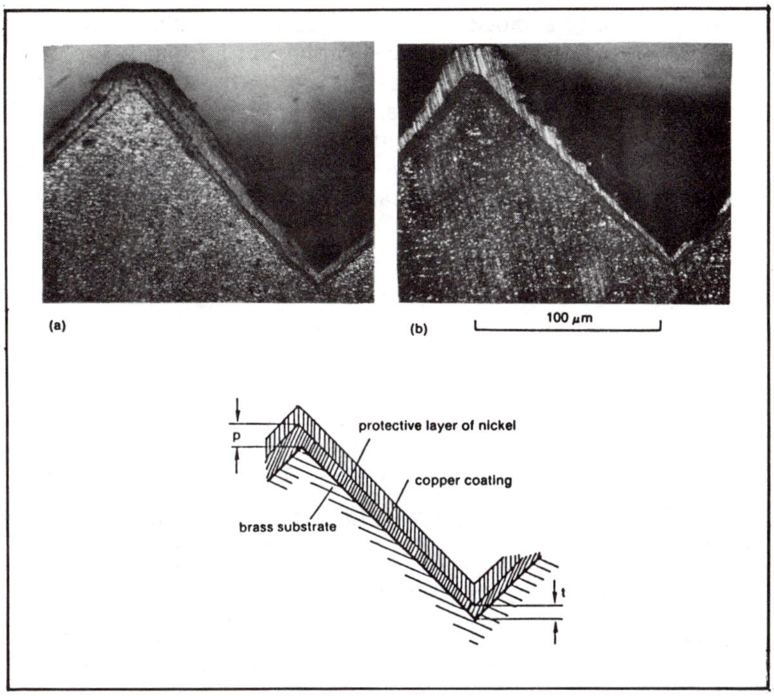

Fig. 4.10—Current distribution on a sawtooth surface.
Bath: 0.5 M $CuSO_4$, 25 ° C;
(a) DC: $j = 3.3$ A/dm^2; $p/t = 4.2$;
(b) Pulsed current: $j_p = 10$ A/dm^2, $T_{on} = 0.2$ μs, $j_m = 3.3$ A/dm^2, $\Delta \cong 1$, $p/t = 2.5$.

Damping of the faradic current can influence current distribution for the following reason. When a faradic current flows between an anode and cathode, the current density on asperities is higher than in recesses due to the additional ohmic drop in the latter. However, when the external current is switched off there is no ohmic drop in the solution although there is still a faradic current due to discharge of the double layer. For a homogeneous distribution of charges over peaks and recesses it is to be expected that the faradic current from the discharge of the double layer would be more uniform over the surface, and therefore the current distribution would be improved by pulsed current with high flattening. Although in the real case the charges are not, in fact, uniformly distributed along the profile of the electrode at the end of a pulse, an improvement in current distribution may still result from the tendency for the charge distribution to equalize during the off-time, a process which may be faster than the charge transfer through the electrode. Each system will behave differently and the characteristic height of asperities will also be a determining factor. For example, for copper deposition from a copper sulfate bath (Fig. 4.10) with Δ near 1, the current distribution is better with pulsed current than with direct current (14), whereas for small values of Δ, which is the most common situation in pulsed current plating, the secondary current distribution is expected to be less uniform than in DC plating.

5. Conclusion

The computations presented in this chapter allow a decision as to whether values of T_{on} and T_{off} in pulse plating have been chosen reasonably in accord with the limitations associated with charge and discharge of the double layer. While precise computation involves a knowledge of electrochemical parameters such as j_0, C, α and n, simpler relationships are available to allow an estimate of the orders of magnitude of t_c, t_d and Δ, which suffice in most practical cases, particularly in avoiding serious disturbances of the pulsed current by capacitive effects at the electrode/solution interface. In practice, especially when working with high currents, other limitations due, for instance, to the length of connecting cables will have to be considered (see Chapter 13).

References

1. J.Cl. Puippe and N. Ibl; *Journal of Applied Electrochemistry,* **10:** 775-784 (1980).

1a. N. Ibl; *Surface Technology,* **10** : 81 (1980)

2. J.O'M. Bockris and A.K.N. Reddy; *Modern Electrochemistry*, Plenum Press, New York : 1195 (1970)
3. M. Braun; *Dissertation ETH 5015,* Zurich (1973)
4. N. Tanaka and R. Tamamushi; *Electrochimica Acta,* **9** : 963 (1964)
5. R. Tamamushi; "Kinetic Parameters of Electrode Reactions", special issue of *Electrochimica Acta* (March, 1972)
6. R. Tamamushi; *Kinetic Parameters of Electrode Reactions of Metallic Compounds,* IUPAC Additional Publication, Butterworths, London (1975)
7. J.J. McMullen and N. Hackerman; *Journal Electrochemical Society,* **106** : 341 (1959)
8. K.J. Vetter; *Elektrochemische Kinetik,* Berlin, Springer-Verlag : 502, 551 (1961)
9. K.J. Vetter; *Elektrochemische Kinetik,* Berlin, Springer-Verlag : 313 (1961)
10. C. Wagner; *Journal Electrochemical Society,* **97** : 72 (1950)
11. E. Robert; *Oberfläche-Surface,* **24,** (12) : 413-420 (1983)
12. A.N. Baraboshkin and V.A. Isaev; *Elektrokhimiya,* **13** : 106 (1977)
13. J.Cl. Puippe and N. Ibl; *Plating and Surface Finishing,* **68,** (6) (1980)
14. B. Sturzenegger; Diploma thesis, ETH Zurich (1980)

Chapter 5

Mass Transport in Pulse Plating

D. Landolt

1. Introduction

Mass transport plays an important role in pulse plating. It limits the maximum rate of deposition, it influences the structure and hence the properties of obtained deposits and it affects the macrothrowing power and the microthrowing power. Under DC conditions the rate of mass transport is determined by prevailing hydrodynamic conditions. In absence of external stirring, free convection phenomena induced by concentration or temperature gradients are the determining factor. If external stirring is applied, e.g. by vibrating or rotating the electrode, by gas stirring or electrolyte pumping, forced convection determines the rate of mass transport at the electrodes. The maximum deposition rate under steady state conditions is given by the limiting current density j_{gg}.

In pulse plating mass transport conditions at the cathode depend on hydrodynamic conditions too, but, in addition, they are strongly influenced by the applied electrical pulse parameters. The theoretical treatment of mass transport in pulse plating, therefore, must take into account nonsteady-state diffusion processes and, for this reason, is mathematically more complicated than for DC. In order to arrive at practically useful predictions, simplified models must be used. It will be shown below that such models are capable of quite accurately describing mass transport in simple metal deposition processes. In other cases, especially in alloy deposition, one may be limited to qualitative predictions of general trends. The most important quantity for mass transport in pulse plating is the pulse limiting current density j_{pg}. It plays a similar role as the limiting current density in DC electrolysis. Its numerical value depends mostly on applied pulse parameters. On the other hand, the maximum average current density in pulse plating is limited by hydrodynamic conditions in a similar way as in DC plating. In the following, we will look at both aspects in more detail.

2. Convective mass transport: limiting current density

Three mechanisms contribute to ionic transport in electrolyte solutions: convection, diffusion and migration (1,2). Convection is bulk movement of the electrolyte and results from the action of external forces such as the mechanical force of a stirrer in forced convection or gravity in free convection. Near solid walls, friction forces cause a slowing down of the electrolyte movement resulting in the establishment of a hydrodynamic boundary layer of thickness δ_v (Fig. 5.1). At the solid surface the electrolyte velocity is zero. Consumption of metal ions at the cathode results in a local decrease of their concentration and thus causes the build-up of a concentration gradient which acts as a driving force for diffusion. At the cathode surface the current

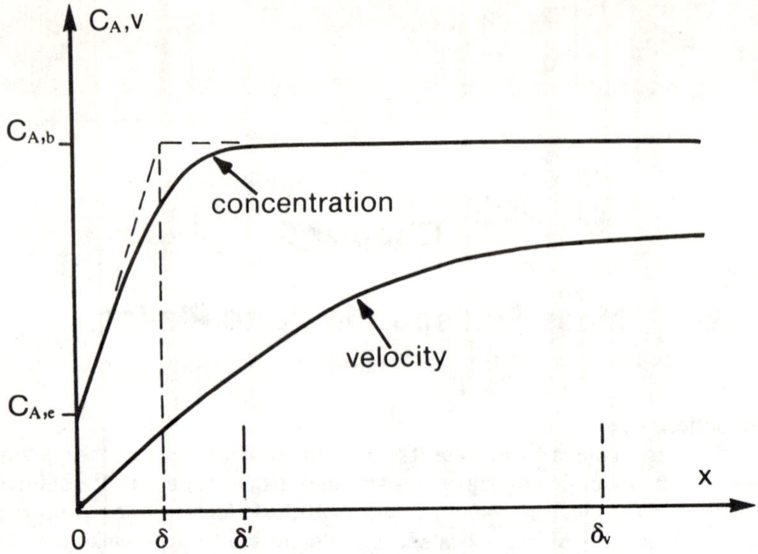

Fig. 5.1—Steady state concentration and velocity profiles at the cathode (schematic).

density is directly proportional to the concentration gradient, the contribution of convection to transport being zero:

$$j = -nF\, D_A \frac{dc_A}{dx} \quad x = 0 \qquad [1]$$

In equation [1] we have neglected the influence of migration which will be discussed below. The charge number n is the number of electrons transferred per reacting metal ion, F is the Faraday constant and D_A is the ionic diffusion coefficient. Further away from the cathode the contribution of convection to transport becomes increasingly important until finally outside the diffusion layer of thickness δ' (Fig. 5.1) it dominates. For practical purposes it is useful to introduce the so-called Nernst diffusion layer thickness δ which is simply the intersection of the linear prolongation of the concentration gradient at x = 0 with the bulk concentration $c_{A,b}$. Equation [1] thus becomes

$$j = -nF\, D_A \frac{c_{A,b} - c_{A,e}}{\delta} \qquad [2]$$

The thickness of the Nernst diffusion layer is always much smaller than that of the hydrodynamic boundary layer δ_v (see Fig. 5.1). It is important to realize that the Nernst diffusion layer thickness is a fictitious quantity which differs in magnitude from that of the real diffusion layer thickness δ'. However, the concept of a Nernst diffusion layer is extremely useful for practical calculations and shall be used here throughout. In physical terms this means we assume that near the cathode a stagnant layer of thickness δ exists in which convection is absent. Outside this layer perfect mixing is achieved through convection and no diffusional effects occur. It may be noted here that under the conditions of interest in electrodeposition, the thickness of the Nernst diffusion layer is always much larger than that of the electrical double layer at the electrode. The latter therefore plays no role for mass transport (double layer charging effects under nonsteady-state conditions are treated in Chapter 4).

We must now briefly consider the influence of the third transport mechanism, ionic migration. It is the result of the existence of an electrical potential gradient in the solution. Depending on the sign of their charge, dissolved ions are attracted towards the anode or the cathode, and the resulting motion contributes to charge transport in the electrolyte. In the bulk solution no concentration gradients exist and the fraction of the total current density carried by an ion A, j_A/j, is equal to the value of the transport number of A, t_A.

$$t_A = \frac{j_A}{j} \qquad [3]$$

The value of t_A depends on the relative concentration of A in the solution and on its mobility (1). Since within the cathodic diffusion layer a concentration gradient of the reacting species exists, the value of the transport number varies with distance from the electrode. This makes it extremely difficult to quantitatively calculate the contribution of migration to overall transport in a general way.

Two special cases can be treated without difficulty, however. The first involves reaction at the electrode of minor ionic species in presence of an excess of supporting electrolyte; for example, deposition of copper from a dilute copper sulfate solution in sulfuric acid. In this case migration contributes negligibly to transport and equations [1] and [2] apply. The second case involves reaction at the electrode of an ionic species present in a binary electrolyte, i.e., an electrolyte consisting of a single salt. Copper deposition from a pure copper sulfate solution is such an example. Here the transport number is constant throughout the diffusion layer and instead of equations [1] and [2] the equations become:

$$j = -\frac{nF\,D_s}{1-t_A}\,\frac{d\,c_A}{d\,x}\bigg|_{x=0} \qquad [4]$$

$$j = -\frac{nF\,D_s}{1-t_A}\,\frac{c_{A,b} - c_{A,e}}{\delta} \qquad [4a]$$

In these equations D_s is the salt diffusion coefficient (1) and t_A the transport number of the reactive ion. In practical electroplating one rarely deals with binary electrolytes. For this reason, and to avoid the difficulty of accounting for migration, one usually works with equations [1] and [2]. In an empirical way one can take care of the influence of migration by defining an effective diffusion coefficient D_{eff} replacing D_A in equations [1] and [2]. The value of D_{eff} can be estimated experimentally from limiting current measurements under well defined hydrodynamic conditions using the same electrolyte. Generally, it may be said that if an anion, for example, a negatively charged metal complex, reacts at the cathode, migration will tend to decrease the value of the limiting current density and if a positively charged ion reacts, it will tend to increase it. In the worst case, the effect of migration on the limiting current density may amount to a factor of two or three (1), but usually in practical plating applications it is much smaller. Following usual practice, in subsequent paragraphs we shall neglect migration effects.

The maximum rate of mass transport in a given system corresponds to the limiting current density j_{gg} at which the surface concentration of the reacting species at the electrode is zero. Therefore, with equation [2]:

$$j_{gg} = -nF\, D_A \frac{c_{A,b}}{\delta} \qquad [5]$$

It follows that the limiting current density is proportional to the bulk concentration of the reacting species and inversely proportional to the Nernst diffusion layer thickness δ. The value of the latter depends on convection conditions. For many electrochemical cell designs and hydrodynamic conditions, the value of the limiting current density can be estimated from dimensionless correlations of the type Sh = f (Re, Sc) where the Sherwoood number

$$Sh = \frac{j_{gg}\, L}{nF\, D_A\, c_{A,b}}$$

is a dimensionless limiting current density, the Reynolds number, $Re = \frac{vL}{\nu}$ is a dimensionless flow velocity and the Schmidt number, $Sc = \nu/D_A$, expresses the ratio between the value of the kinematic viscosity ν and the diffusion coefficient. The characteristic length L characterizes the geometry of the cell. A collection of such correlations can be found in Ref. 3.

3. Single current or potential steps

In this paragraph we are interested in the concentration changes induced at the cathode by the application of a single current or potential step. It is assumed that the Nernst diffusion layer model is applicable and that double layer charging effects are negligible. Figures 5.2 and 5.3 show qualitatively the expected behavior. Let us assume that a constant current step is applied in such a way that the value of the current density exceeds the steady state limiting current density given by equation [2]. According to equation [1] a constant gradient is imposed at the cathode and the surface concentration of the reacting species decreases with time until it reaches zero at $t = \tau$. The time τ is called transition time (4). It manifests itself experimentally by a sudden increase of the electrode potential which leads to the onset of an additional cathodic discharge reaction such as hydrogen evolution. The part of the total current which is carried by reaction of the metal ion A diminishes because the gradient at the electrode surface decreases. Eventually it reaches the steady state value for DC. The behavior is demonstrated by the schematic concentration profiles in Fig. 5.2: For $t < \tau$ the slope of $c = f(x)$ is constant; for $t > \tau$ it diminishes until the steady state concentration profile is reached. If the value of the applied current density step is inferior to the DC limiting current density no transition time is observed, of course, and no second reaction occurs.

The situation is quite different if a potential step is applied (5). The electrode potential fixes the surface concentration of the reacting species which, therefore, is constant. This leads to very high initial concentration gradients at the cathode, but as the region near the surface becomes depleted, the gradient and hence the current density decreases and eventually reaches a steady state. The behavior is illustrated schematically in Fig. 5.3 for a potential step leading to zero surface concentration. If the applied potential were less negative than that necessary to reach zero concentration at the surface, a finite surface concentration value $c_{A,e}$ would be established and the steady state current at infinitely long times would be below the DC limiting current.

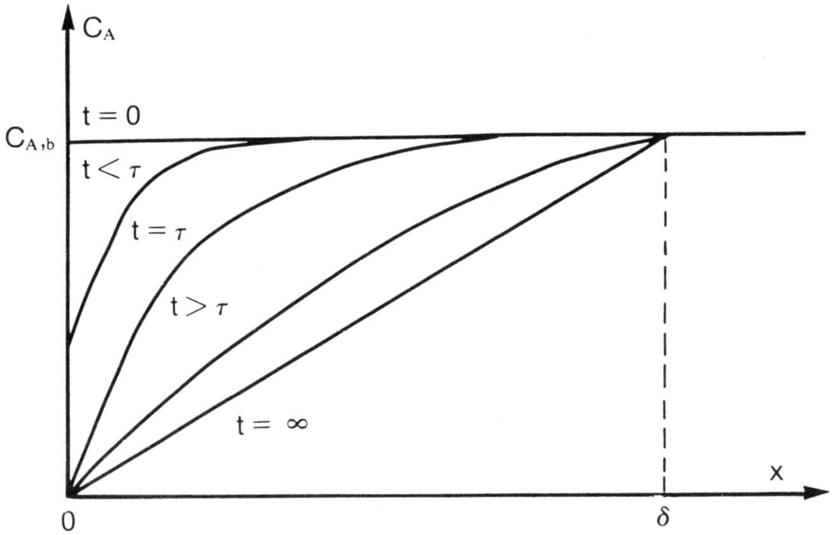

Fig. 5.2—Concentration profiles at the cathode upon application of a constant current step (schematic).

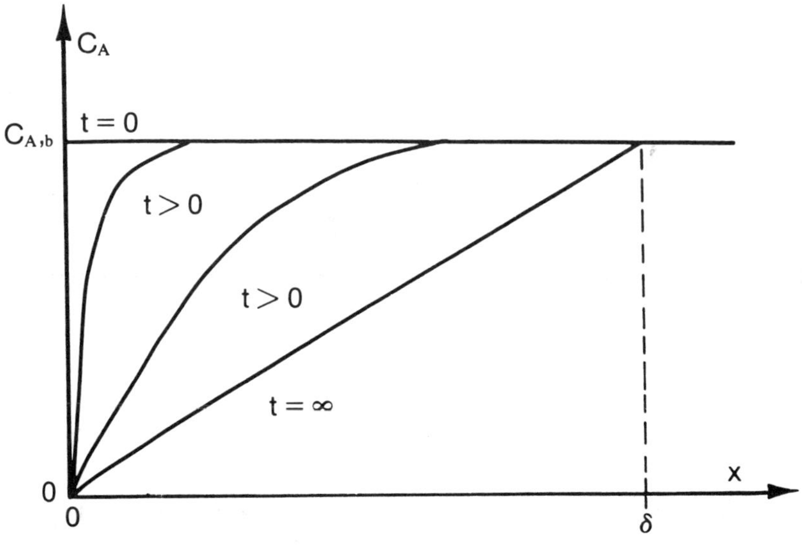

Fig. 5.3—Concentration profiles at the cathode upon application of a constant potential step in the limiting current region (schematic).

The mathematical treatment of the described behavior involves solving Fick's second law of diffusion within the Nernst diffusion layer:

$$\frac{\partial c_A}{\partial t} = D_A \frac{\partial^2 c_A}{\partial x^2} \qquad [6]$$

The initial condition is

$$t = 0 \qquad x \geqslant 0 \qquad c_A = c_{A,b} \qquad [6a]$$

At the outer boundary of the Nernst diffusion layer one has at all times:

$$x = \delta \qquad c_A = c_{A,b} \qquad [6b]$$

The boundary condition at the cathode surface upon application of a current step is

$$x = 0 \qquad \frac{\partial c_A}{\partial x} = - \frac{j}{nF\, D_A} = \text{constant} \qquad [6c]$$

and upon application of a constant potential

$$x = 0 \qquad c_A = c_{A,e} \qquad [6d]$$

The constant current problem was solved by Rosebrugh and Miller in 1910 (6). The decrease of the surface concentration $c_{A,e}$ with time is given by eqn. [7]:

$$c_{A,e} = c_{A,b} - \frac{j\,\delta}{n_A F\, D_A}\, f_{(t)} \qquad [7]$$

where $f_{(t)}$ is a dimensionless time function which for small values of $\frac{Dt}{\delta^2}$ can be approximated (7):

$$f_{(t)} = \frac{2}{\sqrt{\pi}} \sqrt{\frac{Dt}{\delta^2}} \qquad [\frac{Dt}{\delta^2} < 0.1] \qquad [7a]$$

For larger values of $\frac{Dt}{\delta^2}$:

$$f_{(t)} = 1 - \frac{8}{\pi^2} \exp\left(-\frac{\pi\, Dt}{4\,\delta^2}\right) \qquad [\frac{Dt}{\delta^2} > 0.1] \qquad [7b]$$

It is useful to define a dimensionless transition time τ^* and a dimensionless current density j^* :

$$\tau^* = \frac{D\tau}{\delta^2} \qquad [8]$$

$$j^* = \frac{j}{j_{gg}}$$

Introducing these expressions into eqn. [7] [7a] [7b] and setting $c_e = 0$, $t = \tau$ one gets for $\tau^* < 0.1$:

$$\tau^* = \frac{\pi}{4\, j^{*2}} \qquad [9]$$

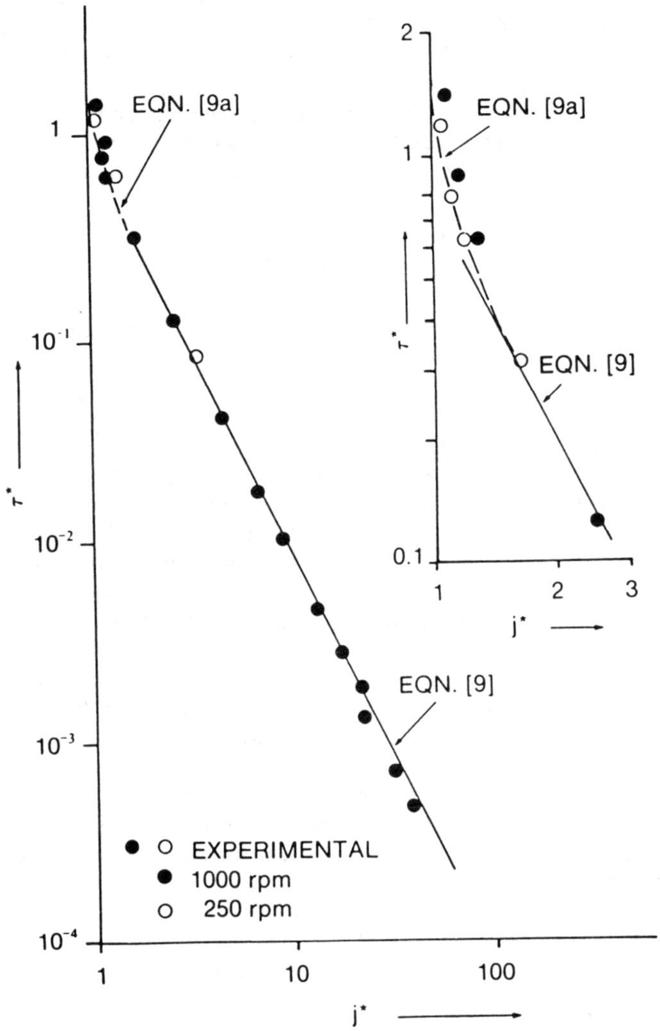

Fig. 5.4—Experimental and theoretical values of the dimensionless transition time as a function of the dimensionless current density $j^* = j/j_{gg}$ for single galvanostatic pulses (8).

and for $\tau^* > 0.1$:

$$\tau^* = -\frac{4}{\pi^2} \quad \ln \left[\frac{\pi^2}{8} \left(1 - \frac{1}{j^*}\right) \right] \qquad [9a]$$

Figure 5.4 shows experimental results for copper deposition by Datta and Landolt (8) which agree well with equations [9] and [9a]. It may be noted that equation [9] is equivalent to the well-known Sand equation for the transition time in a stagnant electrolyte (4).

The mathematical solution of eqn. [6] for the application of a constant potential step has been discussed by Vetter (5). For small times one obtains eqn. [10] which is equivalent to the well known Cottrell equation for stagnant solutions (5).

$$\frac{j}{j_\infty} = \sqrt{\frac{\pi\,\delta^2}{Dt}} \qquad \left[\frac{tD}{\delta^2} < 0.1\right] \tag{10}$$

Equation [10] indicates that the decrease in current density is inversely proportional to the square root of time.

4. Pulse limiting current density

Constant current pulses of the type shown in Fig. 1.1a, have been studied most widely. The basic mathematics describing mass transport for this case have been published by Rosebrugh and Miller (6). Their treatment has been applied by Cheh (9) to the study of pulse plating of gold. A different mathematical approach has been used by Chin (10), while Ibl has given a simple semi-quantitative treatment based on the concept of a duplex diffusion layer with linear concentration profiles (11). More recently, Viswanathan and Cheh (13) published a mathematical solution for arbitrary pulse forms.

The most important quantity to be derived is the pulse limiting current density j_{pg} which is the pulse current density at which the surface concentration reaches zero *at the end of a pulse.* This definition implies that the value of j_{pg} depends on the choice of pulse parameters, in particular the pulse time T_{on}. The longer the pulse time, the smaller is the value of j_{pg}. As will be shown below, the value of j_{pg} also depends on T_{off} and on the ratio j_m/j_{gg}, j_m being the average current density. The mathematical calculation of j_{pg} involves solving equation [6] with the boundary condition [11] where m is the number of pulses applied.

$$\begin{aligned} &(m-1)\,(T_{on} + T_{off}) < t < m\,T_{on},\ x = 0: \quad \frac{\partial c_A}{\partial x} = -\frac{j_p}{nF\,D_A} \\ &mT_{on} < t < m\,(T_{on} + T_{off}),\ x = 0: \frac{\partial c_A}{\partial x} = 0 \end{aligned} \tag{11}$$

Rosebrugh and Miller (6) showed that after only a few current pulses the concentration fluctuations at the cathode become independent of m. This is illustrated by Fig. 5.5, in which the calculated concentration fluctuations of reacting species at the cathode surface are shown in dimensionless form for the first pulses and for stationary conditions.

Using Rosebrugh and Miller's general solution, Cheh (9) has calculated the pulse limiting current density as a function of the dimensionless pulse period for different duty cycles. His results are presented in Fig. 5.6. They show that the value of the pulse limiting current density may far exceed that of the DC limiting current density. The value of j_{pg} decreases with increasing duty cycle and with increasing pulse period. These results have been confirmed by Chin (10). Chin's data are shown in Fig. 5.7 together with experimental results obtained for copper deposition by Datta and Landolt (8). The agreement between theory and experiment is excellent.

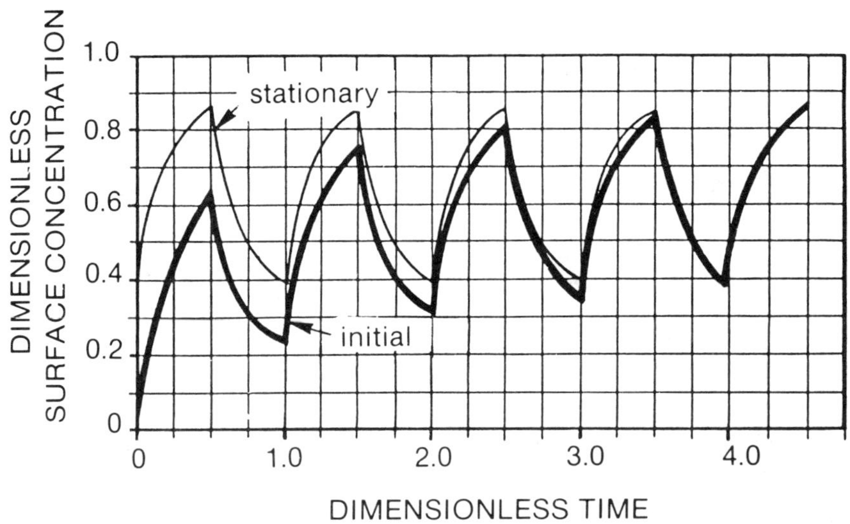

Fig. 5.5—Variation of surface concentration with time at the beginning of a pulse plating experiment calculated by Rosebrugh and Miller [6]. The dimensionless surface concentration shown is

$$c^*_A = \frac{(c_{A,b} - c_{A,e})\, nFD\pi^2}{8\, j_p\, \delta}$$

and the dimensionless time is $t^* = \dfrac{\pi^2\, Dt}{4\, \delta^2}$

A simple approximate model of the mass transport processes in constant current pulse electrolysis was developed by Ibl (11). He considered the diffusion layer to consist of a pulsating and a stationary part as shown in Fig. 1.3. Assuming linear concentration profiles in each part one obtains by simple mass balance arguments:

$$\frac{j_{pg}}{j_{gg}} = [\ \frac{\delta_p}{\delta}(1-\gamma) + \gamma)\ ]^{-1} \qquad [12]$$

where δ_p is the thickness of the pulsating diffusion layer. Introducing the expression for δ_p given by Ibl (11):

$$\delta_p = \sqrt{2\, D\, T_{on}\, (1-\gamma)} \qquad [13]$$

This yields

$$j^*_{pg} = \frac{j_{pg}}{j_{gg}} = [\ \sqrt{\frac{2\, D\, T_{on}}{\delta^2}}\ (1-\gamma)^{1.5} + \gamma]^{-1} \qquad [14]$$

Values of j^*_{pg} calculated from equation [14] are also shown in Fig. 5.7. They lie slightly below those obtained from the more exact solution. A better fit is obtained setting for δ_p (8):

$$\delta_p = \sqrt{\frac{4}{\pi} D\, T_{on} (1-\gamma)} \qquad [15]$$

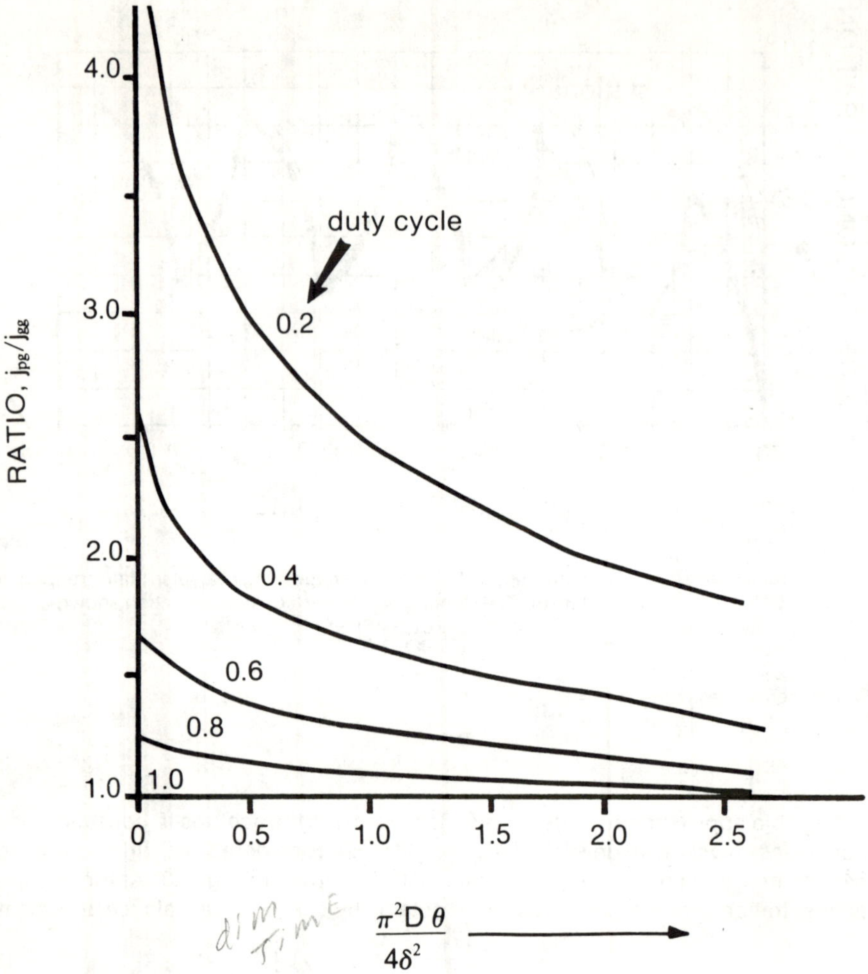

Fig. 5.6—Ratio of pulse limiting current density to DC limiting limiting current density as a function $\frac{\pi^2 D\theta}{4\,\delta^2}$ for different duty cycles calculated by Cheh (9).

yielding

$$j^*_{pg} = [\sqrt{\frac{4}{\pi}\,\frac{DT_{on}}{\delta^2}}\quad (1-\gamma)^{1.5} + \gamma\,]^{-1} \qquad [16]$$

Numerical data obtained with equation [16] are also shown in Fig. 5.7. It is concluded that the use of simple equations is usually adequate for calculating j_{pg} in practical pulse plating applications.

The average limiting current density in pulse plating is given by

$$j_{mg} = \gamma\, j_{pg} \qquad [17]$$

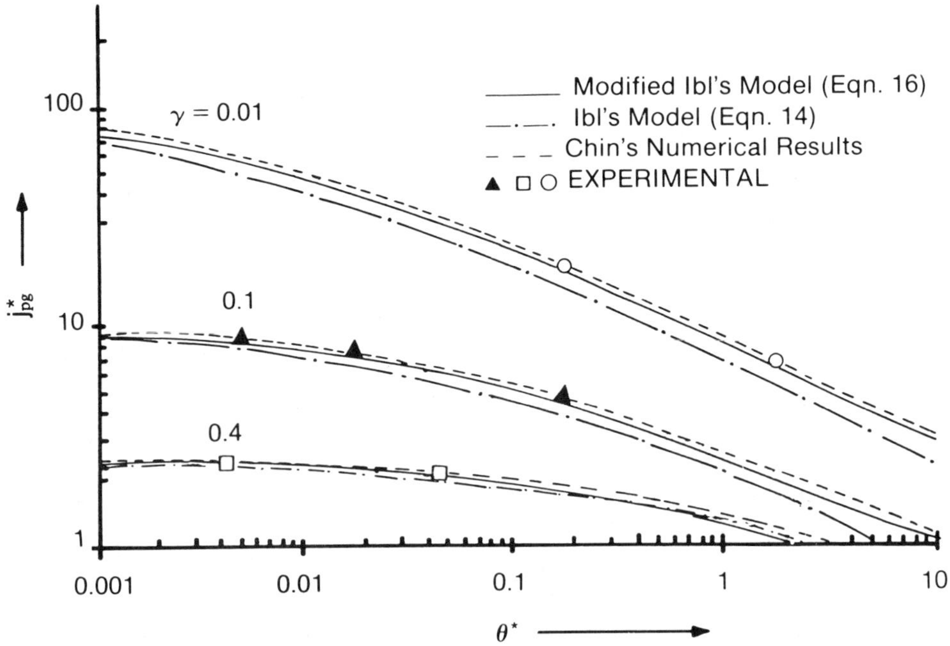

Fig. 5.7—Dimensionless pulse limiting current density $j^*_{pg} = \frac{j_{pg}}{j_{gg}}$ as a function of the dimensionless pulse period $\theta^* = \frac{D\theta}{\delta^2}$, after (8).

Its value is always smaller than that of the DC limiting current density. This is illustrated by Fig. 5.8 (9).

The influence of pulse reversal on the value of j_{pg} has been studied by Cheh (12) and by Chin (10). Chin's results for two different pulse forms are presented in Fig. 5.9 and 5.10. The dimensionless pulse limiting current density is shown as a function of the dimensionless pulse period for different cathodic duty cycles. The anodic pulse current density is kept constant at 1/3 (Fig. 5.9) or 1/10 of the value of j_{gg}. The anodic duty cycle in Fig. 5.10 is 0.1. The data illustrate the influence of pulse reversal on mass transport. During the anodic pulse the concentration of the reacting ions at the electrode surface increases due to dissolution. As a consequence a higher cathodic current density is necessary to deplete the region near the electrode during the cathodic cycle. The value of j_{pg} with pulse reversal is, therefore, higher than that corresponding to the use of cathodic pulses only. This can be seen by comparing the data of Fig. 5.9 and 5.10 with those of Fig. 5.6. On the other hand, the average deposition rate in pulse reversal plating is lower than in normal pulse plating. This is understandable because part of the metal which deposits during the cathodic cycle redissolves during the anodic cycle. In the limit the anodic reaction may even dominate, i.e., more metal is dissolved than deposited. The broken lines in Fig. 5.9 and 5.10 represent the described limit between a net cathodic and a net anodic reaction behavior. In pulse reversal plating the number of electrical parameters that can be chosen freely to optimize deposit properties is higher than in normal pulse plating.

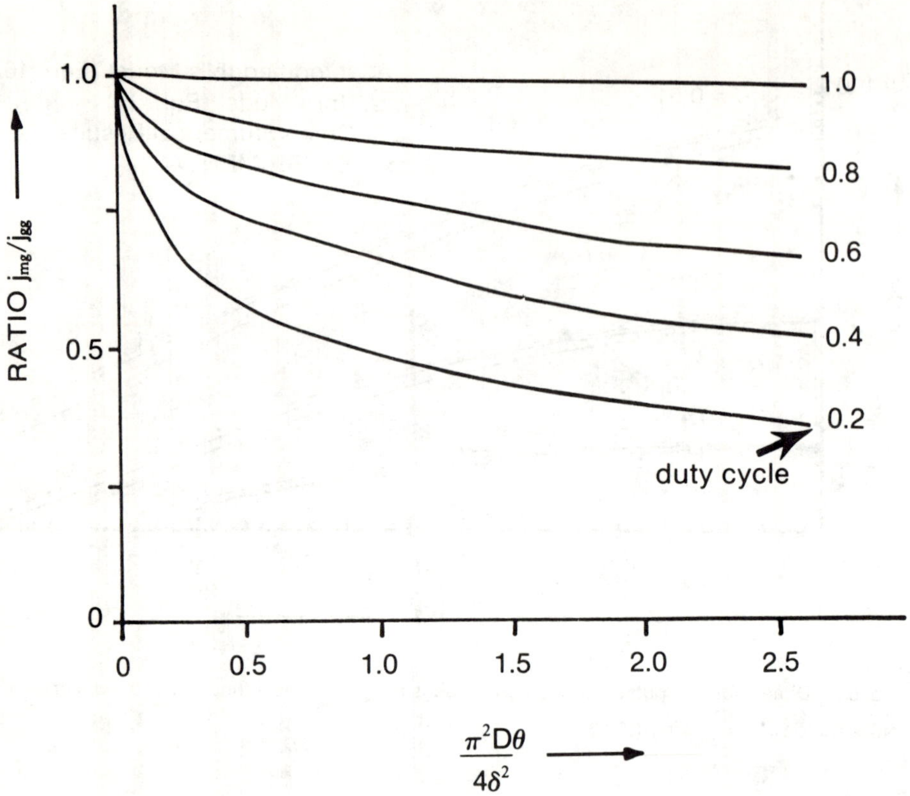

Fig. 5.8—Ratio of average value of limiting current density in pulse plating to DC limiting current density as a function of $\frac{\pi^2 D\theta}{4\delta^2}$ for different duty cycles γ calculated by Cheh (9).

However, the theoretical analysis of prevailing mass transport conditions is greatly complicated and usually requires numerical methods.

5. Practical aspects

5.1 *Deposit structure*

In DC plating dendritic or powdery deposits are obtained if deposition is carried out at or near the limiting current density (16). This is due to the fact that asperities on a surface are more accessible than recesses from a mass transport point of view. The applied current density in DC plating, therefore, usually does not exceed 10-20 percent of the value of j_{gg}. The situation in pulse plating is somewhat different because smooth deposits may be obtained even at the pulse limiting current, provided the pulsating diffusion layer thickness is extremely small (17). Ibl and coworkers verified this behavior for copper (23, 24, 25), silver

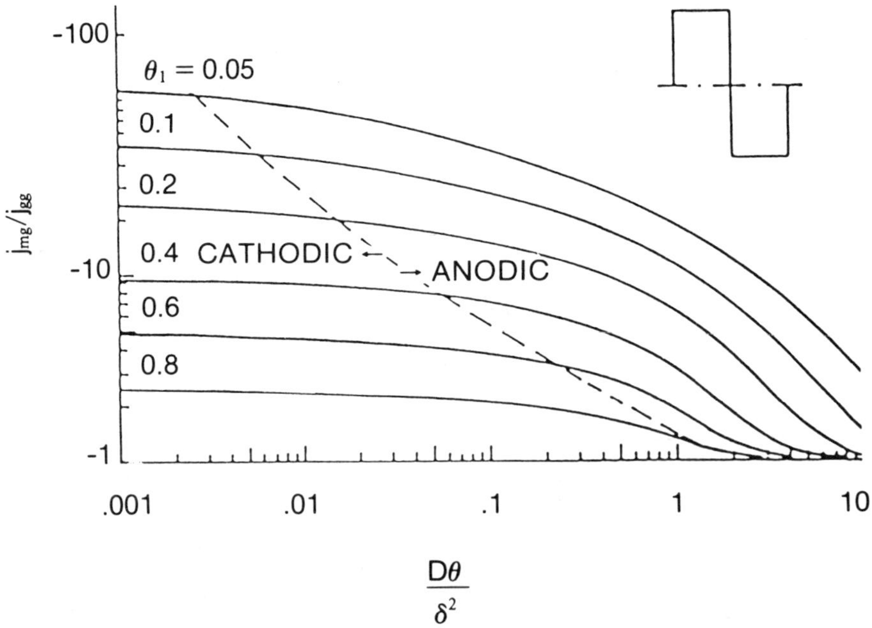

Fig. 5.9—Ratio j_{mg}/j_{gg} as a function of the dimensionless pulse period $\frac{D\theta}{\delta^2}$ for different cathodic duty cycles calculated by Chin (10).

(23, 24) and cadmium (26). If the pulsating diffusion layer is smaller than the original surface rugosity, it espouses the surface profile rendering all points equally accessible (17). Despic, Popov and coworkers (15, 18), using the same reasoning, were able to explain the observed influence of pulse frequency on deposit smoothness. Dendrites can grow only when under low frequency conditions the pulsating diffusion layer can diverge from the surface profile (15). As in DC plating, the average current in pulse plating should stay well below the DC limiting current density j_{gg} if dendrite and powder formation is to be avoided (19). This is illustrated for the case of copper deposition from a copper sulfate/sulfuric acid electrolyte in Fig. 5.11 (20). On the ordinate is shown the ratio j_p/j_{pg} and on the abscissa, the ratio j_m/j_{gg}. It follows that when the latter exceeds a value of ~0.5, powdery noncompact deposits are formed.

5.2 *Current efficiency*

Mass transport may affect current efficiency. In DC plating the current efficiency for metal deposition drops as the limiting current density is exceeded. The same behavior is to be expected in pulse plating. While for pulse plating of pure metals one will usually stay below the limiting current, this is not always possible in alloy plating. For example, during deposition of silver-palladium alloys from an ammonium-based electrolyte, silver deposits under limiting current conditions (21). The influence of pulse parameters on alloy composition

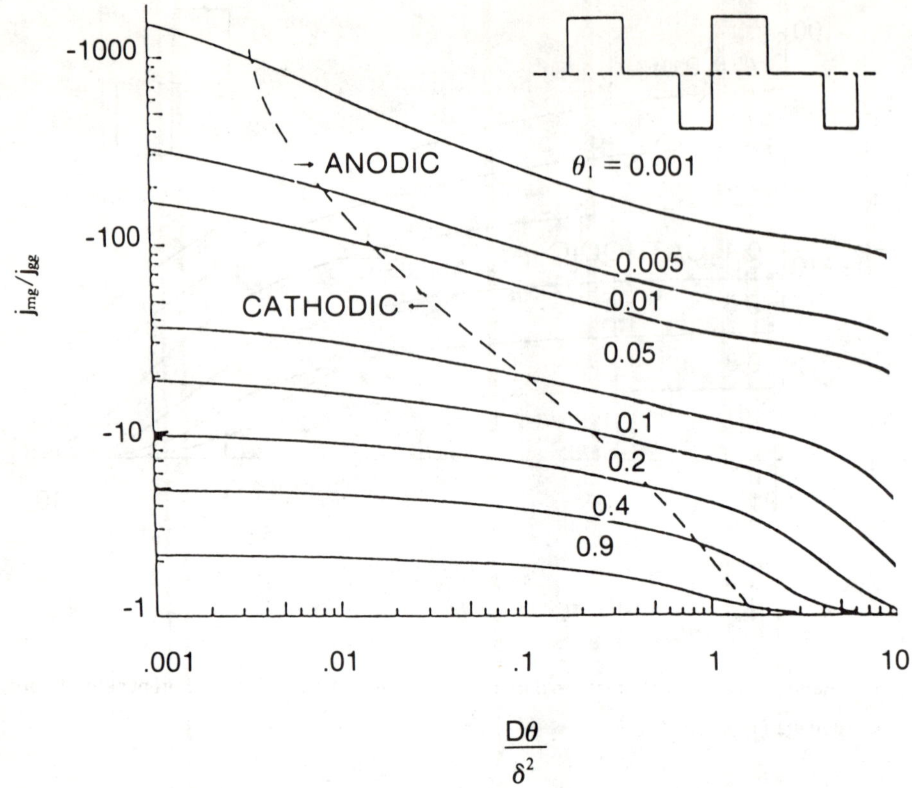

Fig. 5.10—Ratio j_{mg}/j_{gg} as a function of the dimensionless pulse period $\frac{D\theta}{\delta^2}$ for different cathodic duty cycles calculated by Chin (10).

for a system where one component deposits under diffusion control and the other under activation control, has recently been discussed theoretically using the duplex diffusion layer approach (14). Let us assume that deposition of component A of a binary alloy is diffusion controlled at the potential where both A and B deposit and that deposition of B is activation controlled and occurs in the Tafel region. Under DC conditions one therefore has

$$j_{A,gg} = - n_A F D_A \frac{c_A}{\delta} \qquad [18]$$

$$j_B = - k_1 \exp (k_2 E) \qquad [19]$$

Here k_1 and k_2 are constants. The total applied current, provided no other reactions take place, is

$$j = j_{A,gg} + j_B \qquad [20]$$

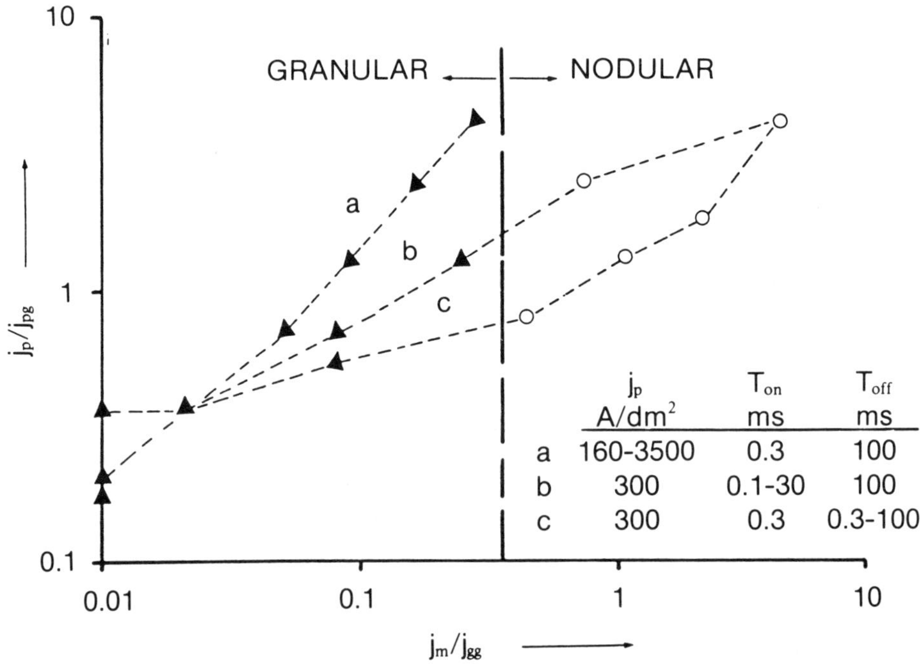

Fig. 5.11—Influence of pulse parameters on deposit morphology for copper deposition from a copper sulfate/sulfuric acid electrolyte (20).

and the current efficiency for the deposition of A is

$$\theta_{A,DC} = \frac{j_{A,gg}}{j} \qquad [21]$$

Note that the concentration of A in the deposited alloy will be directly proportional to θ_A.

In pulse plating the concentration of A in the deposited alloy will depend on the average current efficiency for A, which is given by eqn. [22].

$$\theta_A = \frac{j_{A,m}}{j_m} = \frac{j_{A,m}}{j_{A,m} + j_{B,m}} \qquad [22]$$

Setting

$$j_{A,m} = \gamma\, j_{A,pm} \qquad [23]$$

where the mean pulse current density during the on-time is

$$j_{A,pm} = \frac{1}{T_{on}} \int_0^{T_{on}} j_{A,p}\, dt$$

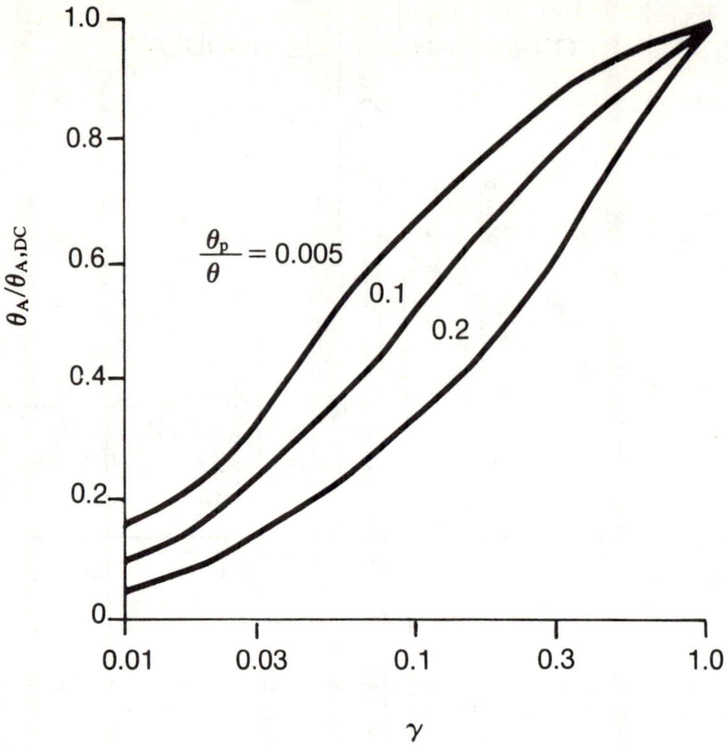

Fig. 5.12—Ratio of current efficiency for the deposition of a substance A in pulse and in DC plating as a function of the duty cycle γ, calculated from equation [26], after reference (14).

One obtains the ratio of the current efficiency in pulse plating to that in DC plating:

$$\frac{\theta_A}{\theta_{A,DC}} = \frac{\gamma\, j_{A,pm}}{j_{A,gg}} \qquad [24]$$

For constant current pulses with amplitudes not very much greater than the pulse limiting current density of A, one may set as a first approximation

$$j_{A,pm} = j_{A,pg} \qquad [25]$$

With eqn. [12] it then follows

$$\frac{\theta_A}{\theta_{A,DC}} = \frac{1}{1 + \frac{\delta_p}{\delta_N}\left(\frac{1}{\gamma} - 1\right)} \qquad [26]$$

The variation of the ratio $\frac{\theta_A}{\theta_{A,DC}}$ calculated according to equation [26] is shown as a function of duty cycle γ for different values of δ_p/δ in Fig. 5.12. It follows from this that the concentration of A in the alloy is always smaller in pulse plating than under DC conditions and depends on applied pulse parameters.

Codeposition of two species both of which are under mass transport control has been investigated theoretically by Viswanathan, Epstein and Cheh (22) for the case where both species are present in equal concentrations. Their results are in qualitative agreement with the above considerations in that the species depositing at less negative potentials is expected to be present in the alloy at lower concentration in pulse plating than in DC plating. If deposition of this species is an undesirable side reaction purer deposits should result from pulse plating compared to DC plating. A general mathematical model for the periodic electrodeposition of alloys has very recently been published by Verbrugge and Tobias (27).

5.3 *Throwing power*

Mass transport may affect macrothrowing power and microthrowing power in DC as well as in pulse plating because it influences the current distribution. As a general rule one would expect that pulse plating at or close to the pulse limiting current improves the macrothrowing power as well as the microthrowing power. Few systematic studies exist, however, in this area. The subject of current distribution in pulse plating will be discussed in more detail in Chapter 6.

References

1. J. Newman; *Electrochemical Systems,* Prentice Hall : 215 (1973)
2. N. Ibl, O. Dossenbach; *Comprehensive Treatise of Electrochemistry,* Vol. 6, E. Yeager Plenum Publ. : 133 (1982)
3. J.R. Selman, Ch. W. Tobias; *Advances in Chemical Engineering,* Vol. 10, T.B. Drew et. al. editors, Academic Press : 211 (1978)
4. K.J. Vetter; *Electrochemical Kinetics,* Academic Press : 205 (1967)
5. K.J. Vetter; *Electrochemical Kinetics,* Academic Press : 213 (1967)
6. T.R. Rosebrugh, W.L. Miller; *Journal of Physical Chemistry,* **14** : 816 (1910)
7. Y.G. Siver; *Russian Journal of Physical Chemistry,* **34** : 273 (1960)
8. M. Datta, D. Landolt; submitted to *Surface Technology.*
9. H.Y. Cheh; *Journal Electrochemical Society,* **118** : 551 (1971)
10. D.T. Chin; *Journal Electrochemical Society,* **130** : 1657 (1983)
11. N. Ibl; *Surface Technology,* **10** : 81 (1980)
12. H.Y. Cheh; *Journal Electrochemical Society,* **118** : 1132 (1971)
13. K. Viswanathan, H.Y. Cheh; *Journal Applied Electrochemistry,* **10** : 37 (1980)
14. D. Landolt; *Oberfläche/Surface,* **25** : 6 (1984)
15. A. Despic, K.I. Popov; *Journal Applied Electrochemistry,* **3** : 161 (1973)
16. N. Ibl; *Advances in Electrochemistry and Electrochemical Engineering,* Vol. 2, Ch. W. Tobias Editor, Interscience Publ. : 49 (1962)
17. N. Ibl, J.Cl. Puippe, H. Angerer; *Surface Technology,* **6** : 287 (1978)
18. K.I. Popov, M.D. Maksimovic, B.M. Ocokoljic; *Surface Technology,* **11** : 99 (1980)
19. D. Landolt; *Proceedings 11th World Congress Metal Finishing* (Jerusalem 1984), p. 1
20. O. Chene, M. Datta, D. Landolt; *Oberfläche/Surface,* **26** : 45 (1985)
21. B. Sturzenegger; *Dissertation ETH 7605,* Zurich (1984)
22. K. Viswanathan, M.A. Farell Epstein, H.Y. Cheh; *Journal Electrochemical Society,* **127** : 2383 (1980)
23. N. Ibl, M. Braun; *Chemie Ingenieur Technik,* **45** : 182 (1973)
24. M. Braun; *Dissertation ETH 5015,* Zurich (1973)
25. H.J. Schenk; *Dissertation ETH 5566,* Zurich (1975)
26. J.Cl. Puippe; *Dissertation ETH 6225,* Zurich (1978)
27. M.W. Verbrugge, Ch. W. Tobias; *Journal Electrochemical Society,* **132** : 1298 (1985)

Chapter 6

Current Distribution in Pulse Plating

O. Dossenbach

1. Introduction

In electrochemical processes the potential difference across the interface solution-electrode as well as the local current density vary in general along the electrode surface. It is common experience that the current density is usually higher on protruding parts of an electrode, or at its edges if the width of the cell is larger than that of the electrode (Fig. 6.1). The reason for this is that at such spots the cross section available for the flow of charges in the electrolyte is on the average larger, and consequently, the resistance to current flow lower.

Potential and current distribution are of decisive importance in applied electrochemistry. They are related to problems such as current efficiency, product selectivity, and utilization of an electrolysis cell. In metal deposition the current distribution determines the variation of the deposit thickness along the cathode surface, and it can have an influence on the roughness of the coating. In anodic shaping (electrochemical machining) and surface treatment (electrochemical polishing) processes, current distribution phenomena are important for surface finish and dimensional control. Current distribution is thus closely associated with concepts of practical importance, such as throwing power or levelling.

Potential and current distribution are governed by phenomena related to charge transport in the electrodes and to charge and mass transport in the electrolyte. Therefore, a general theoretical treatment involves the integration of the fundamental equations of mass and charge transport. However, this leads to very complex mathematical problems, and solutions have been obtained mainly for special cases: In *primary* current distribution, the influence of any kind of overpotential at the electrode is neglected. Current distribution is governed by geometrical factors. In *secondary* current distribution, the influence of activation overpotential is taken into account. Solutions for this situation, and especially for the more general case of *tertiary* current distribution, where both activation and concentration overpotential play a role, have become more easily accessible through the use of computers. Usually the assumption is made that the electrode has a very good conductivity so that it has a uniform electric potential. The reader interested in the fundamentals of current distribution is referred to Newman's book (1) and to Ibl's review articles (2-4).

A theoretical treatment of current distribution will not be given here. Qualitative considerations shall allow us to work out the characteristic features of the different types of current distribution and the specific influences of the relevant parameters. The effects which are expected to be brought about by a pulsating current shall be discussed and compared with the results of experimental investigations. We shall

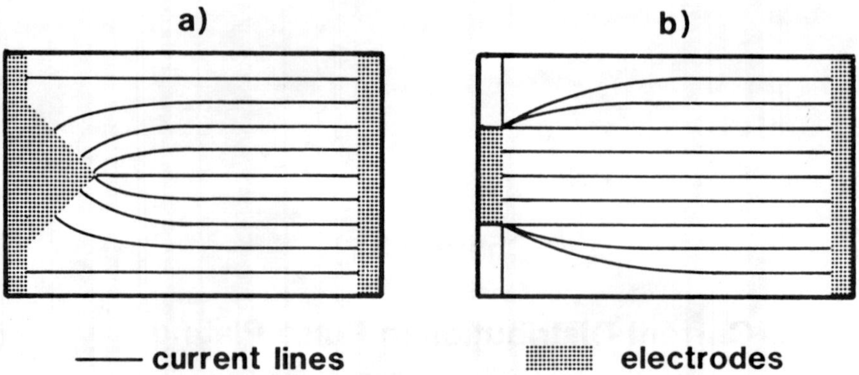

Fig. 6.1—Examples of nonuniform current distribution. High current densities occur a) on protruding parts, b) at electrode edges in cells larger than the electrode.

restrict ourselves to current—and not to potential—distribution since for metal deposition it is the phenomenon of main interest due to its relation to deposit thickness distribution.

2. Qualitative considerations

As has been said in the previous section, current distribution is a matter of mass and charge transport. We can deduce from this that the main factors influencing the distribution of current over an electrode surface are:

- the geometry of the system
- the conductivities of the electrolyte and of the electrodes
- the kinetics of the electrode reaction, i.e., the type and extent of polarization prevailing under the conditions considered. The main kinds of polarization usually taken into account are those due to mass transport and to the charge transfer reaction.

In systems with aqueous electrolytes the conductivity of the electrode is usually much higher than that of the solution. Exceptions are three-dimensional electrodes (porous, fixed-bed, fluidized-bed electrodes) and certain electrode materials with a low conductivity. In metal deposition the resistance of the electrode may not be negligible when a metal is plated on fine wires or in thin layers on a poorly conducting substrate. Such cases will not be considered in the following and the discussion will be restricted to situations where the electrode side of the interface can be assumed to be an equipotential surface.

An illustrative picture of the problem of current distribution can be obtained with a resistance model (Fig. 6.2). Different resistances connected in series oppose the current flow between the electrodes. R_e is the resistance of the electrolyte solution; it is ohmic and depends on geometric factors and on the specific conductivity of the electrolyte. R_p and R_p' are resistances due to the slowness of the electrode reactions. They are given by the slope of the overvoltage-current density curve:

$$R_p = d\eta \,/\, dj \qquad [1]$$

and are called polarization resistances.

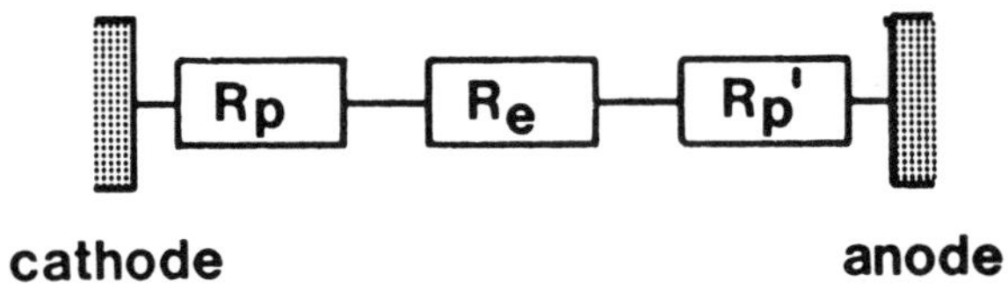

Fig. 6.2—Resistance model of an electrochemical cell.
R_p and R_p' : polarization resistance of cathode and anode, respectively
R_e : electrolyte resistance.

The situation where $R_p << R_e$ and $R_p' << R_e$ corresponds to a primary current distribution which is governed by the geometry of the system. Secondary current distribution occurs when R_p and R'_p are no longer negligible as compared to R_e. R_p is an 'activation' resistance while mass transport effects, i.e., the occurrence of concentration gradients in the solution, can be neglected. Tertiary current distribution prevails if both activation and mass transfer effects contribute to the polarization resistance.

The sum of the different resistances may vary along the surface, as in the case of a cathode whose surface has a saw-tooth profile (Fig. 6.3). If we consider primary current distribution ($R_p = R_p' = 0$) the electrolyte resistance $R_{e,c}$ at the crests of the profile (c) will be smaller than $R_{e,r}$ at the recesses (r) due to the difference in the anode-cathode distances and to the peak effect. The current density at the peak j_c will therefore be larger than j_r in the recess:

$$j_c \,/\, j_r = R_{e,r}/R_{e,c} > 1 \qquad [2]$$

Now, if we add the polarization resistance R_p (for the sake of simplicity the polarization resistance R_p' of the counter electrode is included in R_p) to the electrolyte resistances R_e assuming that $R_{p,r} \approx R_{p,c}$† the ratio of the total resistances will become smaller:

$$(R_{e,r} + R_p) \,/\, (R_{e,c} + R_p) < R_{e,r} \,/\, R_{e,c} \qquad [3]$$

Consequently the current will be more evenly distributed between the peaks and the recesses:

$$j_c' \,/\, j_r' = (R_{e,r} + R_p) \,/\, (R_{e,c} + R_p) < j_c \,/\, j_r \qquad [4]$$

Equation [4] shows that the influence of overpotentials makes the current distribution more uniform.

† Although R_p is a function of current density and varies therefore along the electrode surface, this assumption may be used over a limited range of current density where the current-voltage curve can be approximated by a straight line.

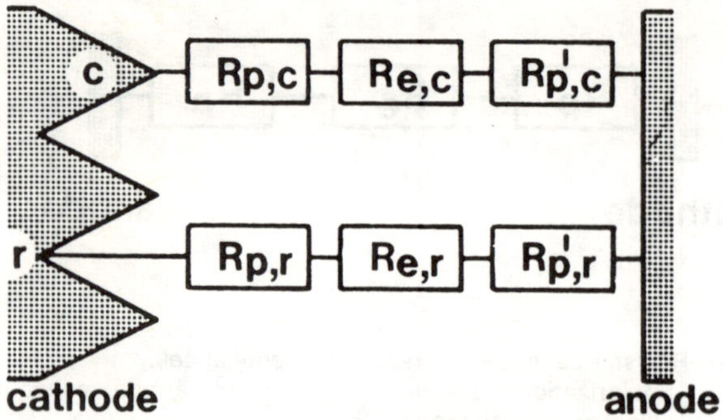

Fig. 6.3—Variation of the resistance along a saw-tooth profile. R_p, R_p' and R_e have the same meaning as in Fig. 6.2. The subscripts c and r refer to the crest and to the recess, respectively.

The equalizing action increases with increasing ratio of the polarization to the electrolyte resistance R_p / R_e which can be expressed as a function of the relevant parameters in the following way:

$$R_p / R_e = \kappa \, (d\eta / dj) / L = Wa \qquad [5]$$

The dimensionless group Wa is called the Wagner number. The larger its numerical value the more uniform is the current distribution, whereas Wa = 0 corresponds to primary current distribution. L is a characteristic length of the system and κ is the specific conductivity of the electrolyte solution.

The theoretical treatment of current distribution problems shows too that the current distribution depends on one or—depending on the number of relevant characteristic lengths—several dimensionless groups of the Wa type.

3. Specific influence of overpotential

For a reaction taking place under activation control, the current-voltage relationship can be described by the Tafel equation if the overpotential is not too small:

$$\eta_a \sim \log(j) \qquad [6]$$

from which we derive that the polarization resistance and therefore Wa decrease with increasing current density:

$$d\eta_a / d_j \sim j^{-1} \qquad [7]$$

The secondary current distribution tends thus to become less uniform at higher current densities.

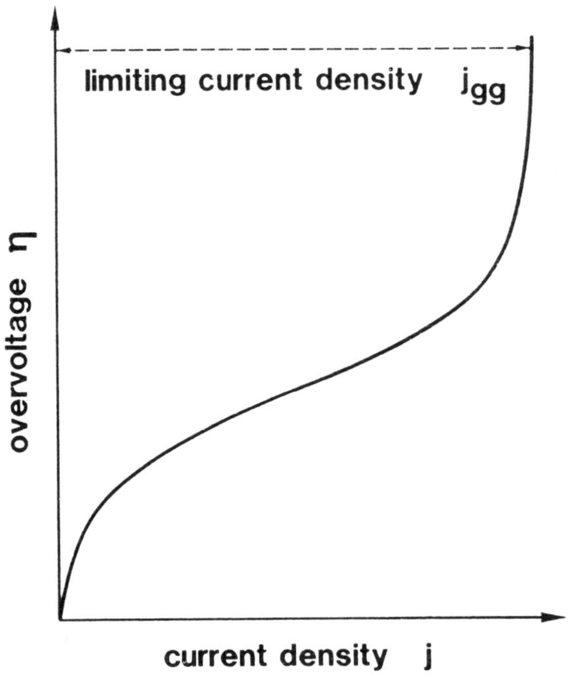

Fig. 6.4—Current-voltage curve (schematic).

However, at higher overpotentials mass transport effects may play a role. The concentration overpotential is related to the current density through:

$$\eta_c \sim \log (1 - j/j_{gg}) \qquad [8]$$

Its derivative increases with increasing j for a given limiting current density j_{gg}:

$$d\eta_c / dj \sim (1 - j/j_{gg})^{-1} \qquad [9]$$

Therefore, we expect that under mass transport controlled conditions the current density is more evenly distributed at higher current densities.

In fact, the schematic current-voltage curve (Fig. 6.4) shows that at lower current densities, where activation overpotential predominates, the slope of the η vs j curve decreases first, then goes through a minimum and increases again in the region where concentration overpotential contributes to the total overpotential. It tends toward infinity at the limiting current where the concentration overpotential becomes overwhelming, i.e., where the deposition reaction is mass transport limited. The Wagner number is then also infinitely large and the current distribution is therefore

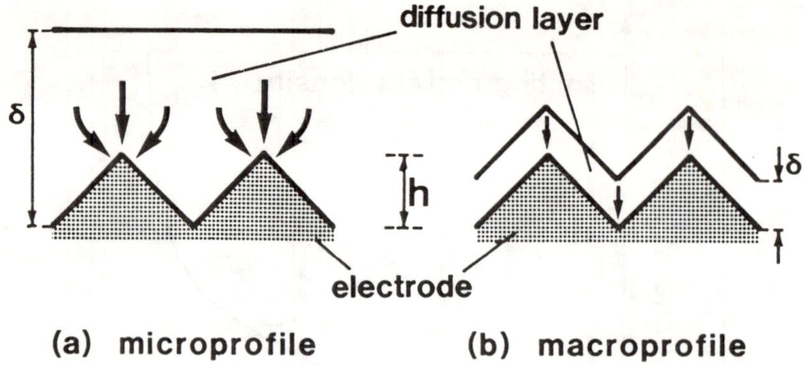

Fig. 6.5—Influence of diffusion layer thickness on tertiary current distribution. a) microprofile: the crests are privileged from the point of view of diffusion, b) macroprofile: the current density is uniformly distributed.

expected to be very uniform. However, in this case the Wagner number is not the only criterion relevant to current distribution. Since the deposition reaction is controlled by mass transport, tertiary current distribution depends mainly upon the uniformity of the diffusion layer thickness. Geometrical influences and hydrodynamics are therefore involved in tertiary current distribution. As far as geometrical factors are concerned two cases are usually distinguished:

(a) A surface profile whose characteristic dimension h is smaller than the thickness of the diffusion layer is called a microprofile (Fig. 6.5a). The peaks are more accessible to diffusion than the recesses due to a peak effect analogous to that in primary distribution. As a consequence the irregularities of the surface are amplified in the course of the deposition; in other words, the microscopic throwing power under these conditions is bad. This explains the general observation that metal deposits obtained at or near the limiting current are rough or powdery (5). This effect is of great importance in metal plating because of its influence on the appearance and the properties of the deposits.

(b) A surface profile with a characteristic length h larger than the diffusion layer thickness is called a macroprofile (Fig. 6.5b). In this figure a constant thickness of the diffusion layer is considered; i.e., each point on the surface is equally accessible for diffusion, which is not affected by geometrical factors. The concentration resistance has here an equalizing effect on current distribution. The macroscopic throwing power at the limiting current is excellent as predicated by the high value of Wa.

It should be noted, however, that the diffusion layer thickness can vary along a macroprofile not for geometrical but for hydrodynamic reasons. Such a variation impairs the uniformity of the tertiary current distribution. Figure 6.6 shows schematically the primary and secondary distribution as well as the distribution at the limiting current for an electrode forming the wall of a rectangular channel with laminar flow of the electrolyte. Secondary distribution is better than the primary one, and also better than that at the limiting current. In the latter case the determining factor is the diffusion layer thickness which in this flow system increases in the downstream direction.

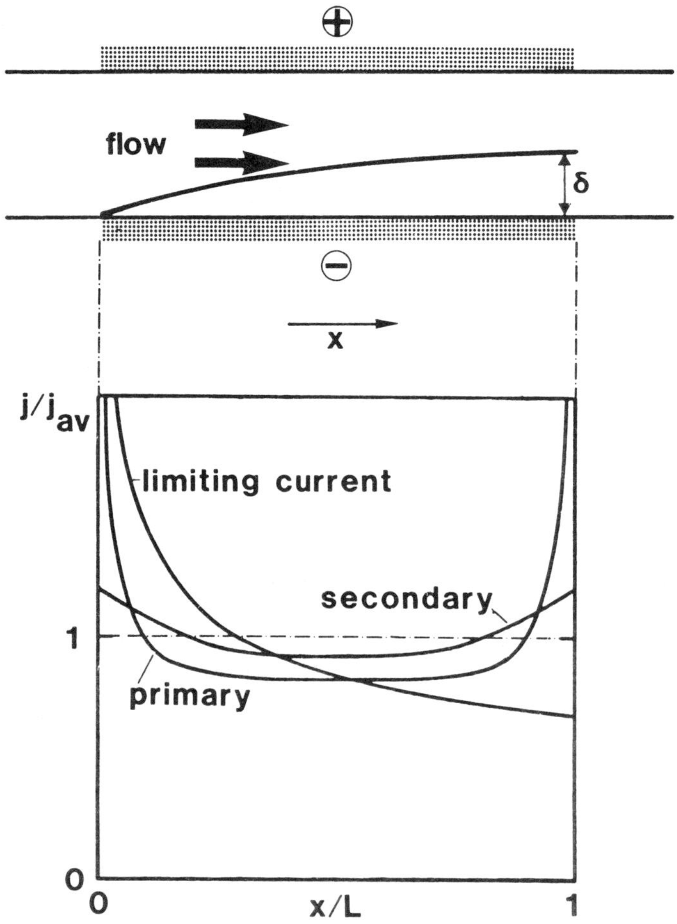

Fig. 6.6—Comparison of different types of current distribution in a channel flow cell (schematic).

4. Current distribution with pulsating current

4.1. *Theoretical aspects*

A quantitative treatment of current distribution in pulse plating would be even more complicated than it is for DC and no attempt has been made so far to solve the pertinent equations for a pulsating current or potential. However, the same general arguments as for DC can be applied accordingly to PC in order to get a qualitative conception:

Primary distribution should not be affected by PC since it does not depend on the current density.

It has been shown in section 3 that the *secondary distribution* becomes less uniform at increasing current densities. The current density which has to be used for the calculation of the Wagner number for a PC is the instantaneous pulse CD. Especially when this is considerably higher than the current densities used in DC, the secondary distribution is expected to be less uniform with PC. The distribution should be the same in both cases if the pulse CD equals the current density in DC. However, in this situation the average deposition rate in PC is necessarily lower. The current distribution in PC may approach the DC distribution or even be improved as compared to the latter if the pulse signal is distorted by double layer capacity effects. This situation—which occurs at short pulse durations and/or low pulse current densities—is discussed in Chapter 4.

In order to obtain a *tertiary distribution,* the pulse duration must be of the order of or larger than the transition time of the reacting species so that the deposition takes place under mass transport controlled conditions (limiting current) for a smaller or larger part of the pulse. For given on- and off-times the transition time is determined by the applied pulse current density j_p (see eqn. 9 of Chapter 5):

$$\tau \sim j_p^{-2} \quad [10]$$

so that the above condition can be fulfilled with a large pulse duration T_{on} and/or a high current density j_p. Obviously a second electrode reaction must set in at $T_{on}/\tau > 1$ with a corresponding drop in current efficiency. The question whether one has a micro- or a macroprofile depends on the length of the pulse. The thickness of the pulsating diffusion layer δ_p—which is relevant for the uniformity of the mass transfer rate—has to be small as compared to the irregularities of the surface profile if a macroprofile is desired. This is the case at short pulses since δ_p is a function of time (see equation 13 of Chapter 5):

$$\delta_p \sim t^{-0.5} \quad [11]$$

irrespective of the current density. With short pulses it is therefore possible to convert a profile which is a microprofile in DC into a macroprofile.

We conclude from the above arguments that a good current distribution is to be expected in PC plating with short pulses (macroprofile) and high pulse current densities (limiting current conditions). Again one has to be aware of the possibility of double layer capacity effects when short pulses are used.

In many investigations of pulse plating reported in the literature, a beneficial or detrimental effect of pulsating current on current distribution, especially on roughness formation, has been observed. However, in most cases the interpretation of the results is difficult because the experimental conditions are not well defined or the range of parameters is too narrow (sometimes a comparison is made with DC on the basis of one single set of pulse parameters). A selection of results of some more systematic studies will be presented in the following in order to test the qualitative predictions made above.

4.2 *Secondary current distribution*

Ibl and coworkers have investigated experimentally the secondary current distribution on rotating disks and on electrodes with a sawtooth profile (6-8). Figures 6.7 and 6.8 show results obtained for nickel deposition from a Watts type electrolyte on a rotating disk with pulse parameters chosen such as to have a negligible mass transport influence. The ratio of the local deposit thickness to the average one is plotted as a function of the reduced radial distance, r_o being the radius of the disk. The figures allow a comparison of the distribution in PC to that in DC at an average pulse current density j_m equal to the current density in the corresponding DC experiment. One can see that the distribution in PC is less uniform than in DC if the comparison is made at equal average rates of metal deposition j_m. Figure 6.7 raises another point of interest: the current distribution for a DC experiment with a CD of 25 A/dm^2 is very similar to that for the PC deposit obtained with j_p = 30 A/dm^2. These results are in agreement with the prediction discussed in section 4.1, according to which the pulse current density is the relevant parameter for current distribution in PC. It had not been possible to determine the Wagner number for the experiments described above from measurements in a Haring-Blum cell nor from the slope of experimental current-voltage curves because the results were too scattered. However, if we assume Tafel kinetics, the ratio of the Wa for the different experiments is equal to the inverse of the ratio of the pulse current densities (Fig. 6.7). The experimental distribution in Figure 6.7 has approximately the shape of the theoretical curves (9). The points corresponding to the DC results at 3 A/dm^2 lie between the curves for Wa = 1.86 and Wa = 0.87, those representing the results obtained at 25 A/dm^2 between the curves for Wa = 0.3 and 0.13. The inverse of the ratio of the current densities is about 8, whereas the ratio of the Wa to which one can assimilate the two sets of experiments is about 6-8. The change in distribution is therefore in fair agreement with the theoretical expectations. The experiments shown in Fig. 6.7 were made with a disk of 2 mm radius, whereas the results in Fig. 6.8 were obtained with a disk of 20 mm radius. Since Wa is inversely proportional to the radius of the electrode, the deposition is expected to be less uniform with the larger disk. This was indeed observed in all experiments. In contrast to PC, the experimental values for DC exhibited a considerable scatter. Moreover, it is interesting to note that the experimental points more or less follow the general shape of the theoretical curves, but that the distribution is in any case less uniform than the primary one which is represented by the curve in Fig. 6.8. This observation apparently conflicts with theory according to which primary current distribution is the most unfavorable case, but it can be understood in terms of current efficiency variations: the overall current efficiency was around 90 percent for the small disk, but as low as 50 percent for the large one. In this case the metal distribution may be substantially different from the current distribution. In fact, the current efficiency for Ni deposition from acidic electrolytes increases with increasing current density (10). The result is that near the edge of the disk where the current density is higher, the current efficiency is larger and the non-uniformity of metal distribution is enhanced. The results of the rotating disk experiments were confirmed by those obtained with the sawtooth profiles. Figure 6.9 shows as an example that for nickel deposition from sulfamate solutions the ratio of the thickness of the deposits on the crest and in the recess was larger in PC than in DC. The comparison is again made at the same average current densities.

Cheh et al. (11) carried out experiments on copper deposition from a sulfate bath in a Haring-Blum (4) cell. They found that PC has relatively little effect on the secondary distribution if the comparison with DC is made at a j_p equal to the DC current density. The somewhat less uniform distribution with PC—the extent of which depended on the on- and off-times—was explained in terms of a current efficiency varying with current density due to the formation of cuprous ions.

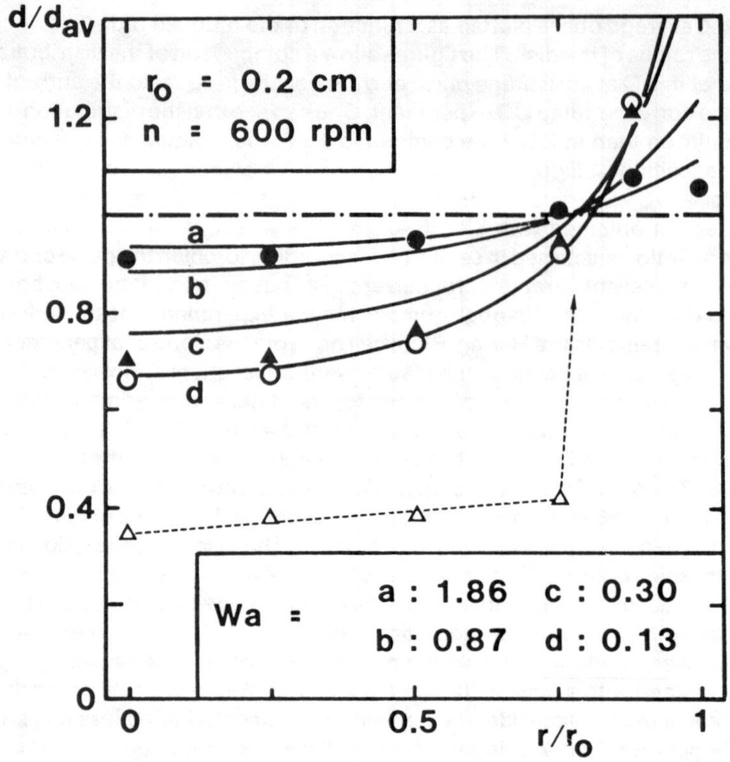

		j_m	j_p	T_{on}	T_{off}
		(A/dm²)		(ms)	
DC	●	3			
	▲	25			
PC	○	3	30	50	450
	△	25	250	0.005	0.045

Fig. 6.7—Thickness distribution of nickel deposit on a small rotating disk electrode. Points: experimental results (6), (8). Curves: Theoretical distribution for different Wagner numbers (9).

Studies on pulsating current copper deposition in a Haring-Blum type cell have also been carried out in this author's laboratory (12). The pulse duration varied between 50 μs and 100 ms; the other pulse parameters were chosen such as to avoid mass transport effects. Figure 6.10 represents a plot of the throwing power S (for a definition of S see, for example, ref. 4) versus the applied overall

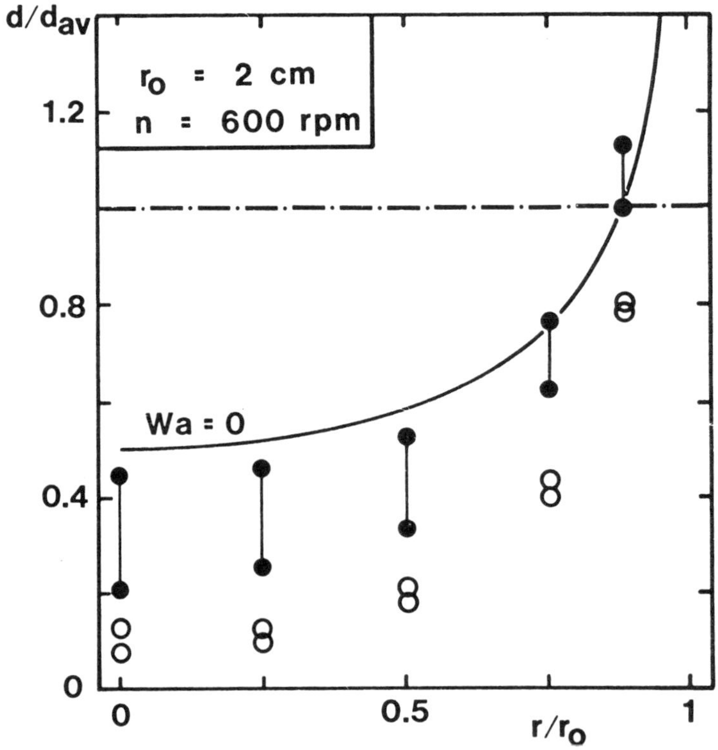

	j_m	j_p	T_{on}	T_{off}
	(A/dm²)		(ms)	
DC ●	3			
PC ○	3	30	50	450

Fig. 6.8—Thickness distribution of nickel deposit on a large rotating disk electrode. Points: experimental results (6), (8); Curve: Theoretical primary distribution (1).

pulse current density. The average current densities varied between 1 and 2.5 A/dm^2 for all experiments and were equal to the current densities of the DC experiments whose results are also shown in the figure (large circles). The results show clearly that the current distribution is mainly a function of pulse current density. The pulse duration has relatively little influence. The throwing power becomes comparable to the DC values if the pulse and the DC current density become equal, with a corresponding decrease of the average current density of the pulses.

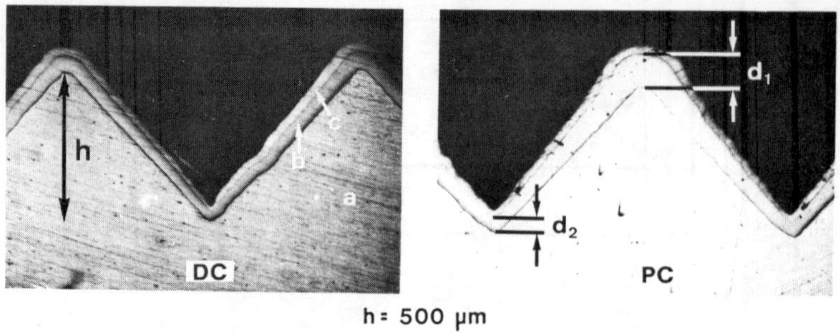

	j_m	j_p	T_{on}	T_{off}	d_1/d_2
	(A/dm²)		(ms)		
DC	5				1.6
PC	5	15	1	2	2.5

Fig. 6.9—Deposit thickness distribution for nickel on a sawtooth profile (6), (7). a) substrate; b) Ni deposit; c) protective Cu layer.

All the results presented above confirm that in secondary distribution the uniformity depends mainly on the actual metal deposition current density rather than on the average one and that therefore, as predicted on the basis of theoretical arguments, a secondary current distribution with a pulsating current is normally less uniform than with DC.

4.3. *Tertiary current distribution*

Tertiary current distribution is usually strongly irregular under DC conditions because the situation of a microprofile normally occurs. Ibl and coworkers were able to convert a microprofile into a macroprofile by using a pulsating current as discussed in section 4.1, therefore making the tertiary current distribution more uniform. They studied the deposition of copper, cadmium and silver in a channel cell with laminar flow (13), (14), (15). The pulses were much longer than the transition time so that the discharge of the metal ions was controlled by their transport through the pulsating diffusion layer and took place under limiting current conditions during the major part of the pulse. The thicknesses of the pulsating diffusion layer were less than half a micrometer. They were realized by using very short pulses; T_{on} was always below 50 μs. In order to make the transition time τ small, pulse current densities as high as 25,000 A/dm² were used. The roughness R of the deposit was measured by a mechanical profilmeter. In Fig. 6.11, R is plotted versus T_{on}/τ for the three metals. We can see that the deposits were remarkably smooth although they were obtained at current densities 1000 to 10,000 times higher than under usual DC conditions.

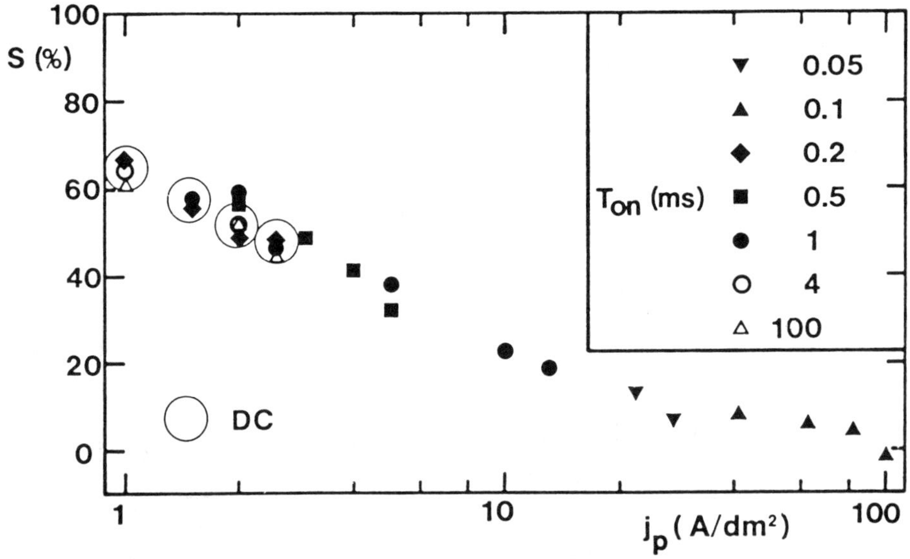

Fig. 6.10—Secondary current distribution in a Haring-Blum cell as a function of pulse current density (12), j_m varied between 1 and 2.5 A/dm^2. Large circles: DC experiments the j values of which can be read on the abscissa.

The thickness of the diffusion layers, also shown in Fig. 6.11, was small enough as to have macroprofiles with constant diffusion layer thicknesses in all cases, and the roughness decreased with increasing T_{on}/τ, i.e., with increasing mass transport influence. For each metal the highest roughness values were observed for pulse durations which were lower than the transition time. In these cases secondary current distribution was predominant and therefore the equalizing action of mass transport on the current distribution was not effective. Figure 6.12 shows for copper deposition that under mass transport controlled conditions ($T_{on}/\tau = 9.1$), the roughness is independent of the deposit thickness, in contrast to the observation in DC (16).

Analogous trends can be found in results obtained by Andricacos et al. (17) (see also Ref. 18) for gold deposition from a cyanide-citrate bath on a rotating disk with a small groove. The authors could show that with PC, throwing power was better than with DC due to the reduced thickness of the diffusion layer.

The results presented above show that a good throwing power on a microscopic scale can be obtained if the diffusion layer is thin as compared to the dimensions of the surface irregularities (macroprofile) and if simultaneously T_{on}/τ has a sufficiently high value (diffusion control). They thus confirm the expectations outlined in section 4.1.

The beneficial effect of a transition from a secondary to a tertiary current distribution with uniform mass transport conditions can also be demonstrated on a macroscopic scale. Landolt et al. (19) have investigated the thickness distribution of copper deposits on a rotating disk. Figure 6.13a shows the secondary distribution for DC and for PC. The average current density was in both cases only about 2 percent of the limiting current density j_{gg}, and therefore the distribution in DC was very uniform. The pulse current density was well

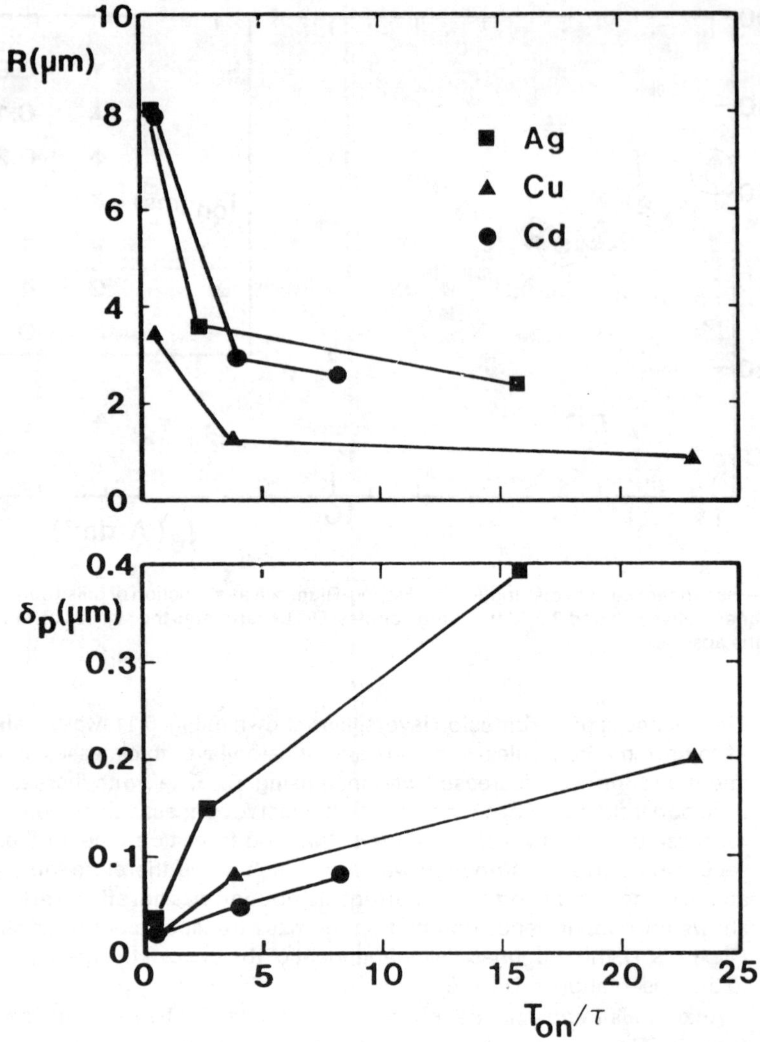

Fig. 6.11—Roughness of electrodeposits on small scale profiles and corresponding diffusion layer thicknesses as a function of the ratio T_{on} / τ for copper (13), (14), silver (13) and cadmium (15).

below the limiting pulse current density j_{pg} so that mass transport effects were negligible, but it was more than 300 times higher than the average current density. As a result the distribution was much less uniform than with DC. The curve has the same shape as the theoretical ones shown in Fig. 6.7. Figure 6.13b shows for PC, as well as for DC, the tertiary current distribution, which is determined by the local mass transfer rate. For the DC experiments, mass transfer limitations occur because the current density is close to the limiting one.

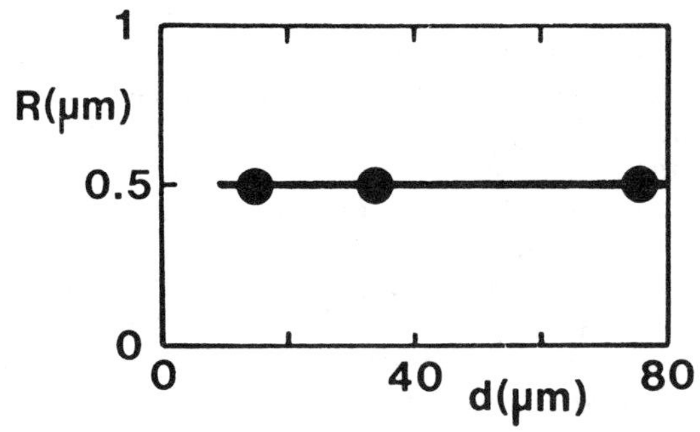

Fig. 6.12—Tertiary current distribution in pulse plating on small scale profiles as a function of deposit thickness (13): No roughness amplification effect.

Since the rotating disk is uniformly accessible for convective mass transport (1), the current distribution under these conditions is very uniform. In PC, mass transport effects occur because the pulse current density is higher than the limiting pulse current density ($j_p > j_{pg}$). The pulsating diffusion layer controlling the transport of the copper ions and thus the deposition rate, has a constant thickness over the disk resulting in a regular deposit thickness distribution. Hence, the improvement due to the transition from a secondary to a tertiary current distribution is evident. In this particular example, the fact that in PC j_m is close to j_{gg} is an additional reason for the uniform current distribution since the stationary diffusion layer, which can be considered to be the same as for DC, has a constant thickness over the disk as mentioned above.

The Haring-Blum experiments presented in section 4.2 have been extended to much higher pulse current densities than those of Fig. 6.10 (12). In Fig. 6.14 the experimentally determined throwing power S is plotted as a function of T_{on} / τ, where τ is a transition time calculated with the total applied current density. However, the partial currents for the two cathodes usually were different and therefore their transition times were different. These were calculated from the measured partial pulse current densities for the individual electrodes in order to assure that for a given experiment T_{on} / τ was larger or smaller than unity, respectively, for both cathodes. It could be shown that this was always the case except for those experiments where $T_{on} / \tau \approx 1$.

The points in the left part of the figure ($T_{on} < \tau$) correspond to results already presented in Fig. 6.10 and pertain to a secondary distribution. The throwing power decreases with increasing T_{on} / τ which is equivalent to an increase in pulse current density. An increase of T_{on} / τ also means that mass transport effects become more and more significant. (Approaching $T_{on} / \tau = 1$ in a pulse experiment is equivalent to approaching the limiting current in DC.) According to what has been discussed in section 3, a minimum of the throwing power should therefore be observed when T_{on} / τ passes from low ($T_{on} < \tau$, secondary distribution) to high values ($T_{on} > \tau$, tertiary distribution). The lowest values for the throwing power were found at $T_{on} \approx \tau$. There was a large scatter in the results, which are represented by the bar in the figure. The average of S for eight measurements was very close to 0 percent, which corresponds to a primary distribution. The occurrence of the minimum in throwing power at $T_{on} \approx \tau$ is explained by the fact that the current distribution is still a secondary one (mass transport limitations just start at the end of the pulse) and, moreover, under the worst conditions since j_p is the highest possible. If the pulse duration is made much larger than the transition time the throwing power does indeed increase again. This is demonstrated by the experimental points in the right hand part of Fig. 6.14. Very high pulse current densities were applied, the average current densities being in the same range as those of the other experiments (1 to 2.5 A/dm^2). The results were rather scattered and no systematic trends could be observed. However, the current distribution was always considerably better than that for the pulse experiments at $T_{on} / \tau < 1$ and for the DC experiments with the same average current density.

With the two examples of Figs. 6.13 and 6.14 it has again been possible to verify that a tertiary current distribution is more uniform than a secondary one provided that the mass transport rate does not vary along the surface. This condition is easily fulfilled in pulse plating with short pulses, whereas in DC where the local diffusion layer thickness depends on hydrodynamic and geometrical factors, only a few special systems have constant mass transport conditions along the surface.

Let us conclude this section with a few remarks regarding roughness formation during deposition. The discussion in this chapter has implied that it is a matter of current distribution and therefore governed by the mass transport and the charge transfer steps of the deposition process. However, other factors may have an influence. It is a general experience (see (20) and Chapter 3) that the grain size of electrodeposits is usually smaller with PC than with DC and that it decreases with increasing j_p. This can be explained in terms of an enhanced nucleation by the high CD and by the resulting high overvoltage. The pulse current exerts thus an effect on the crystallization step. It is to be expected that a deposit with larger grains is more likely to become rough. Other important effects result from the presence of addition agents in the electrolyte. Levelling may be obtained with a substance inhibiting the deposition reaction if its adsorption on the electrolyte surface is mass transport controlled and therefore takes place mainly on the crests of a profile (21). The rate of the charge transfer reaction is then modified locally and the assumption of uniform kinetics along the surface implied in the discussion so far is no longer valid. The adsorption of other species may be able to explain certain roughness effects observed in a deposition process, for example, adsorption of anions (22), or nonuniform passivation by dissolved oxygen (23).

In all these cases it is very difficult to weigh the importance of the different synergistic effects and to judge if and to what extent they are influenced by a pulsating current. The effect of PC is therefore hardly predictable on the basis of simple arguments on current distribution. Much work on the influence of PC on the surface morphology of electrodeposits has been carried out by Popov et al. (24).

4.4. *Periodic current reverse*

An extension of pulsating current electrolysis is periodic current reverse or pulse reverse (PR), where the cathodic pulse is followed by an anodic one whose amplitude and/or duration is such as to have an average current which is cathodic. In fact, PR seems to have been applied to electroplating earlier than PC

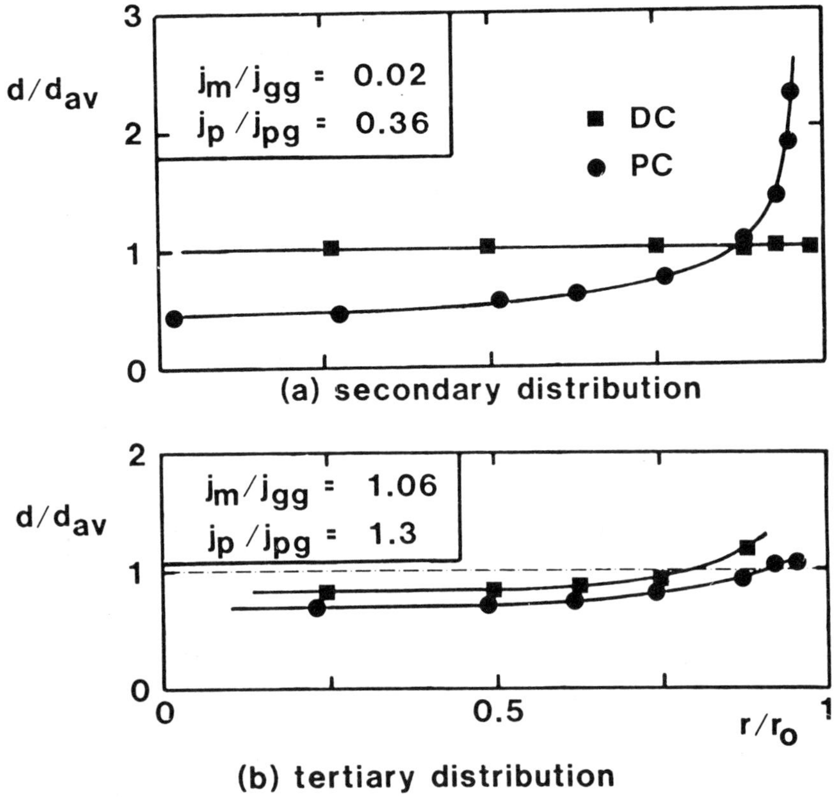

Fig. 6.13—Comparison of secondary and tertiary current distribution with DC and PC on a rotating disk electrode (19).

(25, 26). The investigations reported in the literature are mainly empirical and not systematic. Also, in early work the cycle times were quite large, typically 10 to 20 seconds, sometimes in the order of minutes, so that the characteristic features of nonsteady state electrolysis were probably lost. One has then rather a sequence of alternating steady state cathodic and anodic periods. Deposits with less rugosity than with DC are usually obtained with such currents (27, 28). The results obtained by Hickling et al. (28) may serve as an example: The authors deposited copper from a cyanide solution using PR with constant cathodic and anodic current densities of 10 A/dm^2. The total period was always 15 s, the ratio of cathodic to anodic pulse time T_c / T_a varied between 1.4 and 5. In experiments with originally smooth electrodes, the roughness increased first and then attained a constant value which did not change with further increase of deposition time, but which was larger the higher T_c / T_a.

When the initial roughness was larger than the steady state value, levelling was observed and the same constant roughness as with smooth cathodes was approached. These findings can be explained easily in terms of the different

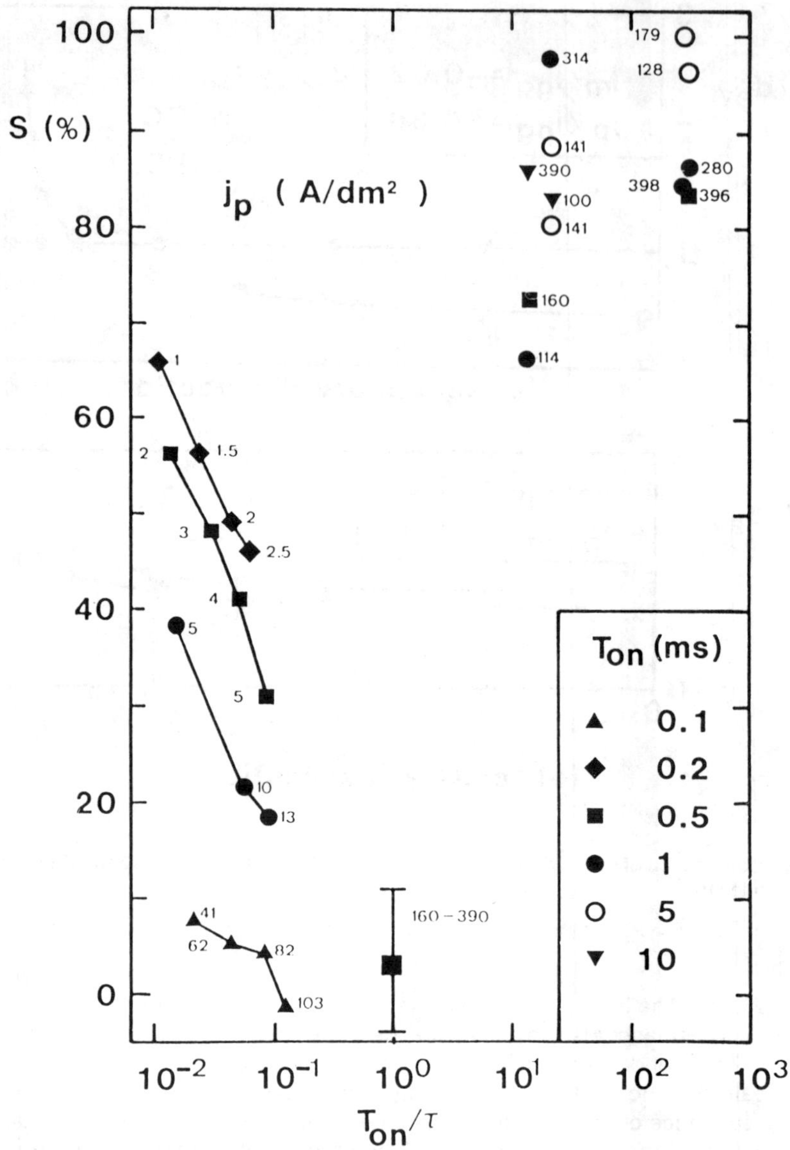

Fig. 6.14—Throwing power as a function of the ratio T_{on} / τ showing the influence of the transition from secondary to tertiary current distribution for copper deposition in a Haring-Blum cell (12).

effects exerted by a nonuniform current distribution on the cathodic and in anodic processes: A nonuniform CD leads to a roughening during deposition whereas it tends to level out surface irregularities in a dissolution process (29). In PR there is a competition between these two effects, and after a certain time a steady situation is established with a roughness depending on the ratio of the cathodic to the anodic cycle.

More recently Popov et al. (30) found that with PC and with PR smoother copper deposits could be obtained than with DC, but the roughness was higher with PR than with PC. One difference between a zero-current period and an anodic current period is that in the latter case etching may occur during dissolution which can contribute to an increase in roughness. Crystallographic factors may thus have an influence here too. The same authors (31) could show that the macroscopic throwing power was considerably better with a periodic reverse than with a pulse current. Analogous arguments as for Hickling's results can be used to explain this behavior.

The potentialities of periodic current reverse in terms of plating thickness distribution have barely been exploited yet. Further advantage could be taken by using PR with cycles in the millisecond rather than in the traditional range of seconds. A substantial improvement in deposit thickness distribution with PR is to be expected from highly irregular current distributions resulting from short anodic pulses with large amplitudes.

5. Concluding remarks

Theoretical considerations predict that current distribution with PC is usually more irregular than with DC if the comparison is made for equal average current densities. It is common practice to distinguish between primary, secondary and tertiary current distributions. Pulse plating does not affect primary current distribution, while it renders the secondary one less uniform than with DC. Tertiary current distribution can be improved by PC as compared to DC if the pulsating diffusion layer has a constant thickness along the electrode. This condition is fulfilled in cases where it is thin with respect to the characteristic dimensions of the surface profile, i.e., when short pulses are applied. Tertiary current distribution prevails when the pulse current densities are higher than the limiting ones, which inevitably leads to considerable losses in current efficiency. Moreover, a limiting factor for the high pulse currents required will often be presented by pulse generator performance.

In practical applications the current distribution is not clearly primary, secondary or tertiary, but usually lies between a secondary and a tertiary one. PC is therefore expected to render the current distribution less uniform. However, it has to be kept in mind that this statement implies Tafel kinetics for the charge transfer step at the electrode which are not necessarily applicable when inhibitors, brighteners and other additives are present in the plating bath. It is therefore easily conceivable why trends opposite to the above mentioned predictions are sometimes reported. Since the results obtained with practical systems are usually of proprietary nature, it is difficult to know the precise values of the plating parameters, and it is hardly possible to extend the approach presented in this chapter to such cases.

The influence of pulse plating on the current distribution on poorly-conducting cathodes like wires or thin metallic substrates has not been discussed here. Qualitatively it can be expected that in these situations PC makes the current distribution less uniform.

Significant improvements in plating thickness distribution can be expected from pulse reverse plating. They are due to a compensation of non-uniformities in current distribution occurring in the cathodic and the anodic cycles, respectively. Full advantage of this plating technique can be expected from short and high amplitude current pulses. While PR with cycles in the range of 10 s or more has very often been applied, few applications were found with PR cycles in the ms range. The latter is still an exploratory range of PR plating conditions with great possibilities in terms of uniform plating thickness distribution.

References

1. J. Newman; *Electrochemical Systems,* Prentice Hall, Englewood Cliffs NJ, (1973)
2. N. Ibl; *Oberfläche-Surface,* **16** : 23 (1975)
3. N. Ibl; *Techniques de l' Ingenieur,* **D902,** Paris (1976)
4. N. Ibl; in: *Comprehensive Treatise of Electrochemistry,* Vol. 6, ed. by E. Yeager, J. O'M. Bockris, B.E. Conway, S. Sarangapani, Plenum Press, New York (1983)
5. N. Ibl; in: *Advances in Electrochemistry and Electrochemical Engineering,* Vol. 2, ed. by C.W. Tobias and P. Delahay, J. Wiley, New York (1962)
6. N. Ibl; *2nd International AES Symposium on Pulse Plating,* Chicago (1981)
7. B. Sturzenegger; Diploma thesis ETH, Zurich (1980)
8. C. Iten; Diploma thesis ETH, Zurich (1980)
9. J. Newman; *Journal Electrochemical Society,* **113** : 501, 1235 (1966)
10. H.W. Dettner; J. Elze; *Handbuch der Galvanotechnik* Vol. II, C. Hanser, Muenchen, p. 99 (1966)
11. H.Y. Cheh; P.C. Andricacos; H.B. Linford; *Plating and Surface Finishing* **44,** (7) : 42 (1977)
12. P. Oswald; Diploma thesis ETH, Zurich (1983)
13. M. Braun; N. Ibl; *Oberfläche-Surface,* **14** : 49 (1973)
14. H.J. Schenk; *Thesis 5566 ETH,* Zurich (1975)
15. J.C. Puippe; *Thesis 6225 ETH,* Zurich (1978)
16. N. Ibl; D. Bindschedler; H.J. Schenck; *Oberfläche-Surface,* **16** : 278 (1975)
17. P.C. Andricacos; H.Y. Cheh; H.B. Linford; *Plating and Surface Finishing,* **44,** (9) (1977)
18. D. Landolt; *Oberfläche-Surface,* **25** : 6 (1984)
19. O. Chene; M. Datta; D. Landolt; *Oberfläche-Surface* **26**, (2) : 45 (1985)
20. N. Ibl; *Proc. 10th World Congress of Metal Finishing "Interfinish",* Kyoto (1980)
21. J. Osterwald; *Oberfläche-Surface,* **17** : 89 (1976)
22. N. Ibl; J.C. Puippe; H. Angerer; *Surface Technology,* **6** : 287 (1978)
23. N.Y. Kovarskii; L.A. Kuznetsova; B.B. Chernov; *Zhurnal Prikladnoi Khimii,* **48** : 826 (1975)
24. K.I. Popov et al.; "Fundamental Aspects of Pulsating Current Metal Electrodeposition," *Surface Technology* **11** : 99 (1980); **11** : 111 (1980); **14** : 323 (1981); **15** : 161 (1982); **17** : 3 (1982); **17** : 125 (1982); **19** : 181 (1983); **22** : 155 (1984); **22** : 159 (1984)
25. P. Baeyens; *Transactions Institute of Metal Finishing,* **31** : 429 (1954)
26. J.W.Dini; *Metal Finishing,* **6,** (7) : 53 (1963)
27. G.W. Jernstedt; *Annual Proceedings American Electroplaters' Society,* **37** : 151 (1950)
28. A. Hickling; H.P. Rothbaum; *Transactions Institute of Metal Finishing,* **34** : 199 (1957)
29. D. Landolt; *Techniques de l'Ingenieur,* Paris, **D950** (1979)
30. K.I. Popov; M.D. Maksimovic; D.C. Totovski; *Surface Technology,* **17** : 125 (1982)
31. K.I. Popov; D.C. Totovski; M.D. Maksimovic; *Surface Technology,* **19** : 181 (1983)

Chapter 7

Pulsed Electrodeposition of Zinc and Cadmium

W. Paatsch

1. Zinc

Although electroplated zinc is widely used for corrosion protection, especially of iron base metals, very little work has been done so far in the field of pulsed electrodeposition. A similar situation exists for cadmium.

The main reason seems to be that the best known advantage of pulse plating, a decreased requirement for metal by reducing the layer thickness, is not as important for unnoble metals as it is for precious metals. Nevertheless, some other properties of zinc and cadmium layers on steel substrates can be more nearly optimized by pulse plating, as will be discussed in this chapter.

1.1. *Electrolytes*

The influence of pulse plating on the deposition of bright zinc layers on steel substrates using commercial high cyanide (100 g/l NaCN) as well as low acid, ammonium chloride-containing solutions was investigated by Paatsch (1).

Further work on low acid bright zinc electrolytes (60 g/l $ZnCl_2$, 300 g/l NH_4Cl, pH = 4-7) was done by Krivtsov et al. (2,3) and Gadsov (4). Some information has appeared in the literature on the use of pulsed current for electroplating porous zinc layers for battery operations using concentrated alkaline zincate solutions (35 percent by weight KOH, 60 g/l ZnO) (5-8).

1.2. *Plating rate*

Very little information has been given in the literature concerning the deposition current efficiency using pulse plating. In agreement with pulse plating theory (9) according to Ref. 3, the limiting overall plating rate of zinc generally does not exceed the values for direct current plating conditions.

The pulse current density influences the concentration at the electrode, but not the deposition mechanism. This, however, is possible if two or more competing reactions take place in the deposition process; e.g., the metal ion discharge and hydrogen evolution. As both reactions show different kinetics and current efficiency values, it can be possible to exceed the direct current plating rate using suitable pulse plating parameters. Using a cyanide bright zinc electrolyte (131 g/l NaCN, 130 g/l NaOH, 52 g/l Zn) and an average current density of $j_m = 1$ A/dm^2, the metal deposition current density exceeds the value for direct current conditions by about 10 percent with increasing pulse duration and duty factor.

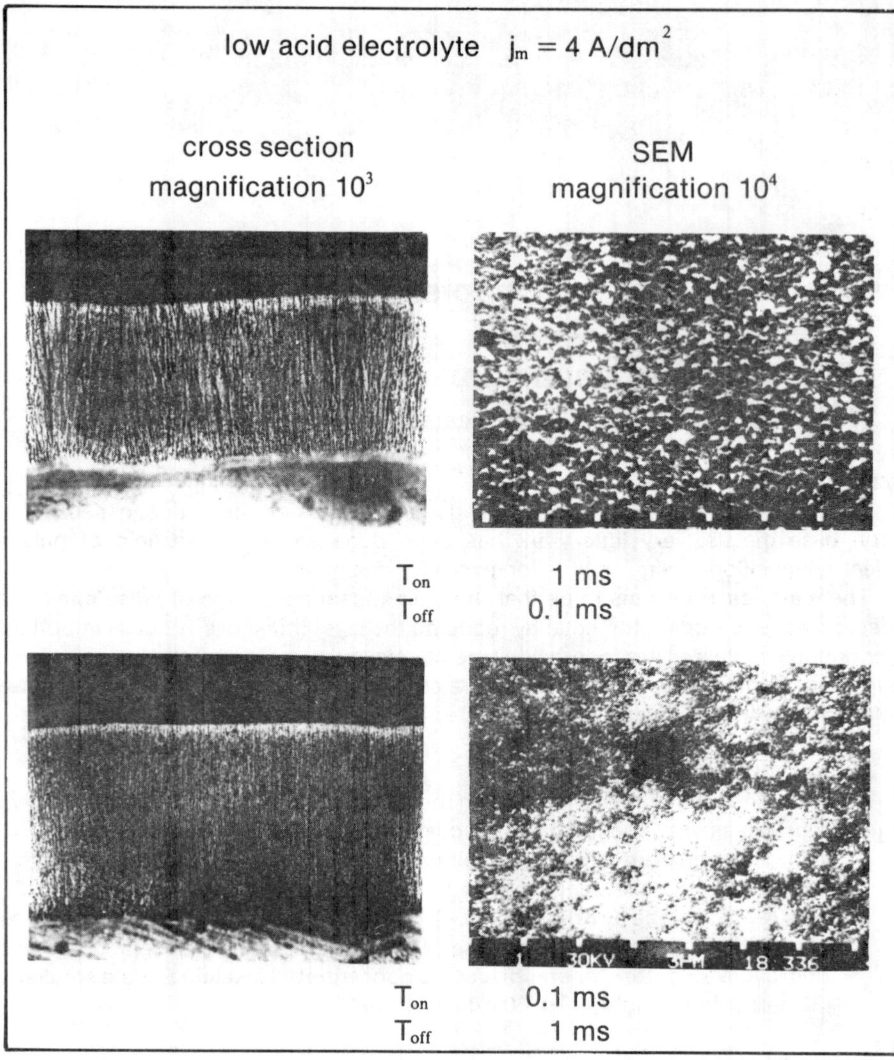

Fig. 7.1—Layer structure and surface morphology of pulse-plated zinc layers.

Choosing $j_m = 8\ A/dm^2$, a maximum plating rate about 30 percent higher as compared to the direct current value can be obtained for $T_{on} = 0.2$ ms and $\gamma = 0.7$ (1). This result is in general agreement with hydrogen permeation experiments.

1.3. *Structure*

The most interesting advantage obtained with pulsed current zinc deposition is the formation of a finer grain sized layer structure, resulting in a more homogeneous surface appearance (1-3). Figure 7.1 shows the layer structure and surface morphology of pulse plated bright zinc layers using a low acid electrolyte (27 g/l zinc, 110 g/l chloride, 22 g/l boric acid, pH = 4-5) (1).

The deposits show a finer grain size and a leveled surface if a shorter pulse duration (e.g., T_{on} = 0.1 ms) and longer break times (e.g., T_{off} = 1 ms) are used. This effect is due to an enhancement of the nucleation rate by the high pulse current density and the resulting high overvoltage (4). Krivtsov et al. (2) are of the opinion that the zinc surface will be partly deactivated during the break time by the adsorption of impurities. This results in a reduced growth of existing crystallization centers and new ones are formed during the following pulse duration, followed by the formation of a more microcrystalline and uniform deposit. According to Ref. 2, the most appreciable improvement in grain size fineness occurs up to a pulse duration of T_{on} = 2.5 ms and duty cycle $\gamma = 1.25\ 10^{-4}$. At average current densities of j_m = 2 and 3 A/dm^2 and pulse durations of T_{on} = 1.25 ms and 0.6 ms however, dendrites are observed to grow at edges.

High levelling zinc coatings of 80-85 percent reflectivity can be deposited from a low acid electrolyte without brightening agents (60 g/l $ZnCl_2$, 300 g/l NH_4Cl, pH = 4-7, T = 20-25° C) if a pulse current density j_p = 300 to 400 A/dm^2, a pulse duration T_{on} = 0.2 to 0.35 ms and a frequency of 50 to 100 Hz are used (3). It should be mentioned that plating baths with no organic additives are easier to handle and less expensive.

As the layer structure and the hardness are closely correlated, the microhardness of zinc deposits was found to be higher for pulse plating conditions (10). The Vickers-microhardness was increased from 124 HV 0.1 for direct current plating to 140 HV 0.1 for T_{on} = T_{off} = 0.1 ms and to 183 HV 0.1 for T_{on} = 0.1 ms and T_{off} = 1 ms. This change is due to the different grain sizes and possibly to electrolytic reaction products which can be preferentially incorporated into the zinc layers during the break time.

1.4. *Corrosion behavior*

Corrosion tests of pulse plated and direct current plated zinc using commercial cyanide and low acid bright zinc electrolytes were performed in standardized artificial climates (1). No significant difference in corrosion behavior was found in these tests, as well as in outdoor field tests under industrial and marine conditions. Corrosion potential measurements in different solutions did not give differing values. The composition, and therefore the electrochemical activity of the zinc deposits investigated, were about the same regardless of deposition procedure. However, Krivtsov et al. (3) found an enhancement of the corrosion behavior of pulse plated-layers as compared to direct current conditions, by a factor of 1.8 to 2. This effect is believed to be due to the reduced porosity of the layers formed under optimum pulse plating conditions.

1.5. *Hydrogen embrittlement*

Especially for low-alloyed-high-strength steel with a tensile strength higher than 1200 N/mm^2 being used, for example, in the building and aircraft industries, the phenomenon of hydrogen-induced delayed cracking is well known. Electrochemical hydrogen permeation measurements clearly show that the time period for the first zinc nucleation is most important with respect to penetration of the codeposited hydrogen into the base metal causing embrittlement (11). Once a very thin but continuous zinc layer is built up, a barrier is formed preventing the hydrogen atoms from further diffusion into the base metal. Figure 7.2 shows the formal hydrogen coefficient D_H^* for a 50 μm thick steel membrane shortly after the zinc plating process was started. In comparison to the diffusion behavior

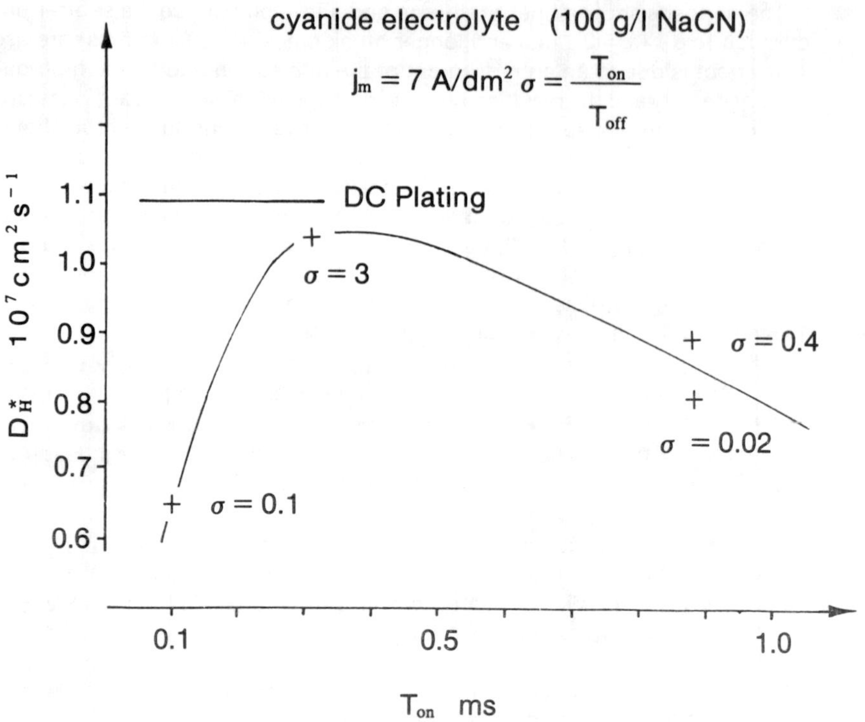

Fig. 7.2—Formal hydrogen diffusion coefficient D_H^* of in-situ direct- and pulse-plated zinc layers on a steel membrane.

under direct current plating conditions, optimum reduction of hydrogen diffusion is obtained by pulse plating (12) with short current pulses (e.g., $T_{on} = 0.1$ ms) and small duty factors (e.g., $\gamma = 0.09$). The effect is due, as already mentioned, to smaller grain size and therefore, decreased porosity of the deposited thin barrier layer as well as a change of the electrode kinetics. For high current density values, hydrogen evolution is more inhibited than the metal deposition reaction; also, the overvoltage for hydrogen evolution at zinc is much higher than at the iron base metal.

Additional tensile test experiments with high-strength steel specimens prove that hydrogen can be avoided to some extent by pulsed electrodeposition of bright zinc (12).

1.6. *Porous zinc layers from alkaline solution*

Zinc electrodes in storage batteries (e.g. nickel-zinc batteries) are subject to the problem of formation of dendritic deposits during the discharge and charging processes. With regard to cycle life and efficiency, good adherent deposits of high porosity are needed. Compact spongy and dendritic layers can be deposited, however, depending on the current density (13). Zinc sponge consisting of very few dendrites is deposited at low current densities and a relatively high zinc concentration in the electrolyte, occuring during electrolysis of zincate solutions. This phenomenon may be caused by the incorporation of colloidal particles of zinc hydroxide or very fine zinc particles into the cathode deposit, resulting in dendrites at a later growth stage. The dendritic growth was found to be mass-transfer controlled.

Despic and Popov (8) showed that compact deposits can be achieved by pulse plating using the same average current density at which dendrites are formed under direct current plating conditions. In these experiments zinc was deposited from a 1.5 M zincate solution in 5 M KOH on copper wires. In pulse plating, potentiostatic symmetric square wave pulses of different frequencies up to 10^4 Hz were used. With increasing pulse frequency, dendrite formation was reduced and compact deposits resulted. According to pulse plating theory (14), the thickness of the pulsating part of the diffusion layer is a function the pulse duration. If the thickness is small compared to the surface roughness of the base metal, which is possible for sufficiently short pulse duration, compact deposits are formed. Similar results were reported by Romanov (7).

Further work was done using a three-component impulse current consisting of high cathode and small anode current pulses followed by a break time (6). For practical reasons, the cathodic pulse duration and the break time were equally chosen to be 5 ms and the anodic pulse duration to be 10 ms. Even for prolonged times of electrolysis, the formation of 1.5 mm thick, adherent deposits of about 50 percent porosity was possible.

It is suggested that the anodic current pulses result in a dissolution of the peaks of the deposited zinc, increasing the zincate concentration at the electrode surface. Chin et al. (5) investigated the morphology of a zinc rotating electrode charged with superimposed square-wave alternating voltage in a 0.74 M zincate solution in 8.4 M KOH. The study covered a range of alternating voltages up to 200 mV rms and of frequencies from 1 to 100 Hz. At the low cathodic direct current overpotentials normally encountered in battery operations, the alternating voltage greatly improved the current distribution and changed the nonadherent mossy zinc deposit into a compact epitaxial deposit. At higher cathodic direct current-overpotentials of 75 to 200 mV, (with respect to the rest potential value of the zinc electrode), the effect of an alternating voltage diminished. However, superimposition of alternating voltage seemed to widen the potential region for the growth of epitaxial deposits.

2. Zinc alloys

The influence of pulse plating on the anomalous alloy deposition of zinc-cobalt alloys has been investigated by Paatsch (15). Using a low acid electrolyte (150 g/l $ZnSO_4$, 14.5 g/l $CoSO_4$, 50 g/l $(NH_4)_2SO_4$, 25 g/l H_2BO_3, T = 20° C) the cobalt content of the zinc deposits at a constant average current density of j_m = 2.5 A/dm^2 can be increased from 0.02 percent for direct current plating conditions to 0.2 percent for T_{on} = 1 ms and T_{off} = 10 ms. With pulse plating, the layers are more compact and show a more homogeneous surface appearance if compared to direct current conditions. Salt spray tests according to ASTM B-117 show a slightly improved corrosion resistance of the pulse plated deposits.

Nenov et al. (16) investigated the influence of pulse plating on the composition and structure of zinc-nickel alloys using an ammonium chloride electrolyte. Using a duration time of T_{on} = 0.25 ms and a duty factor γ = 0.05, the average current density can be increased up to j_m = 8 A/dm^2.

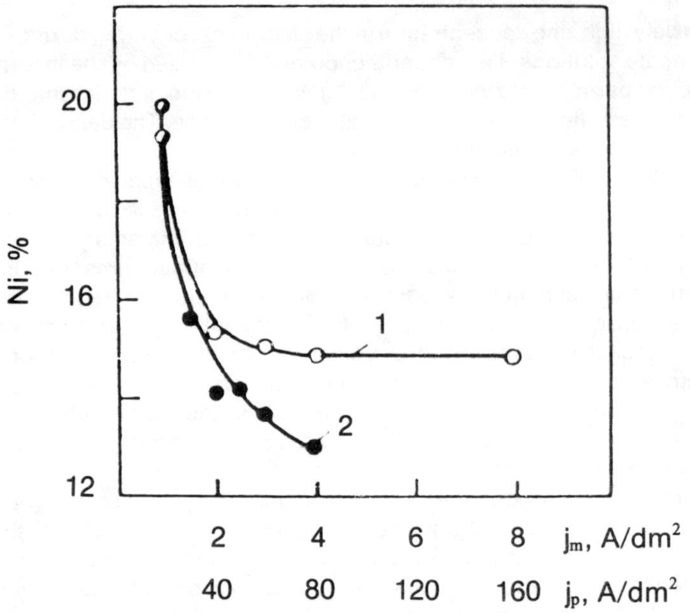

Fig. 7.3—Influence of current density on nickel content.
1: pulse plating, $T_{on} = 0.25$ ms, $\delta = 0.05$
2: direct plating

Under direct current conditions, a limiting value of $j = 4$ A/dm^2 was found. The change of the nickel content with the current density is shown in Fig. 7.3. The nickel content of pulse plating layers is enhanced if compared to deposits plated under direct current conditions of the same average current density. All electrodeposited zinc-nickel alloys consist of cubic γ- and tetragonal δ-phases, but the proportion is changed by pulse plating conditions and new textures are found. In addition, a smaller grain size is received if the deposits are formed using a duration time of $T_{on} = 0.25$ ms and break time of $T_{off} = 4.75$ ms if compared to direct current conditions or longer duration times of the same average current density $j_m = 2$ A/dm^2. Similar results were found by Paatsch (15). Further on, in comparison to direct current plating conditions, an increase of the macrothrowing power can be observed if a duration time of $T_{on} = 1$ ms, a duty factor of $\gamma = 0.09$ and an average current density of $j_m = 5.2$ A/dm^2 is used (electrolyte: 75 gl $ZnSO_4$, 70 g/l $NiSO_4$, pH = 12-12.5, T = 20° C).

3. Cadmium

The use of cadmium is very restricted because of its toxicity. Therefore, very little information has appeared in the literature on the use of pulsed current and the results reported are very similar to those reported for zinc in Ref. 1. Ibl and Puippe (17) found using an acid electrolyte (0.75 M $CdSO_4$, 1 M H_2SO_4) as well as a cyanide bath (0.23 M Cd^{2+}, 170 g/l NaCN, 18 g/l NaOH, organic additives) that the covering power was much better and the grains finer if the layers were pulse plated instead of deposited under comparable direct current conditions. The quality of the deposits was nearly as good as that obtained from commercial baths with organic additives. This means that pulse plating allows the use of plating baths with little or no organic additives, which are relatively expensive, or at least with a reduced amount. Further investigations on the influence of the break time on the deposit morphology (18) show that unlike copper and gold deposition, cadmium layers show a reduced grain size with longer T_{off} periods.

It is suggested that corrosion during the break time deactivates the surface and inhibits surface diffusion, resulting in an increased nucleation rate during the following pulse period. Further work on the influence of pulse plating on the properties of deposited cadmium using commercial low acid and cyanide solutions show no significant differences with respect to the corrosion behavior and the wear resistance (19). Yahalom (20) finally reported that as in pulse plating of zinc, hydrogen embrittlement of high strength steel can be reduced by pulsed electrodeposition of cadmium.

4. Summary

The pulsed current deposition of zinc and cadmium as well as zinc-cobalt and zinc-nickel alloys has been established as superior to that of conventional direct current deposits. The major advantage is a comparably smaller grain size resulting in more compact and less porous layers of bright appearance. Therefore, pulse plating allows the use of plating baths without organic additives (or at least with a reduced amount), being easier to handle and less expensive. Furthermore, pulse plating of zinc and cadmium reduces the risk of hydrogen embrittlement when plating high strength steel. When plating zinc alloys, especially zinc-nickel layers, the plating rate can be doubled and the nickel content can be changed to a large extent.

References

1. W. Paatsch; "Electrodeposition of Zinc by Pulse Plating," *Galvanotechnik*, **71** : 107 (1980)
2. R.I. Gurovich and A.K. Krivtsov; "Copper and Zinc Plating by Pulsating Current," *Zhurnal Prikaldnoi Khimii*, **41** : 1227 (1968)
3. N.A. Kostu and A.K. Krivtsov; "Application of Pulsating Current for Bright Zinc Deposition," *Zaschita Metallov*, **18** : 634 (1983)
4. I. Gadshov and I. Nenov; "Potential Distribution for Zinc Deposition by Direct and Pulsating Current," *Metalloberfläche*, **38** : 199 (1984)
5. D.T. Chin, R. Sethi and J. McBreen; "Zinc Electrode Morphology in Alkaline Solutions," *Journal Electrochemical Society*, **129** : 2677 (1982)
6. K. Appelt and K. Jurewicz; "Zinc Deposition from Alkaline Solutions by Three-Component Impulse Current," *Electrochimica Acta*, **24** : 253 (1979)
7. V. Romanov; "An Investigation of the Causes of Formation of Zinc Sponge During Electrolysis of Zincate Solutions," *Zhurnal Prikladnoi Khimii*, **36** : 1057 (1963)
8. A.R. Despic and K.I. Popov; *Journal Applied Electrochemistry*, **3** : 161 (1973)
9. H.Y. Cheh; "Electrodeposition of Gold by Pulsed Current"; "The Limiting Rate of Deposition by P-R Plating," *Journal Electrochemical Society*, **118** : 551 and 1132 (1971)
10. W. Paatsch; "Electrodeposition of Zinc by Pulse Plating," *1st AES International Pulse Plating Symposium, Ext. Abstracts*, Boston (1979)
11. W. Paatsch; "Investigation of Hydrogen Embrittlement of High Tensile Strength Steel During Electrodeposition Processes. Part I: Hydrogen Permeation Experiments," *Metalloberfläche*, **33** : 546 (1979)
12. W. Paatsch; "Investigation of Hydrogen Embrittlement of High Tensile Strength Steel During Electroplating Processes, Part II: Tensile Tests," *Metalloberfläche*, **34**: 174 (1980)
13. M. Froment and G. Maurin; *Electrodeposition and Surface Treatment*, **3** : 245 (1975)
14. N. Ibl; "Pulsed Electrodeposition of Metals," *Metalloberfläche*, **33** : 51 (1979)

15. W. Paatsch, in preparation
16. I. Nenov, I. Gadshov and K. Pangarov; "Composition and Structure of Direct Current and Pulse Plated Zinc-Nickel Deposits," *Galvanotechnik,* **75** :1107 (1984)
17. N. Ibl and J.Cl. Puippe; "The Morphology of Pulse-Plated Deposits," *Plating and Surface Finishing,* **68** (Oct., 1980)
18. J.Cl. Puippe and N. Ibl; "Influence of Charge and Discharge of Electrical Double Layer in Pulse Plating," *Journal of Applied Electrochemistry,* **10** : 775 (1980)
19. R. Sanders and W. Bayer; "Pulse Plating of Cadmium," Diploma thesis, Technical High School, Aachen, W. Germany (1980)
20. J. Yahalom; "Properties of Pulse-Plated Cadmium," *1st AES International Pulse Plating Symposium, Ext. Abstract* Boston (1979)

Chapter 8

Pulsed Electrodeposition of Nickel and Chromium

B. Sutter

1. Nickel

The multifarious applications of nickel plating in fields ranging from electronics to heavy engineering call for a wide range of properties, the relative importance of any of which depends on the particular application involved. For example, ductility is of prime importance in electroforming, or in the plating of parts which are subsequently deformed in manufacturing or assembly operations; low porosity is the main requirement when the coating, either alone or in conjunction with top coatings such as gold or chromium, is to function as a decorative/protective coating or as a diffusion barrier; internal stress is important as affecting ductility and porosity or cracking of deposits, particularly when thick coatings are required; hardness and frictional properties play an important part in wear-resistant applications.

As already indicated, these properties are to a considerable extent interrelated, and modification of coating characteristics in various directions can be achieved by the use of various types of bath, operational conditions and additive systems, sometimes in conjunction with modulation of the current supply; an early example of which is the superimposition of AC on DC as a means of reducing stress in deposits. In this context, the use of various forms of pulsed current plating offers interesting possibilities, and this aspect has been widely studied particularly in recent years, by a number of workers.

The results obtained, insofar as they affect the structure and resulting properties of coatings, often apply to baths without the use of additives, and, although varying widely, show the same general tendencies with respect to the effect of pulsed current plating, in the frequency range from 10 to several hundred Hz, on properties such as grain size, brightness, hardness, tensile stress and porosity. In the following, significant results are reviewed under the headings of individual properties.

1.1. *Structure and appearance*

Pulsed current plating clearly influences on the structure and appearance of deposits, the effects being in general more marked with higher peak current density. Results pertaining to various types of bath are discussed in the following sections.

1.1.1. Chloride baths

Kendrick (1) produced bright and semi-bright deposits from a chloride bath without additions by pulsed reverse current plating under the following conditions:

cycle frequency	10 to 100 Hz
$j_{p(a)} / j_{p(c)}$	1/2 to 3/4
$T_{on(a)} / T_{on(c)}$	0.10 to 0.25
cathodic current density $j_{p(c)}$	40 A/dm^2

There was, however, an accompanying increase in tensile stress and brittleness. At higher frequencies ($\geqslant$ 400 Hz), with high anodic current density ($j_{p(a)} / j_{p(c)} = 0.5$ to 1.5) and anodic pulse lengths ($T_{on(a)} / \theta = 0.15$ to 0.30) deposits were mat and far more ductile.

Under similar conditions, Crossley, Kendrick and Mitchell (2) observed a spectacular modification in the internal microstructure of deposits. Whereas direct current plating produced coatings with no preferred orientation, heavily distorted grains, and high dislocation density, pulsed reverse current plating at low frequency (10 Hz) resulted in a smooth surface, with fine grains and low dislocation density under the following conditions:

$j_{p(c)}$	20 A/dm^2
$j_{p(a)} / j_{p(c)}$	0.5
$T_{on(a)} / \theta$	0.25

When the frequency was increased to 400-500 Hz, the structure became coarse-grained and the surface rough.

At intermediate frequencies (e.g., 75 Hz) a mixed structure of fine and coarse grains was observed. This corresponds to the break in the curve of Fig. 8.1 relating the number of grains per unit area to the frequency, from which it may also be noted that an increase in grain size coincides with a decrease in tensile stress.

In the above cases, deposits obtained by pulsed reverse current plating showed a preferred orientation, in contrast to those produced by direct current plating.

1.1.2. Watts bath

Investigations into the effect of pulsed current plating on the structure of deposits from a Watts type solution without additives have generally confirmed an increase in brightness, reduction in grain size, and a progressive transition of structure from columnar to lamellar.

Using a pulsed current density of 9900 A/dm^2, with T_{on} 0.3 ms and T_{off} 10 ms, Viswanathan and Raub (5) reported increasing grain refinement with increasing pulsed current density, with attenuation of pyramid-shaped grains and of the columnar structure, whereas in a bright nickel bath, pulsed current produced no significant modification of structure. W. Kleinekathöfer, Ch. Raub and E. Raub (4) likewise found finer grain size, with evolution from a columnar to a lamellar structure by the application of pulsed current with $j_p = 104$ A/dm^2, $T_{on} = 1$ ms, and $T_{off} = 20$ ms.

Results obtained by Kleinekathöfer, Muramaki and Raub (3) relating to the structural features of pulsed current deposits over a range of parameters are shown in Table 8.1.

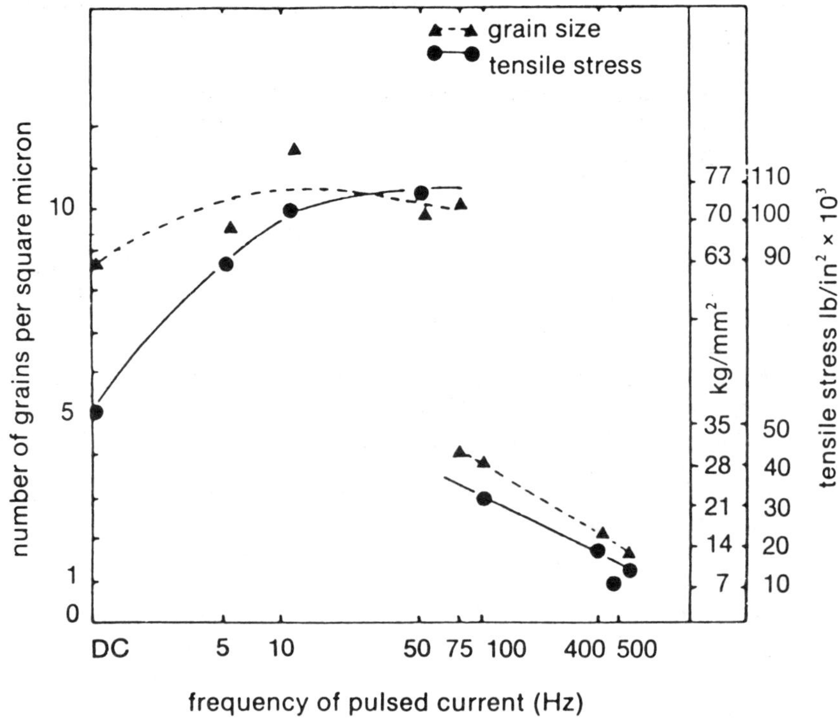

Fig. 8.1—Relationship between grain size, tensile stress and pulse reverse frequency (2).

TABLE 8.1

Structure and Microhardness of Nickel Deposits Made from a Watts Bath as a Function of Pulse Parameters

pulsed current density $j_{p(c)}$ (A/dm^2)	frequency (Hz)	$T_{on(c)}$ / θ %	structure	microhardness (HV 0.01)
direct current	—	—	columnar	191
10	500	50	columnar	223
10	50	50	fine-grained	279
25	400	20	fine-grained columnar	259
25	80	20	fine-grained columnar	367
55	910	9	fine-grained	261
55	91	9	fine-grained	385
100	480	5	fine-grained columnar	262
100	48	5 %	fine-grained columnar	389

The effect of pulsed current plating on deposits from baths containing small quantities of brightening additives was studied by Robert (6) under the following conditions:

average current density j_m	5 A/dm^2
T_{on} / θ	10 to 25 %
Frequency θ^{-1}	500 Hz ($\theta = 2$ ms)

With brightener concentrations from 0 to 1 percent, coatings produced were light and bright, whereas direct current plating gave only dull, dark deposits.

Sutter (7) also found an improvement in appearance of deposits from an additive-free Watts bath, from mat with direct current to a satin finish with pulsed current with average current density 5 A/dm^2, T_{on} 1 ms, T_{off} 19 ms.

The structure of the corresponding DC and pulse plated deposits is shown on micrographic cross-sections in Figure 8.2. The columnar structure which results from the use of DC becomes a combination of columnar and lamellar when using pulsed current.

Grain refinement was also observed in the work of Tai-Ping Sun et al. (8). Using pulse plating with average current density 8 A/dm^2, T_{on} / θ ratios of 0.75, 0.5 and 0.25 and frequency of 12.5 Hz, the effect increased with reduction in the pulse ratio, i.e., increase in pulse current density. Still finer grain structure resulted from the application of pulsed reverse plating, using the same average current density, but with $j_{p(c)} = j_{p(a)}$; $T_{on(c)}$ 60 ms, $T_{on(a)}$ 20 ms, and frequency 12.5 Hz ($\theta = 80$ ms).

1.1.3. Sulfamate bath

The effects of pulsed current plating in sulfamate baths are similar to those reported for Watts baths, i.e., increased fineness of grain, and production of bright deposits from additive-free solutions (9,10). Typical conditions for such results are the following:

j_p	100 A/dm^2
T_{on} / θ	0.1
Frequency θ^{-1}	100 Hz ($\theta = 10$ ms)

1.1.4. Woods bath

A significant increase in brightness, together with a reduction in the fissuring that is characteristic of this type of deposit in direct current plating, has been reported by Robert (6) under the following pulsed current conditions:

Average current density, j_m	3 A/dm^2
T_{on} / θ	0.15 to 0.2
Frequency, θ^{-1}	100 Hz ($\theta = 10$ ms)

1.1.5. Nickel-phosphorus electrolytes

Studies on the pulsed current deposition of nickel-phosphorus alloys from a bath containing nickel sulfate, nickel chloride, phosphoric and phosphorous acids, have been made by Lashmore and Weinroth (11,12). As compared with direct current plating, pulsed current deposits showed a finer and a more

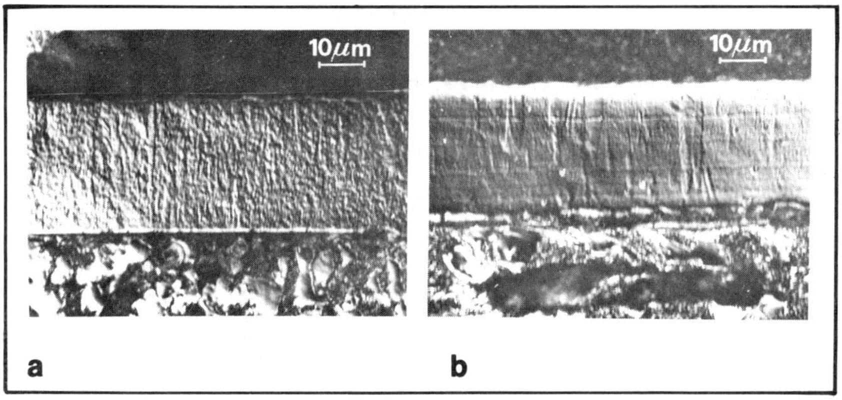

Fig. 8.2—Micrographic cross-sections (a) of DC plated nickel from an additive-free Watts bath ($j = 5$ A/dm^2) and (b) of pulse plated nickel from an additive-free Watts bath ($j_m = 5$ A/dm^2, $T_{on} = 1$ ms, $T_{off} =$ 19 ms) (7).

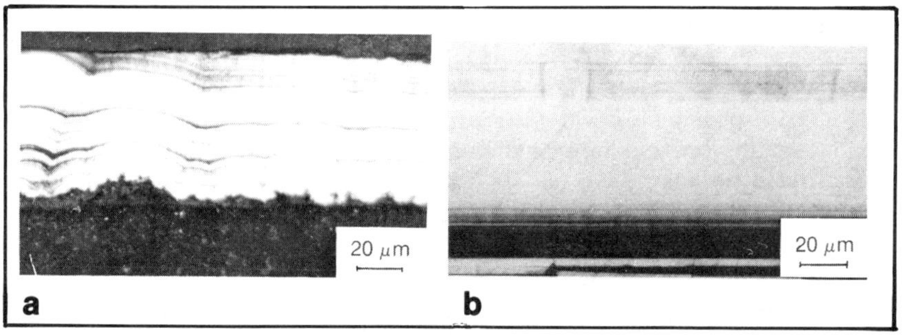

Fig. 8.3—Etched cross sections of (a) DC plated and (b) pulse plated nickel alloys containing 12 percent phosphorus by weight (12).

uniform layer spacing in the lamellar structure (Fig. 8.3). Furthermore, the deposits had an increased phosphorus content, and the local phosphorus atom environment more closely resembled that of electroless nickel.

1.2. *Hardness*

The effects of pulsed current plating on the hardness of deposits is directly related to structural modifications. Deposition parameters that bring about a reduction in grain size and increase in brightness usually increase hardness.

1.2.1. Chloride bath

In tests with pulse reverse current, Kendrick (1) obtained deposits from chloride baths which, depending on the frequency, were either harder or softer than those produced by direct current plating at the same average current density (Table 8.2).

1.2.2. Watts bath

Results obtained concerning the effect of a range of pulsed current parameters on the hardness of deposits from a Watts bath, with constant average current

TABLE 8.2

Microhardness of Nickel Deposits Made from a Chloride Bath as a Function of Pulse Reverse Plating Parameters (1)

frequency (Hz)	cathode current density, $j_{p(c)}$ (A/dm^2)	anode current density, $j_{p(a)}$ (A/dm^2)	$T_{on(a)} / \theta$ (%)	microhardness (HV 0.05)
direct current	20	—	—	290
10	20	10	25	368
50	20	5	25	345
100	20	30	25	249
400	20	15	25	171
500	20	20	25	145

density of 5 A/dm^2 (4) are included in Table 8.1. There is a general increase in hardness over that of DC coatings, but the maximum effect occurs with high pulsed current densities and low frequencies. Sutter et al. (13) have demonstrated the possibility of varying the hardness of deposits at will by adjustment of pulsed plating parameters, as shown in Fig. 8.4, where average current density and pulse frequency are constant and hardness is controlled by the pulsed current density. Similar observations have been made by Olson (14) and Tei Ping Sun et al. (8).

1.2.3. Sulfamate bath

Effects here are similar to those observed with the Watts bath. A significant increase in hardness is obtained at intermediate frequencies and high pulsed current densities (10) as illustrated in Fig. 8.5.

1.2.4. Nickel-Phosphorus bath

Results obtained by Lashmore and Weinroth (12) during the work described in section 1.1.5. above are shown in Fig. 8.6. With phosphorus contents of over 4 percent, alloy deposits produced by pulsed plating are harder than DC coatings, the effect increasing with increasing phosphorus content.

1.3. *Porosity*

Although relatively little work has been reported on this aspect (5,6,10), it seems fairly well established that the use of pulsed current plating parameters that result in grain refinement will also lead to a reduction in porosity of deposits as compared with direct current plating.

This effect is illustrated in Fig. 8.7 for a Watts solution (13), and in Fig. 8.8 for a sulfamate bath (15).

1.4. *Ductility*

While very ductile but soft deposits have been obtained by pulsed reverse current plating in additive-free chloride baths at frequencies in the order of 500

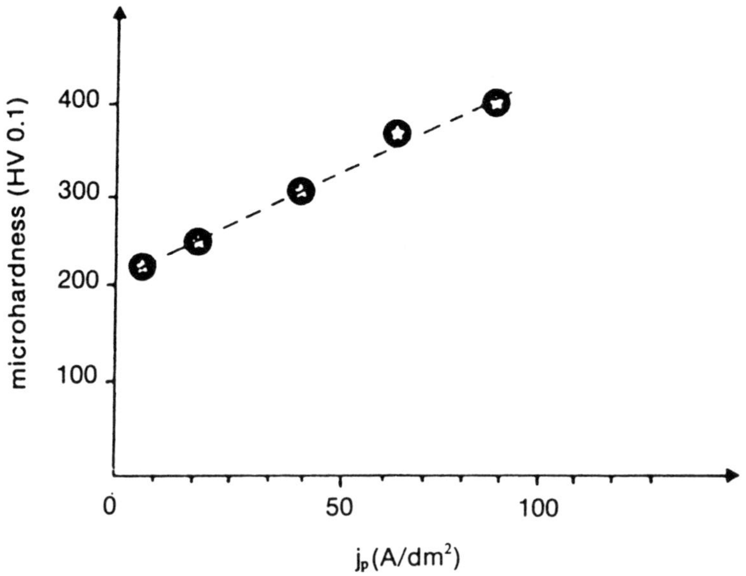

Fig. 8.4—Vickers hardness of pulse plated Watts nickel versus pulsed current density with $j_m = 5$ A/dm^2 and $\theta^{-1} = 50$ Hz (13).

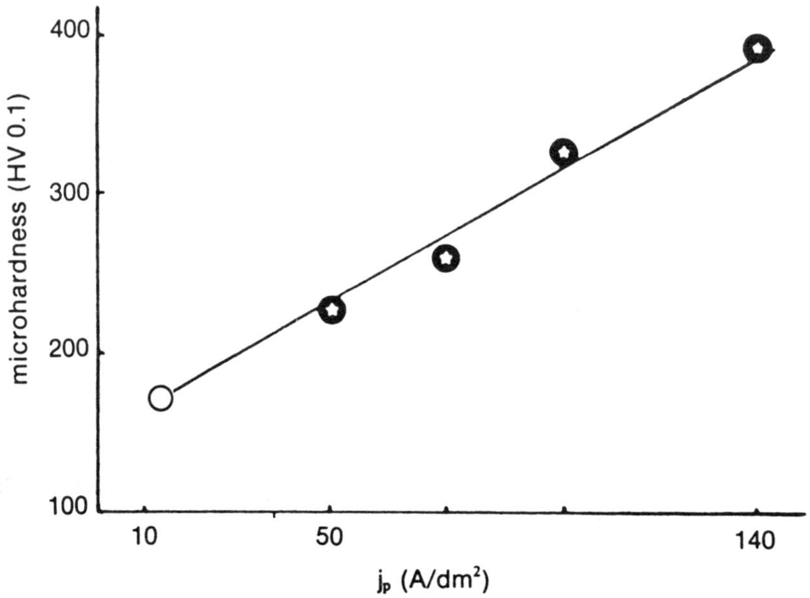

Fig. 8.5—Vickers hardness of pulse plated sulfamate nickel deposits versus pulsed current density with $j_m = 10$ A/dm^2 and $\theta^{-1} = 100$ Hz, O : DC (10).

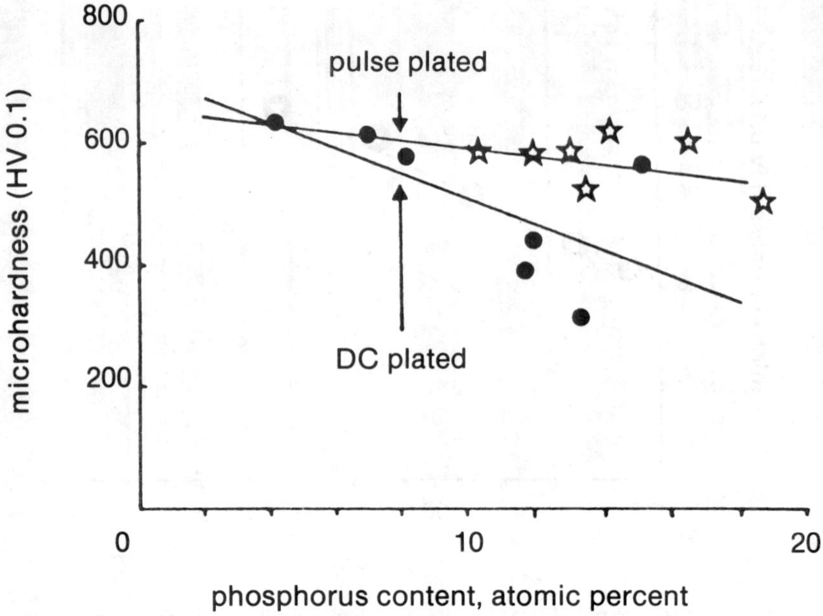

Fig. 8.6—Microhardness as a function of the phosphorus content of pulse plated and DC plated Ni-P alloys (12).

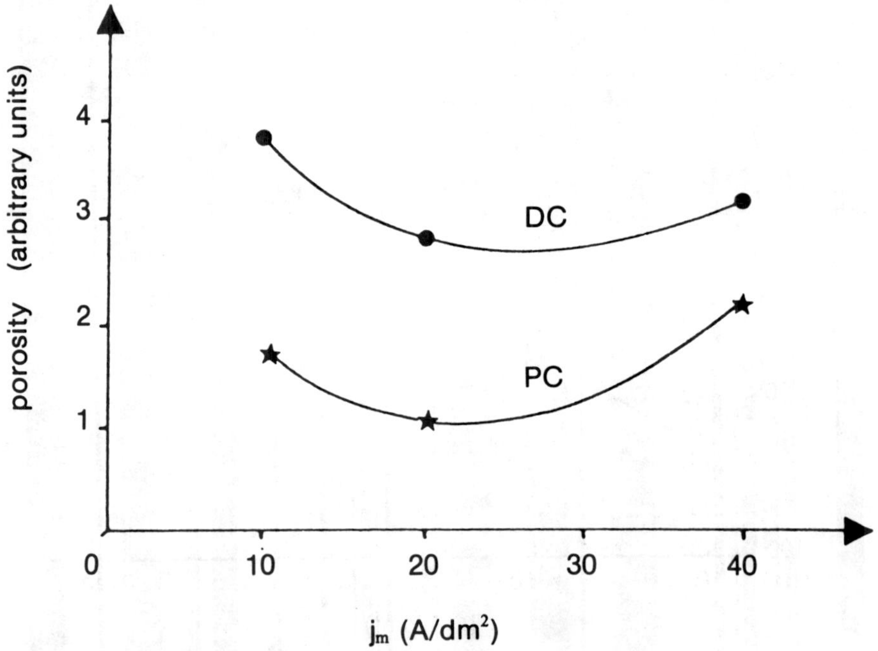

Fig. 8.7—Porosity of pulse plated and DC plated Watts nickel deposits as a function of average current density j_m and j respectively. T_{on} = 1 ms, T_{off} = 9 ms, deposit thickness = 2 μm (13).

Hz (1,2), it appears that pulse plating has only a limited effect in the case of bright nickel baths. Nevertheless, it offers some improvement in bright nickel deposits by allowing a reduction in the concentration of additives (6).

1.5. *Internal stress*

Internal stress in nickel deposits can be modified by the type of electrolysis current used (14). With Watts and chloride baths an increase has been observed in pulse plating with low duty cycles (3).

However, as shown in Fig. 8.1, pulsed reverse current plating under certain conditions can lead to a substantial decrease in internal stress of coatings from a chloride bath (1,2), and it is also possible to reduce stress in deposits from a Watts bath by superimposing pulsed current of low frequency (20 Hz), as indicated in Fig. 8.9 (16).

In the case of the sulfamate bath, internal stress is strongly influenced by the current form (3), increasing with decreases in the duty cycle (9) and providing the possibility of controlling stress levels without the use of additives.

1.6. *Throwing power*

Macrothrowing power is marginally reduced in pulsed current plating in chloride, Watts, and sulfamate type baths, a reduction in duty cycle bringing about a decrease (3). This observation is in accord with theoretical considerations by Ibl (17) (see also Chap. 6), which predict an adverse effect on secondary current distribution in pulse current plating. Microthrowing power, on the other hand, may be either increased or decreased relative to direct current plating, depending on electrolysis conditions (3).

This difference in response of secondary and tertiary throwing power to pulse plating (17) explains the difficulty in formulating a general rule, particularly in view of the marginal nature of the effect. For example, an in-depth study using a Watts bath (6) "did not give conclusive evidence either or an improvement or a worsening directly following the application of pulsed current." It nevertheless appears that there is a tendency for reduction in macrothrowing power and an improvement in microthrowing power in pulse plating under conditions in which the structure of the deposit is significantly modified.

2. Chromium

2.1. *Introduction*

With few exceptions, chromium is deposited from aqueous solutions of chromic acid. These electrolytes operate with low current efficiency (10-20 percent) (18) as compared with most electroplating baths and require the use of high current densities (20-50 A/dm^2) in direct current plating. Hence the cost of chromium plating rectifiers is higher than that of equipment used in other types of plating, representing a substantial part of overall processing costs.

Thus pulsed current plating, involving still more expensive current generators, is, on a first view, at some disadvantage, unless the improvements which may result are spectacular and/or the electrical power requirements do not too greatly exceed those required in conventional direct current plating.

For these reasons, applications of pulsed current to chromium plating are still relatively limited, normally being restricted to cases in which some improvement can be obtained without the use of excessive pulsed current densities.

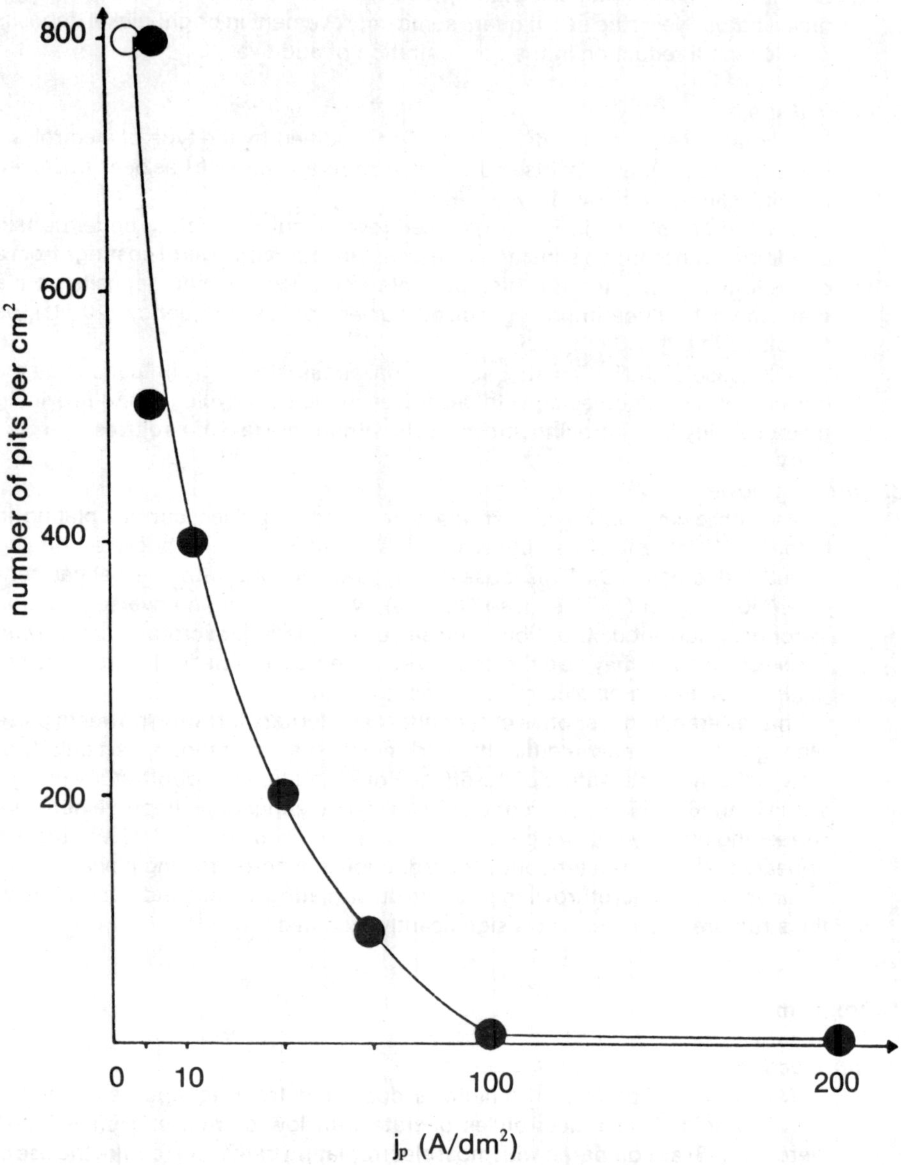

Fig. 8.8—Porosity of pulse plated sulfamate nickel deposits as a function of pulsed current density, $j_m = 10$ A/dm^2, $\theta = 10$ ms, deposit thickness = 0.5 μm, • DC (10 A/dm^2); o PC (15).

These applications fall into two main categories:

— modification of structure, and especially the realization of crack-free deposits;

— production of alloy deposits which cannot be obtained by direct current plating.

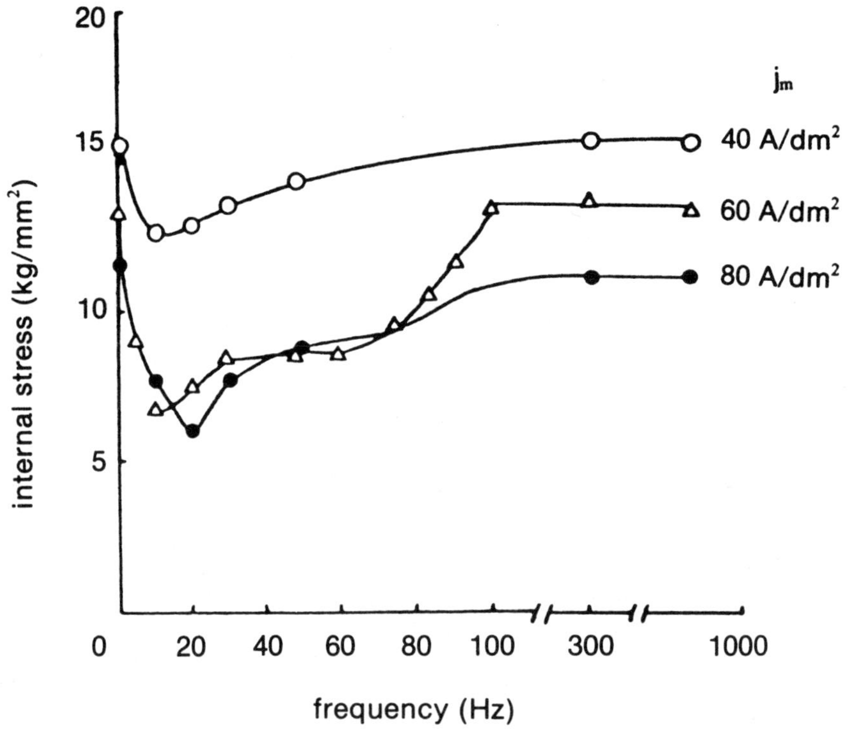

Fig. 8.9—Internal stress of Watts nickel deposits as a function of frequency of superimposed PC ($j_p = 1.6\ j_m$, $T_{on}/\theta = 0.5$) (16).

2.2 *Modification of Structure*

In the electrodeposited condition, the properties of chromium differ radically from those applicable to the metal as produced by conventional metallurgical processes, particularly in respect to a very high hardness and interesting frictional properties. A further characteristic feature of chromium plate is its tendency to spontaneous cracking. Thin deposits ($< 0.5\ \mu m$) are usually crack-free, but those with thicker coatings, i.e., hard chromium deposits in general, develop cracks.

According to Snavely (19), cracking is related to the formation of an unstable chromium hydride during the electrolytic process, which decomposes spontaneously, even at ambient temperatures, resulting in a contraction of the coating by about 15 percent, which in turn leads to crack formation, with a reduction in the protective value of the coating.

With this in mind, Faust et al. (20) in 1961 investigated the possibility of pulse plating as a means of realizing crack-free coatings. It was assumed that the contraction due to the decomposition of each hydride layer caused cracking of subsequent layers. This being so, the on-time of a pulse must be restricted to produce only a monolayer of deposit, followed by a relatively long off-time to allow for decomposition of the deposited layer before continuing build-up of the coating. The result of such a process should be a crack-free chromium deposit in which no unstable layer is covered by another layer.

On this principle it was, in fact, possible to obtain crack-free chromium deposits directly on aluminum, using a conventional chromium bath under the following conditions (20):

T_{on}	0.5 to 5 ms
T_{off}	0.5 to 2.5 ms
j_p	15.5-23 A/dm^2
Temperature	60° C

There was a reduction in cathode efficiency and a loss of some 50 percent in hardness. Deposits produced in this way have a columnar structure and remain uncracked even after heating to redness. The coating is semi-bright in appearance, but can be polished, and its protective value far exceeds that of conventional cracked chromium.

Crack-free, low-stress deposits have also been produced by Schreiner and Raub (21) with pulsed current plating using a bath containing silicofluorides. This type of finish, despite of its low stress, offers no advantage over conventional chromium plating with respect to its effect in reducing the fatigue strength of a steel substrate (22). The latter aspect has also been studied by Knödler (27) and Logan (28), both authors noting a reduction in fatigue strength of steel when plated with crack-free chromium produced by pulsed current. Knödler attributed the behavior in fatigue tests to stress concentration resulting from the formation of major cracks penetrating to the substrate, and offered a number of suggestions relating to the choice of electrolysis parameters to reduce stress concentration in crack-free deposits, including:

— low concentration of chromic acid in bath
— low current density
— high plating temperature
— short on- and off-times

Conditions used by Schreiner and Raub (23) for the deposition of crack-free deposits were as follows:

Electrolyte	
Chromic acid, CrO_3	250 g/l
Sulfuric acid	2.5 g/l
Potassium silicofluoride, K_2SiF_6	2.0 g/l

Pulse plating parameters	
j_p	40 A/dm^2
T_{on}	1 ms
T_{off}	2 ms

The deposits produced showed much better anodic behavior than that of conventional cracked chromium, particularly in the presence of chloride, and demonstrated a clear improvement in corrosion resistance.

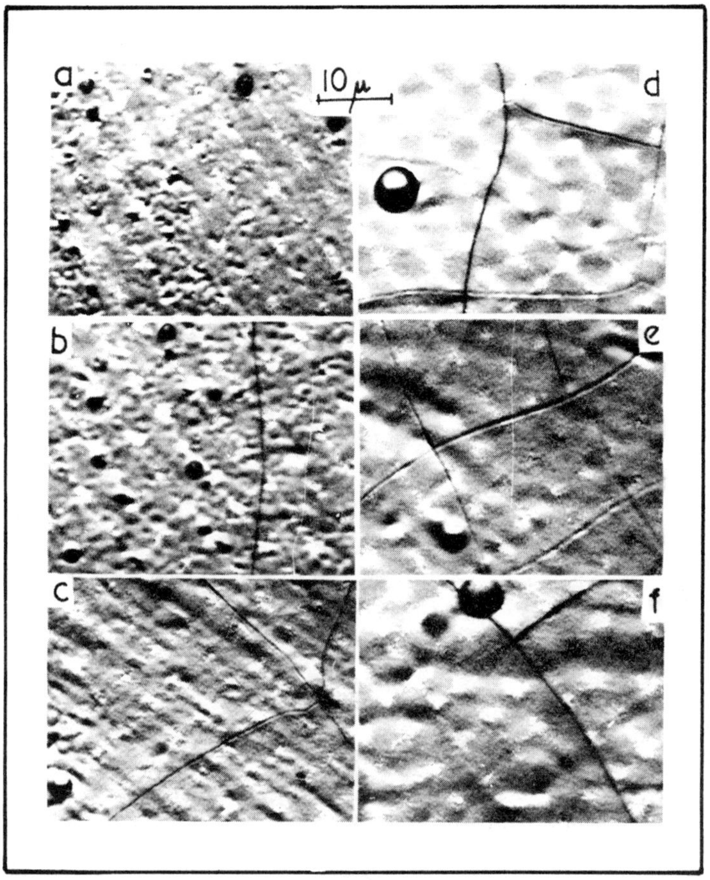

Fig. 8.10—Growth of a conventional Cr deposit on mild steel. Total thickness in μm (a) 1.0, (b) 1.5, (c) 3.0, (d) 4.5, (e) 6.0, (f) 12.5 (25).

Working with a conventional chromium bath (250 g/l CrO_3 : 2.5 g/l H_2SO_4) at 40° C, with current density of 22 A/dm^2, Saiddington and Hoey (24,25) observed that the formation of cracks was associated with incremental increases of deposit thickness (Fig. 8.10) and hence sought to realize crack-free coatings by a similar principle to that used by Faust (20), i.e., by interruption of electrolysis before individual layers reached the critical thickness. Several types of interruption were tried:

— switching off current for periods of 1 second to 2 minutes;
— withdrawing the cathode from the electrolyte for similar periods;
— reversing polarity, giving an anode current density of 4.4 A/dm^2, or by imposing an anode current density of 5 A/dm^2 for a fraction of a second.

Conclusions from this study were as follows:

— The incidence of crack formation is considerably reduced when the frequency of interruptions is such as to produce individual layers of approximately 0.5 μm;

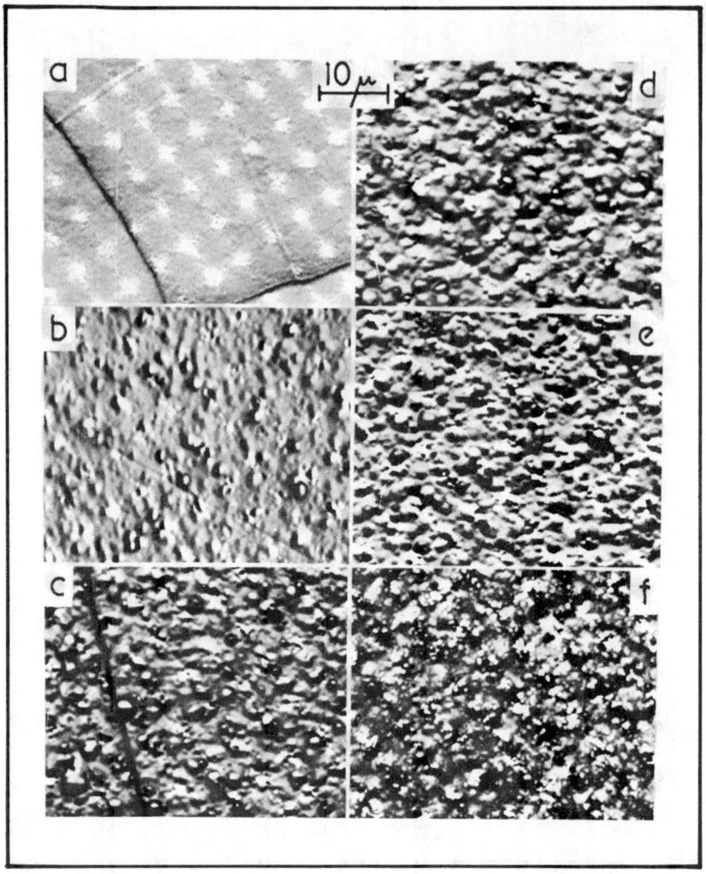

Fig. 8.11—Effect of reversed current interruption at decreasing intervals. 3 μm of Cr on bright Ni. Number of interruptions and thickness levels in μm: (a) none, (b) one (1.5 + 1.5), (c) two (1.0 + 1.0 + 1.0), (d) three (0.75 + 0.75 + 0.75 + 0.75), (e) five (6 x 0.5), (f) eleven (12 x 0.25) (25).

— Simple current interruptions for 1 second with or without withdrawal of the cathode have no effect;

— Interruptions are effective only when they result in discontinuity in the growth of deposits, illustrating the necessity of dissipating the layer of electrolyte in contact with the cathode. This may be achieved by withdrawals of the cathode for long periods, which reduces cracking, but most effectively by reversal of the direction of current flow, when the cathode film is dispersed as a result of the sudden increase in potential.

Using the latter process, crack-free plating was possible by manual reversal of current for a fraction of a second at intervals corresponding to the deposition of a layer of 0.35 μm. Similar results were obtained using a more highly concentrated bath, the average rate of plating not being appreciably affected. The elimination of cracks considerably increased the protective value of deposits, but there was a reduction in brightness which was more marked with

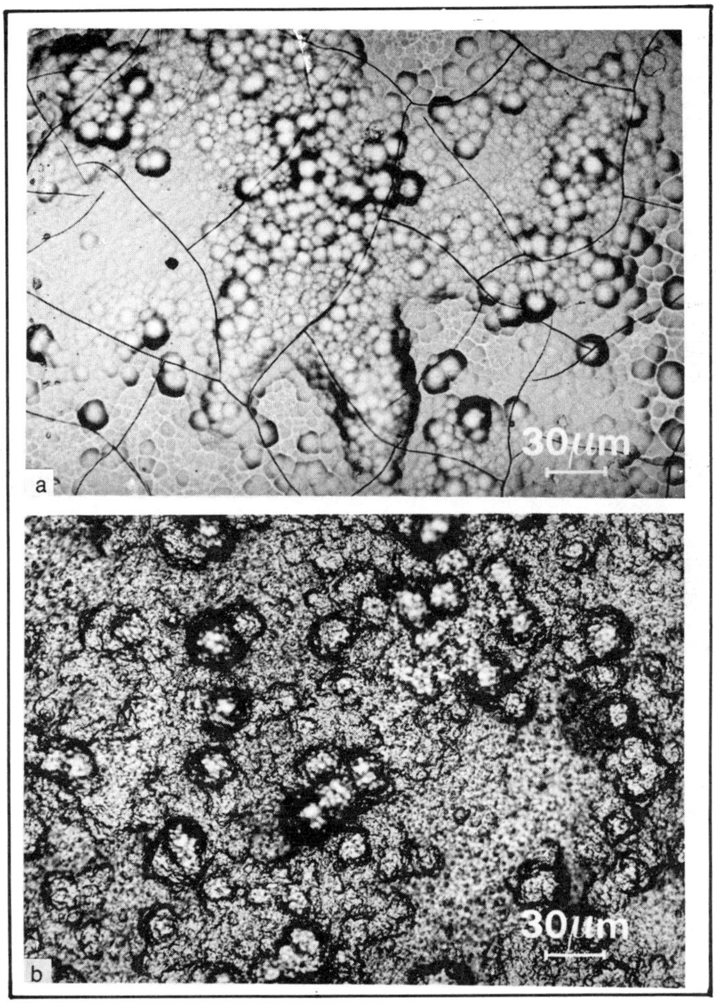

Fig. 8.12—Micrographic pictures of Cr surfaces obtained (a) by DC, $j = 50$ A/dm^2; (b) by PR, $j_{p(c)} = 50$ A/dm^2, $j_{p(a)} = 50$ A/dm^2, $T_{on(c)} = 5$ s, $T_{on(a)} = 5$ ms. The plating bath consists of CrO_3 250 g/l and H_2SO_4 2.5 g/l at 50° C (29).

increase in the frequency of interruption. Surface examination indicated that this was due to the formation of numerous nucleation sites following each interruption, which favored the growth of tiny nodules, causing dispersion of light and reduction in surface reflectivity (25). The effect of reverse current interruptions at various intervals on the surface appearance and cracking of deposits is illustrated in Fig. 8.11. It was considered that the numerous nucleation sites could also contribute to reduction in cracking by interfering with cumulative tensile stress.

Following a somewhat similar procedure, crack-free chromium deposits have also been produced by Mortier et al. (29) and Sutter (7) by pulsed reverse current

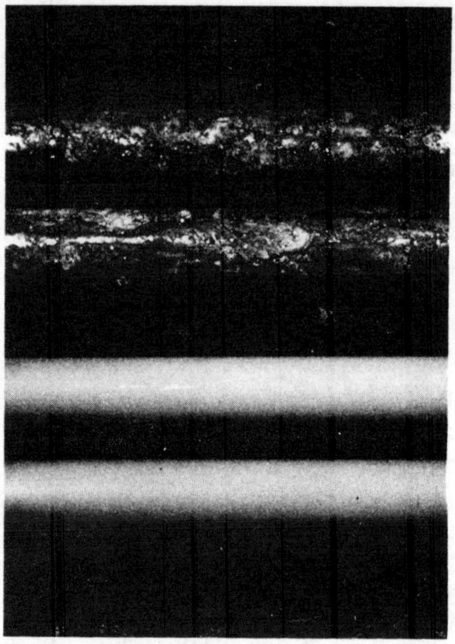

Fig. 8.13—Rods coated with 15 μm of chromium after 72 hours exposure to salt spray. (a) and (b) correspond to experimental conditions a) and b) described in caption of Fig. 8.12 respectively. (a) DC: the entire surface of the rod is covered with rust; (b) PR: grey mat appearance with no trace of corrosion. (Rod diameter 16 mm).

plating from a conventional chromium bath, operated at 50° C, and using the following parameters:

$j_{p(c)}$	30 - 50 A/dm^2
$j_{p(a)}$	5 - 50 A/dm^2
$T_{on(c)}$	5 - 15 s
$T_{on(a)}$	30 - 50 ms

The deposits produced were grey, mat in appearance, but readily polished. Hardness was reduced as compared with conventional chromium plate (850 HV, cf. 1000 HV), but nevertheless retained a useful level. Surfaces of DC and pulsed reverse current deposits are shown in Fig. 8.12 and the improvement in corrosion resistance of crack-free coatings is illustrated by the results of a salt spray corrosion test shown in Fig. 8.13.

Nesnidal (26) has studied the production of crack-free chromium deposits from a conventional chromium bath by the application of pulse plating over a range of conditions, with duty cycles varying from 25 to 75 percent and cycle times from fractions of a millisecond to over 100 milliseconds. She concluded that essential features in permitting the production of crack-free coatings by pulsed current were the escape of hydrogen from coatings during the off periods and the fact that the deposit cannot be coherent and hence requires no stress relief.

2.3. *Concluding remarks*

The foregoing review establishes that production of crack-free chromium is possible using two types of pulsed current:

— pulse plating with on- and off-times of approximately 1 millisecond;

— pulsed reversed current plating with short anodic periods and very low frequencies (< 1 Hz).

In either case, crack-free deposits are grey mat in appearance, with a significant reduction in hardness but greatly increased protective value as compared with conventional coatings.

References

1. R.J. Kendrick; *Transactions of the Institute of Metal Finishing,* **44** : 78-83 (1966)
2. J. A. Crossley, R.J. Kendrick, W.I. Mitchell; *Transactions of the Institute of Metal Finishing,* **45** : 58-63 (1967)
3. W. Kleinekathöfer, T. Muramaki, Ch. J. Raub; *AES International Pulse Plating Symposium,* American Electroplaters Society, Boston (1979)
4. W. Kleinekathofer, Ch. J. Raub, E. Raub; *Metalloberfläche,* **36** (9): 411-420 (1982)
5. M. Viswanathan, Ch. J. Raub; *Galvanotechnik* **66,** (4): 277-282 (1975)
6. E. Robert; *Oberfläche Surface,* **24,** (12): 413-420 (1983)
7. Sutter, B.; *Oberfläche Surface,* **25,** (1): 16-19 (1984)
8. Tai Ping Sun, C.C. Wan, Y.M. Shy; *Metal Finishing,* **72:** 33-38 (1979)
9. W. Kleinekathöfer, Ch. J. Raub; *Surface Technology,* **7:** 23-34 (1978)
10. M.A. Laroche, F. Mortier, B. Sutter, G. Vasseur; *Cetim-Informations,* **86:** 38-42 (1984)
11. D.S. Lashmore, J. Weinroth; *AES Second International Pulse Plating Symposium,* American Electroplaters Society (1981)
12. D.S. Lashmore, J. Weinroth; *Plating and Surface Finishing,* **69,**(8): 72-76 (1982)
13. B. Sutter, F. Mortier, M.A. Laroche, J. Reby, G. Vasseur; *Interfinish '84* (October 21-26, 1984)
14. R. Olson; *Plating and Surface Finishing,* **68,** (2): 38-39 (1981)
15. F. Mortier; *Cetim-Informations,* **72,** 21 (1981)
16. M. Yamamoto, T. Sato; *Bulletin—Japan Society of Precision Engineering,* **14** (2): 119-120 (1980)
17. N. Ibl; *AES International Pulse Plating Symposium,* American Electroplaters Society (1981)
18. G. Dubpernell; *Modern Electroplating (Third Edition),* John Wiley and Sons Inc.
19. C.A. Snavely; *Transactions Electrochemical Society,* **92** : 537 (1947)
20. G.L. Faust, G.R. Schaer, D.E. Semones; *Plating,* **48** (6): 605-612 (June 1961)
21. H. Schreiner, Ch.J. Raub; *Z.f Werkstofftechnik/J. of Materials Technology,* **6,** (12) : 414-416 (1975)
22. G. Nover, Ch.J. Raub, H. Speckhardt; *Metalloberfläche,* **34,** (4) : 169-173 (1980); G. Nover, Ch. J. Raub; *Metalloberfläche,* **35,** (9) : 319-329 (1981)
23. H. Schreiner, Ch. J. Raub; *Metalloberfläche,* **34**, (9) : 357-361 (1980)
24. J.C. Saiddington, G.R. Hoey; *Plating,* **61,** (10) : 923-930 (1974)
25. J.C. Saiddington; *Plating and Surface Finishing,* **65,** (1) : 45-49 (1978).
26. V.L. Nesnidal; *AES International Pulse Plating Symposium,* American Electroplaters Society, Boston (1979)
27. A. Knödler; *AES International Pulse Plating Symposium,* American Electroplaters Society, Boston (1979)
28. H.L. Logan, *J. Research. Nat. Bur. Standards,* **43** (1964)
29. F. Mortier, J. Reby, B. Sutter, G. Vasseur; *Cetim Report* 12 F 39 0 (Unpublished) (1984)

Chapter 9

Pulsed Electrodeposition of Precious Metals

A. Knödler

1. Introduction

The term "precious metals" comprises gold and silver, together with the metals of the platinum group: platinum, palladium, ruthenium, iridium, rhodium and osmium. Of these, gold and silver are the best-known and most widely used in the electrodeposited condition. Gold plating, in particular, has probably received more attention than any other electrodeposition process in terms of investigational and applicational studies of pulse plating. This followed from the relatively early observation that the use of pulses in the millisecond range resulted in substantial improvements in density and other properties, such as conductivity, contact resistance and porosity, both in the case of the pure metal and particularly gold alloyed with such metals as cobalt or nickel, widely used in the electronics industry. On this basis, further investigation was stimulated by the promise of considerable economy in usage of the metal at a time of escalating prime metal cost. In the absence of such a potent spur, investigations concerned with pulse plating of silver and the platinum metals have been relatively limited. In the case of silver, some work has been directed to alloy deposition with metals such as tin and antimony, aimed at general improvement of hardness and other physical properties, as well as the continuing search for a non-tarnishing finish.

Of the platinum metals, only palladium, rhodium, and more recently ruthenium, are of major interest as electrodeposits. Palladium has been the subject of a fair number of pulse plating studies, in view of its potential application as an economic substitute for gold, coupled with an intrinsic interest in the deposition process arising from the well-known characteristics of the metal in respect to hydrogen adsorption-desorption and hydride formation, which is significantly affected by the application of the pulsed current technique, with possible beneficial effects on deposit properties. Rhodium, probably the longest established member of the group in the decorative and industrial electrodeposition field, appears to have been completely neglected in the present context, but some work on pulsed current deposition of ruthenium has been reported. This metal has some interesting properties in terms of a high hardness, intrinsic corrosion resistance, and good erosion resistance in electrical contact applications. A major limitation in state-of-the-art electrodeposits produced by direct current is porosity, related to microcracking, and preliminary indications are that pulse plating may result in significant improvement in this area.

2. Gold

2.1. *Pure gold*

One of the earliest examples of studies on the electrodeposition of gold by pulsed current is that by Cheh (1), who developed a diffusion model to calculate the deposition rate of metals by a periodic pulsed current. Although the magnitude of the instantaneous applied current, that is, the pulsed current density, can be considerably higher than the current density of direct current plating, the limiting overall plating rate from baths containing the $Au(CN)_2^-$ ion was generally lower. Experimental investigations were carried out in a phosphate, a citrate and a cyanide gold solution. It was concluded that the advantage of applying pulsed current during electrodeposition is not to increase the limiting overall plating rate but to raise the pulsed current density, thereby possibly improving the physical properties through structural modifications of the electrodeposits.

In a later study, Cheh (2) extended this model developed for PC plating to calculate the limiting rate of deposition by pulse reverse plating, the diffusion model of PR plating being a complete analogy to that of PC plating. Satisfactory agreement was observed between theory and experimental results for large values of T_{on}/θ.

Macrothrowing power measurements in a Haring cell with a citrate-phosphate gold solution under both DC and PC conditions indicate that a combination of fast pulses and lower peak current densities improve the macrothrowing power of gold. However, this improvement diminishes rapidly at increasing peak current densities. Current efficiency is lower in PC plating as compared to DC plating (3).

Results of experiments with gold electrodeposition on a rotating-disk electrode indicate that under the conditions used in this work microthrowing power can be improved considerably by applying pulsed current and also by increasing agitation of the solution. Furthermore, it was found that the applied current density has a significant influence on the microthrowing power, whereas pulse characteristics such as duty cycle and cycle time seem to have relatively little effect on microthrowing power (4).

Rehrig (5) showed porosity of gold deposits to be lowest for pulse plated deposits, followed by asymmetric AC and DC plated deposits, while it is highest for evaporated gold layers (Fig. 9.1).

In a later paper, Rehrig and coworkers (6) investigated the effect of 30 different forms of pulsed current on the morphology of soft gold. These conditions ranged from 0.1 to 9 msec on-time and from 9 to 90 percent duty cycle. At low peak current densities, the crystallographic orientation of the gold deposits was immutable and the grain size increased with the duration of the pulse. With higher peak currents, $\leqslant$ 50 percent duty cycle, crystallographic orientation underwent a transition as the pulse duration increased. In these cases, the grain diameter was inversely proportional to the peak current and the duration of the pulse. Orientation and grain growth can be governed by charge transfer and/or surface diffusion of adions and adatoms.

During studies on the porosity of gold electrodeposits, R.J. Morrissey and A.M. Weisberg (7) found that at very low thicknesses the crystallography and the porosity of the deposits are mainly dependent on the surface characteristics of the underlying substrate. At greater thicknesses the slope of the porosity/thickness curve is controlled by parameters relevant to the deposit itself. Between these two extremes there is a sharp, well defined transition region in which the porosity of the deposit falls extremely rapidly as shown in Fig. 9.2. In this figure, porosity is expressed as the exposed basis-metal area fraction of the specimen which is related to the corrosion potential of gold-plated copper, thus affording a convenient and sensitive means for determining electroplate

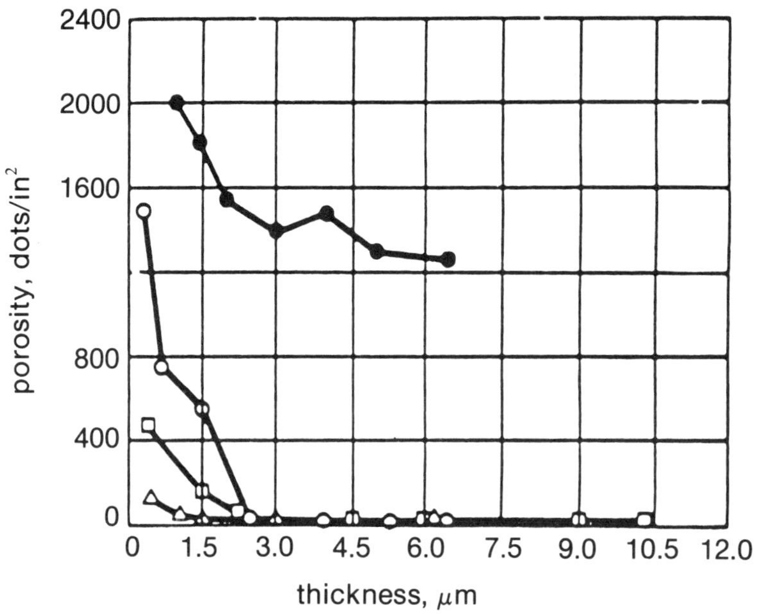

Fig. 9.1—Relationship between porosity and thickness of gold films deposited by • electron gun evaporation, □ asymmetric AC plating, o DC plating and △ pulse plating (5).

porosity. Numbers at curves indicate on- and off-time of the pulsed current in msec. The curves in Fig. 9.2 are all of the same form but are displaced downward and to the left, that is, to lower porosity values, as the duty cycle is decreased. Lowest porosity is found in PC plated gold deposits at 1 msec on-time and 19 msec off-time.

J.W. Dini and H.R. Johnson (8) studied the use of pulsed current or asymmetric A.C. current instead of direct current for the electrodeposition of pure gold from a cyanide-citrate-sulfate electrolyte. Table 9.1, which is taken from their paper, compares yield strength, ultimate tensile strength, hardness and elongation for 250 μm thick gold foils produced by each of the three deposition modes.

The following conclusions can be drawn from Table 9.1:

(a) In the as-deposited condition, gold foils produced by asymmetric AC plating have a higher hardness and tensile strength than those produced using DC or PC, but are less ductile.

(b) In the as-deposited condition, PC plated gold foils were slightly stronger than DC plated foils, but elongation was similar.

(c) DC plating produced large, coarse-grained deposits. PC plated deposits were also coarse-grained but the individual grains were more elongated and more uniform in size. Use of asymmetric AC resulted in a much finer grain size and a higher tensile strength.

(d) Heat treatment at 300° C for two hours reduced the yield strength, the ultimate tensile strength and hardness of all deposits while it increased their elongation.

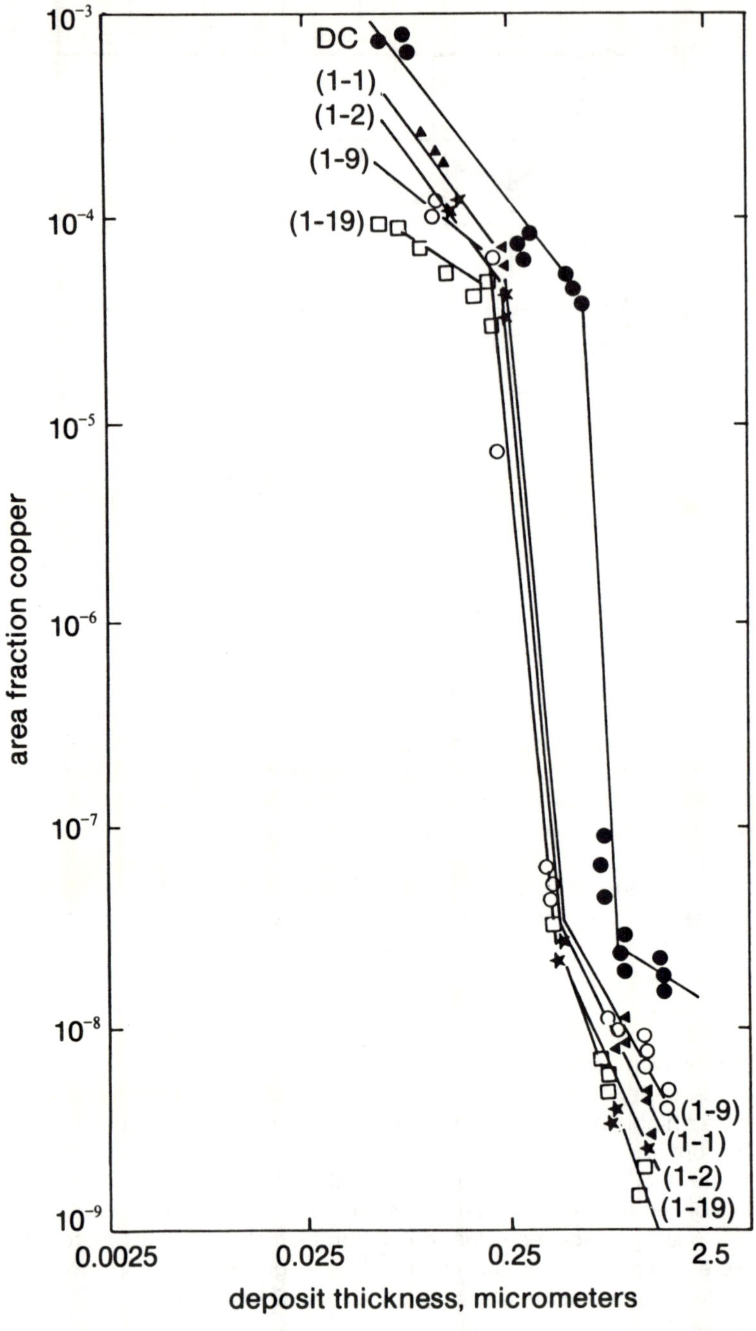

Fig. 9.2—Porosity vs. thickness of DC and PC plated gold on copper. Numbers at curves indicate on-off-time in msec (7).

TABLE 9.1
Properties of As-Deposited and Heat Treated 250 μm Thick Gold Foils Plated from a Citrate Bath

	Yield strength*, MPa			Ultimate tensile strength, MPa			Elongation, % in 25 mm			Knoop hardness (25 g)	
Method of deposition	as-deposited	2 h at 300° C	4 h at 300° C	as-deposited	2 h at 300° C	4 h at 300° C	as-deposited	2 h at 300° C	4 h at 300° C	as-deposited	2 h at 300° C
DC	60	59	52	141	126	123	26.3	29.1	30.2	69	68
AC	101	21	33	203	106	120	16.5	27.8	25.7	98	49
PC	79	35	29	161	116	108	26.4	41.8	43.3	70	57

*0.2 per cent offset.

In another investigation, Dini and Johnson (9) used electrochemical porosity tests and Auger electron spectroscopy to evaluate two gold electroplating formulations (citrate and phosphate) at varying current densities and deposition modes (DC, PC, asymmetric AC) for eventual use on hybrid microcircuits. A comparison of the deposition modes did not lead to clearcut conclusions at current densities of 0.108 and 0.324 A/dm^2. At the more realistic current density of 0.756 A/dm^2, DC deposits show the steepest slope of the porosity vs. thickness curves and, therefore, have the highest porosity. This is evident from Fig. 9.3, taken from the work of Dini and Johnson, when the curves are extrapolated to a 1 μm thickness. The results confirm those by Rehrig (5).

During their research work on the effect of PC on the electrodeposition of pure gold from weakly acidic gold cyanide baths, Wang, Tung and Wan (10) found that PC decreases the current efficiency and increases microhardness of the deposits. Contact resistance is lower and less affected by external pressure than the equivalent DC deposits. Porosity data were too erratic to reach a general conclusion. DC plated deposits between 2.7 and 10 μm showed the least porosity. At thicknesses below 2.7 μm a pulsed current density of 0.5 A/dm^2 and an on/off ratio of 1 produced the least porous deposits.

A study of the mechanical properties of gold deposits from a phosphate bath by Ch. J. Raub and J. Lendvay (11) revealed that PC plated deposits had a tensile strength 25 percent higher than DC plated deposits while elongation increased by a factor of two (Fig. 9.4). The difference is explained by the finer grain size and the higher purity of the PC plated deposits. DC plated deposits show a slight increase in tensile strength after heat treatment at 100° C, followed by a continuous decrease at higher annealing temperatures. Pulse plated deposits, which have lower impurity contents, do not show this hardening effect. As one might expect of a purer metal, its tensile strength also decreases much more rapidly.

There is only one reference in the literature to pulse plating from a non-cyanide gold electrolyte. J. Horkans and L.T. Romankiw (12) studied the effect of PC deposition on the properties of gold films from Au(I) sulfite solutions. Internal stress, grain size, purity, and resistivity of PC plated gold showed no apparent relationship to the pulse frequency. This can be explained in terms of the irreversibility of the Au/Au(I) couple in sulfite solutions. Those properties that were influenced by the conditions of deposition were the thickness distribution of gold deposited through a mask, the crystallographic texture, and surface morphology. Gold deposited through a photoresist mask from sulfite solutions shows large thickness variations both across any given line and from line to line. Use of PC plating at a high frequency (500 Hz) was effective in eliminating nonuniformity in the thickness of a given line.

2.2. *Gold alloys*

Viswanathan and Raub (13) studied the effect of PC plating of gold-nickel alloys from an alkaline bath. A slight increase in the nickel contents of PC plated deposits was observed, as compared to DC plated deposits. Nickel contents in DC plated deposits were much more sensitive to current density variations than in PC plated layers. As shown by SEM micrographs, grain size in PC plated gold-nickel alloys was higher than in those plated using DC. In neither case

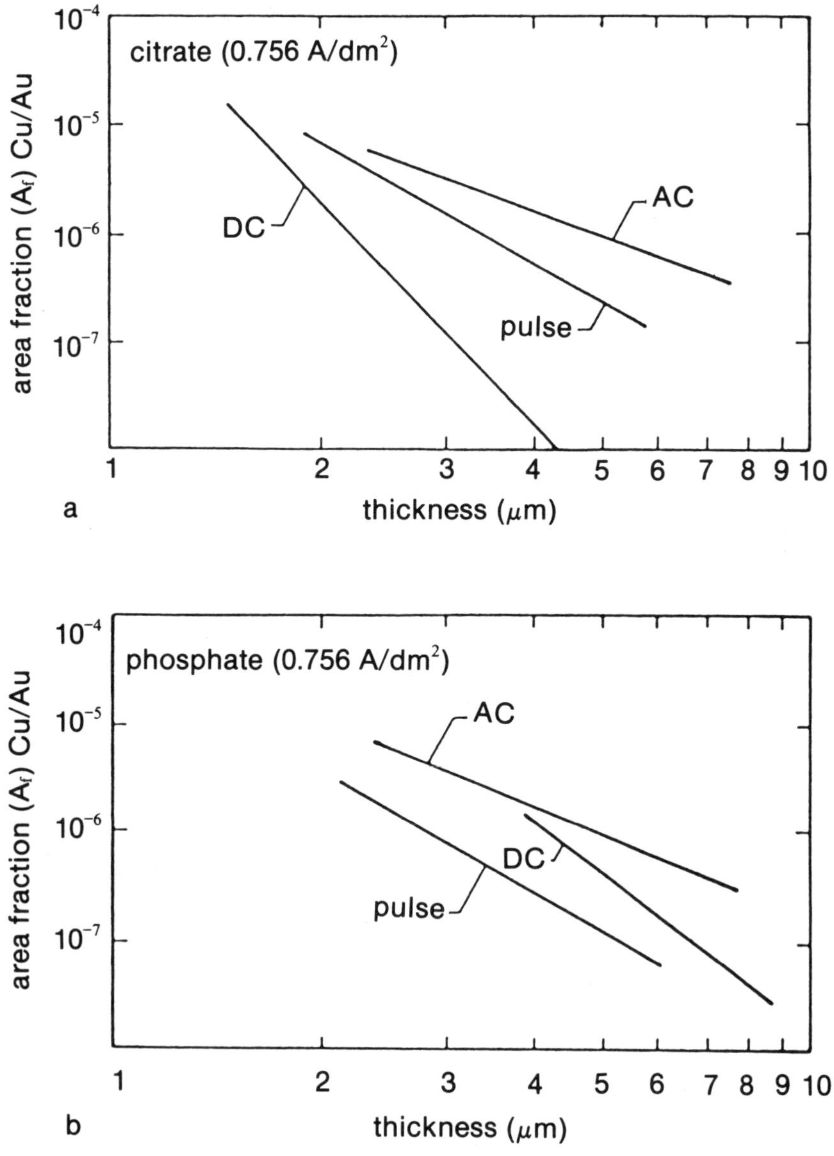

Fig. 9.3—Porosity vs. thickness of gold deposits plated from a citrate (a) and a phosphate (b) bath (9).

could typical crystal faces be detected but instead, "cauliflower"-like grains with a coarse substructure.

Much of the later work on pulse plating of gold alloys has been done with acidic electrolytes containing potassium gold cyanide and an organic buffer (for example, citric acid) in the presence of small amounts of cobalt or nickel as a brightener. Despite their $\geqslant$ 98 purity, gold deposits from these electrolytes have

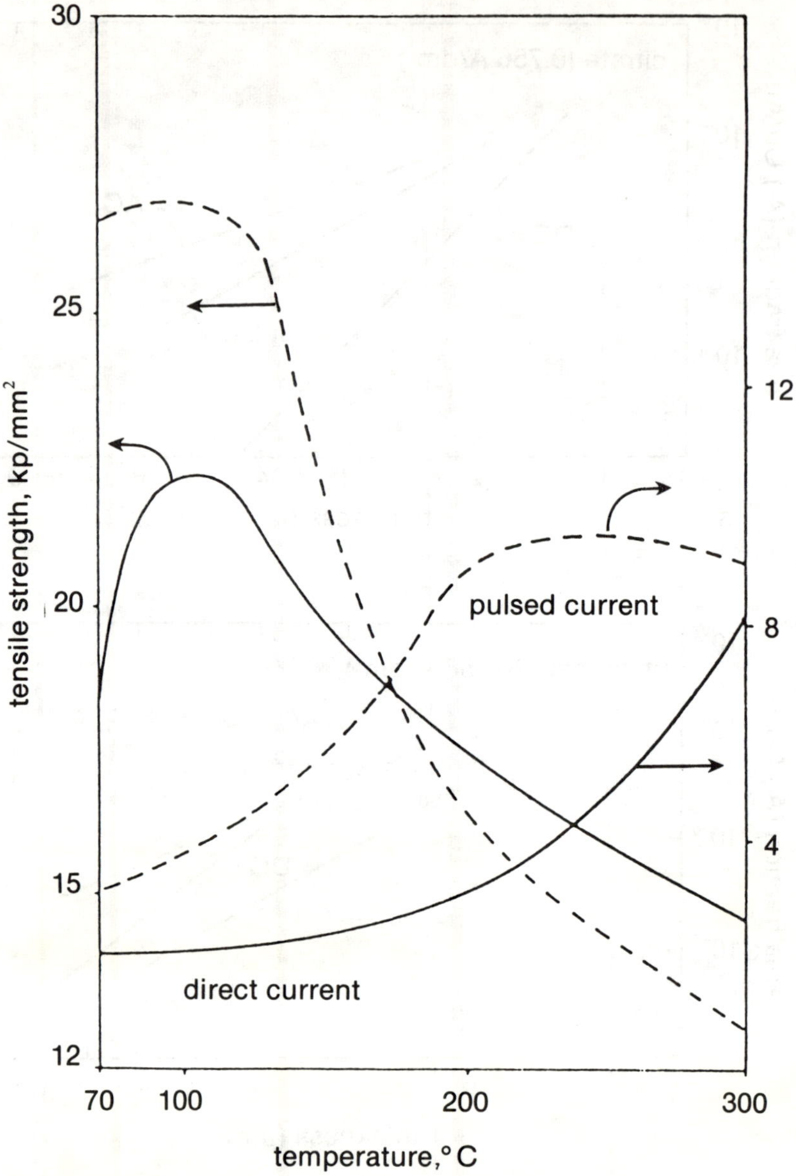

Fig. 9.4—Tensile strength and elongation of PC and DC plated gold deposits from a phosphate solution (20 μm) (11).

high hardness and wear resistance. For this reason such deposits are used widely for gold plating of electrical contacts.

Large amounts of carbonaceous material were detected in such deposits by Munier (14). Considerable interest has been stimulated regarding the nature of these inclusions and the mechanism by which the material is included in the

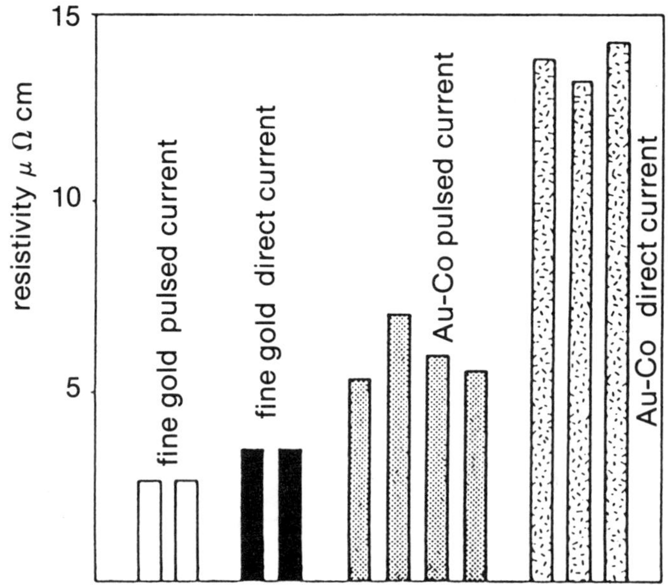

Fig. 9.5—Influence of pulse plating on the electrical resistivity of pure gold and of gold cobalt alloy deposits (18)

gold alloy deposits. It was soon realized that the source of carbon is the cyanide ion, which is liberated at the cathode when gold is deposited:

$$Au(CN)_2 + e^- = Au + 2\ CN^-$$

Munier suggested that the carbonaceous inclusions were polymeric hydrocyanic acid, $(HCN)_x$. It is now fairly conclusively established that the included carbon-containing substance consists, at least to a substantial degree, of a cobalt complex, KCo $[Au(CN)_2]_3$, which is precipitated at the cathode during gold deposition (15-17).

The effect of pulse plating on the electrodeposition of gold-cobalt alloys from weakly acidic cyanide solutions was thoroughly studied by F.H. Reid (18). Reid succeeded in isolating the polymer, thus confirming the findings of Munier. PC plated gold alloys had much smaller amounts of polymeric inclusions and a lower electrical resistivity. Figure 9.5 shows the influence of pulse plating on the electrical resistivity of pure gold and of gold-cobalt alloy deposits according to Reid. The density of PC plated gold-cobalt deposits was very near to the theoretical density of fine gold (19.3 g/cm^3), in contrast to DC plated alloy deposits which had a density of only 16.6-17.2 g/cm^3, despite their high gold content of $\geqslant$ 98 percent. The higher density of PC plated gold-cobalt alloys can be explained by their higher purity.

Since gold-cobalt and gold-nickel electrodeposits are used widely as a finish on electrical contacts, the questions arose as to the effects these polymeric inclusions have on wear behavior, solderability and ultrasonic bonding. According to Souter (19), the polymeric inclusions reduce friction and wear by serving

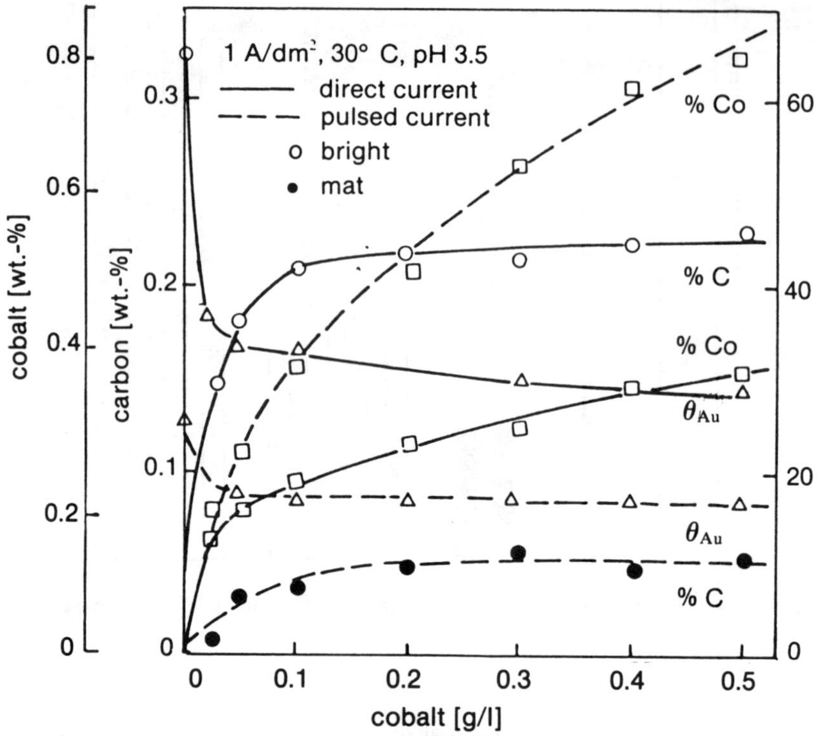

Fig. 9.6—Cobalt and carbon content, current efficiency of DC and PC plated gold-cobalt alloy deposits. On-time 2.3 msec, off-time 20 msec (23, 27).

as lubricants. This view was contradicted by M. Antler (20-22) who suggested that adventitious organic contaminants from the air can accumulate on the surface and may provide effective lubrication for both polymer-containing and polymer-free gold alloy deposits.

In the following years much effort went into clarifying the role of polymeric inclusions. One possibile way of reducing their codeposition with gold was the application of pulse plating, as suggested by Reid (18). The aim of these efforts was to produce gold alloy deposits with higher ductility, yet still possessing high hardness and wear resistance. A detailed study of the influence of pulse plating on the composition and the properties of gold-cobalt and gold-nickel deposits was carried out by Knödler (23-25). The results will be discussed in more detail.

For this study an acidic citrate plating solution with small additions of either cobalt or nickel was used. Figure 9.6 shows the contents of cobalt and carbon and the current efficiency of DC and PC plated gold-cobalt deposits as a function of the cobalt concentration of the plating bath. Cobalt contents of PC plated deposits are twice as high as those of DC plated deposits while carbon contents are reduced to one fifth. Current efficiency of PC plating is lower by about 40 percent.

The corresponding results for gold-nickel alloy deposition are shown in Fig. 9.7. The increase of the alloy metal content by use of pulse plating is even more

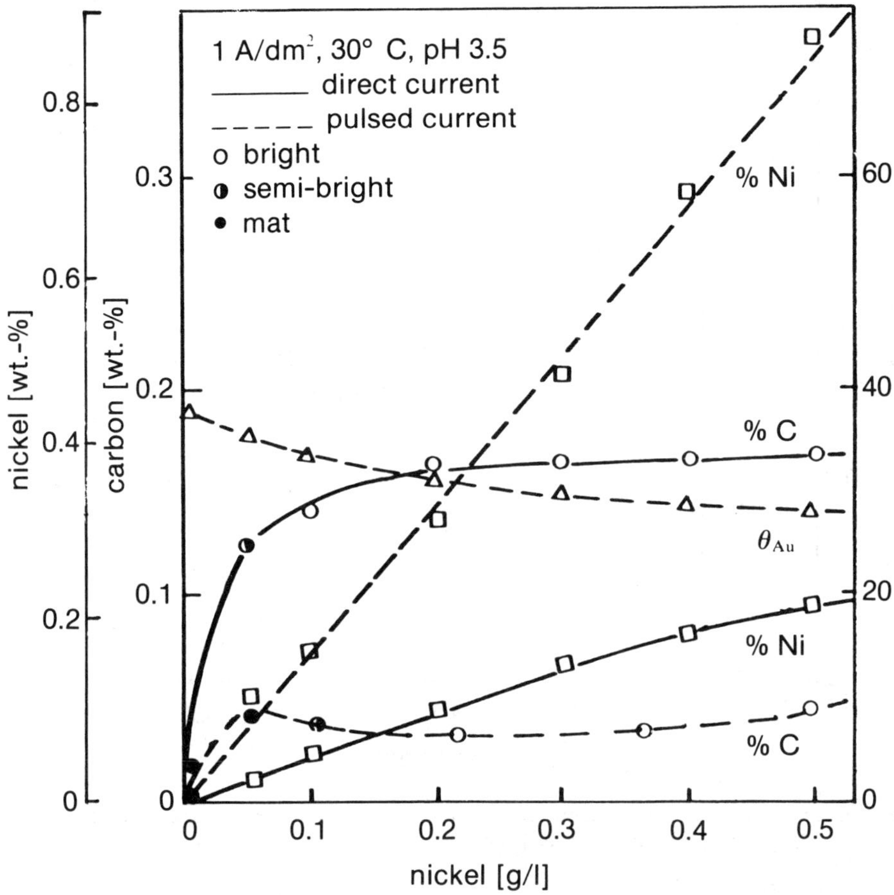

Fig. 9.7—Nickel and carbon content, current efficiency vs. nickel concentration of DC and PC plated gold-nickel alloy deposits. On-time 2.3 msec, off-time 20 msec (23, 27).

apparent than in gold-cobalt alloys. Pulse plated gold-nickel alloys are bright, in contrast to gold-cobalt deposits.

In Fig. 9.8 the effect of varying the off-time on the composition of gold-cobalt alloys is shown. On-time was held constant during this series of experiments, so pulsed current density had to be increased in order to maintain the average current density constant at 1 A/dm^2. As the off-time is increased, carbon contents fall rapidly to low values characteristic of PC plated deposits, becoming constant at off-times of about 40 msec. Cobalt contents increase with higher off-times, also becoming constant at about 40 msec off-time.

A similar behavior is exhibited by gold-nickel alloy deposits (Fig. 9.9). As off-time is increased, carbon contents of the deposits fall even more rapidly than in gold-cobalt deposits, becoming constant at about 10 msec off-time. In contrast to gold-cobalt deposition, nickel contents continue to rise with increasing off-time.

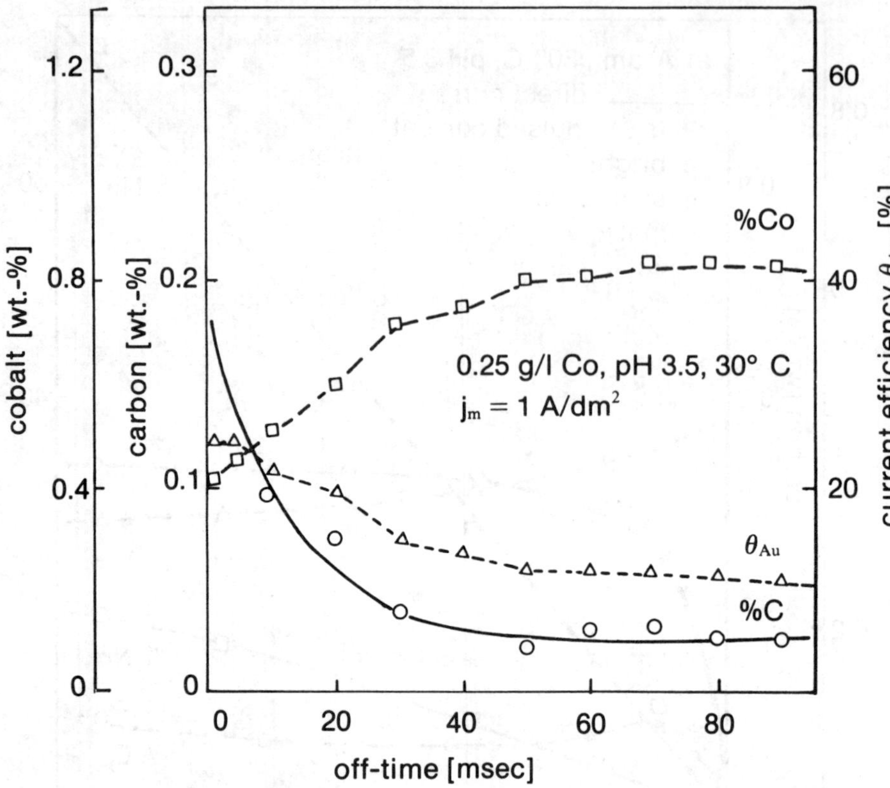

Fig. 9.8—Cobalt and carbon content, current efficiency vs. off-time of gold-cobalt alloy deposits. On-time 2.6 msec, average current density 1 A/dm^2 (23, 27).

Since it was not clear whether this effect is caused by the concomitant rise of pulsed current density, another series of tests was carried out in which pulsed current density was held constant, while both on- and off-time, i.e., cycle time, were increased. This resulted in a constant average current density of 1 A/dm^2 and a constant duty cycle of 9.9 percent. Results of this series of experiments are shown in Fig. 9.10 for gold-cobalt and in Fig. 9.11 for gold-nickel alloys. With increase in cycle time carbon contents fall rapidly to low values, becoming constant at about 10 msec cycle time. Cobalt and nickel contents continue to rise and a constant value at about 30 msec.

These results may be interpreted following a diffusion model developed by Cheh (1). During on-times the cathodic diffusion layer is depleted of cobalt and nickel ions respectively, and restored during the off-periods by diffusion of these ions from the bulk of the solution. Thus the surface concentration of cobalt or nickel at the beginning of the next current pulse is higher, which results in higher cobalt or nickel contents in the deposits. Fig. 9.10 and Fig. 9.11 reveal that cobalt or nickel contents become constant at off-times of about 30 msec. Obviously, this time is sufficient for the alloy metal diffusion to achieve a stationary value. On the other hand, carbon-containing reaction products formed during the

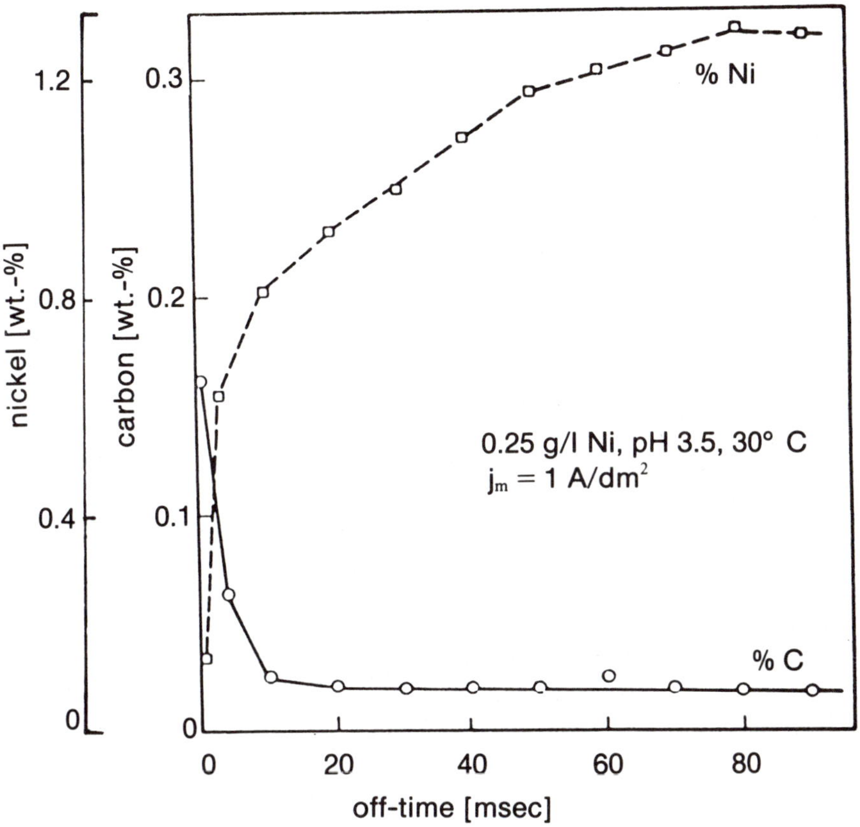

Fig. 9.9—Nickel and carbon content vs. off-time of gold-nickel alloy deposits. On-time 2.6 msec, average current density 1 A/dm^2 (23, 27).

cathodic current pulse may diffuse away from the cathode into the bulk of the solution, thus reducing the possibility of their codeposition in the deposit. It is evident from Fig. 9.10 and Fig. 9.11 that a cycle time of about 10 msec is sufficient for the reaction products to attain a stationary concentration in the diffusion layer. This period is distinctly lower than that for the cobalt of nickel ions to attain an equilibrium state.

Figure 9.12 shows the effect of temperature on gold-nickel alloy deposition under pulse plating conditions. As temperature is increased, the nickel contents rise rapidly to about 2.5 wt-% at 60° C. A similar but less marked increased was observed under DC plating conditions in an earlier investigation (26). Carbon contents go through a minimum at 40° C. This minimum is not observed with DC plating. Gold-nickel alloys deposited at higher temperatures are only semi-bright as shown in Fig. 9.12.

There is a distinct difference between gold-nickel and gold-cobalt electrodeposits produced by pulse plating: In the latter case, as the temperature rises, cobalt contents remain almost constant, while carbon contents decrease continuously.

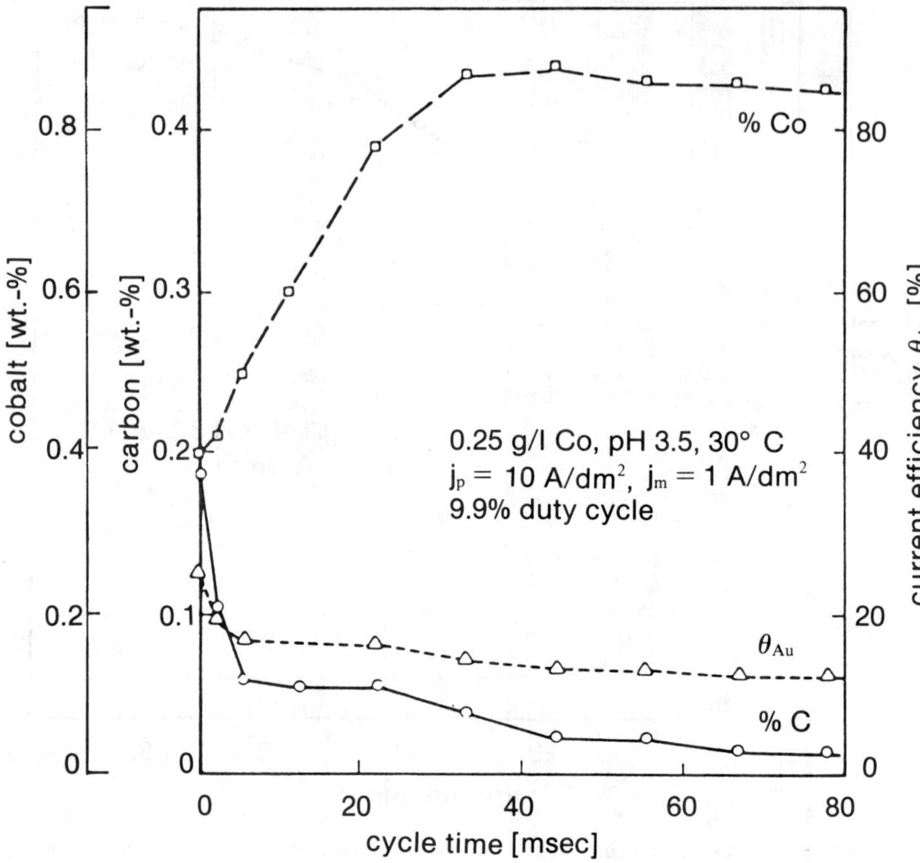

Fig. 9.10—Cobalt and carbon content, current efficiency vs. cycle time of gold-cobalt alloy deposits (23, 27).

In another series of tests the average current density was increased while pulsed current density and off-time were held constant. This implies an increase of on-time. Results of gold-cobalt deposition show that, as the average current density is increased, cobalt contents decrease linearly while carbon contents rise steadily to high values characteristic of DC plated deposits. Increasing the on-time at constant pulsed current obviously means an approach to DC plating conditions. This result is also consistent with the diffusion model described above. The difference between DC and PC plated alloy deposits becomes less obvious at high cobalt concentrations in the plating bath as shown in Fig. 9.13. The explanation is that at high cobalt concentrations the composition of the deposit is no longer controlled by diffusion processes in the cathode layer.

Knödler also measured the gas content (hydrogen, nitrogen, oxygen) of DC and PC plated gold-cobalt and gold-nickel deposits. As an example, the results of gas analysis of gold-cobalt deposits are presented in Table 9.2. As will be seen from the table, carbon and gas contents of deposits from addition-free solutions are very low. Addition of only small amounts of cobalt to the

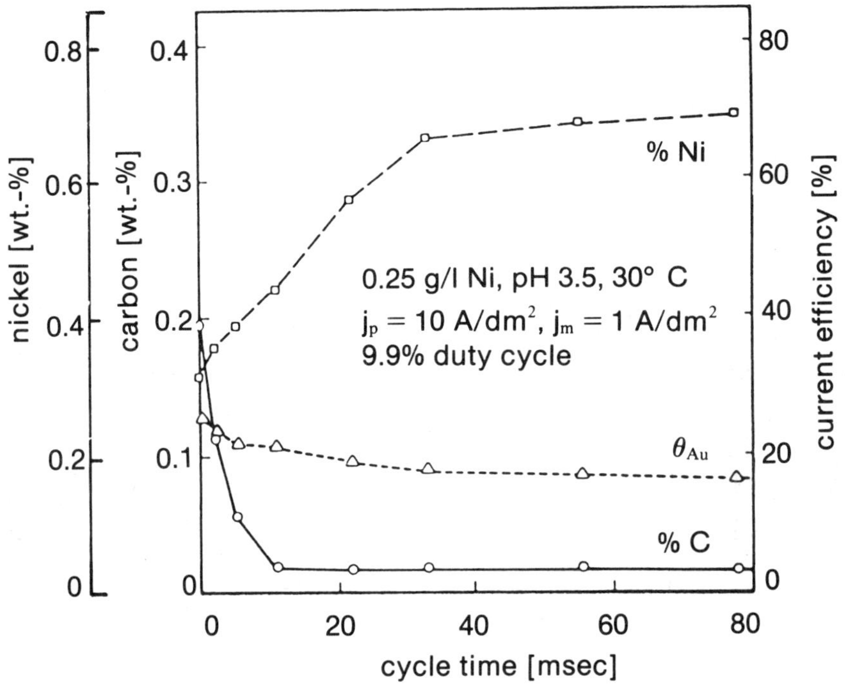

Fig. 9.11—Nickel and carbon content, current efficiency vs. cycle time of gold-nickel alloy deposits (23, 27).

plating solution causes the carbon and gas contents to rise steeply. It will also be noted that by applying pulsed current, the carbon and the gas contents are reduced by a factor of about 3 to 5. There is evidently a close relationship between the amount of cobalt included in the deposit and the concentration of non-metallic material (C , H, N, O).

The atomic ratio of H/C in DC and in PC plated alloy deposits is far in excess of what would be expected if carbon were present as polymeric hydrocyanic acid $(HCN)_x$. Also, the ratio of N/C in DC plated gold-cobalt deposits is much lower than 1:1. This makes the assumption by Munier of a polymeric inclusion unlikely. Similar measurements were carried out with gold-nickel deposits.

In addition, Knödler measured the mechanical properties of DC and PC plated alloy deposits. DC deposits are known to be highly stressed. Failure of contacts plated with gold alloys from acidic plating solutions has been attributed to crack formation due to high tensile stresses. By using pulse plating, stress is reduced drastically as compared to DC plating (Fig. 9.14). Tensile stress decreases as deposition temperature is raised. The lower tensile stress of PC plated deposits can be attributed to the lower codeposition of nonmetallic material. However, the decrease of tensile stress is accompanied by a partial loss of brightness. High tensile stresses, codeposition of large amounts of nonmetallic elements and brightness seem to be interconnected.

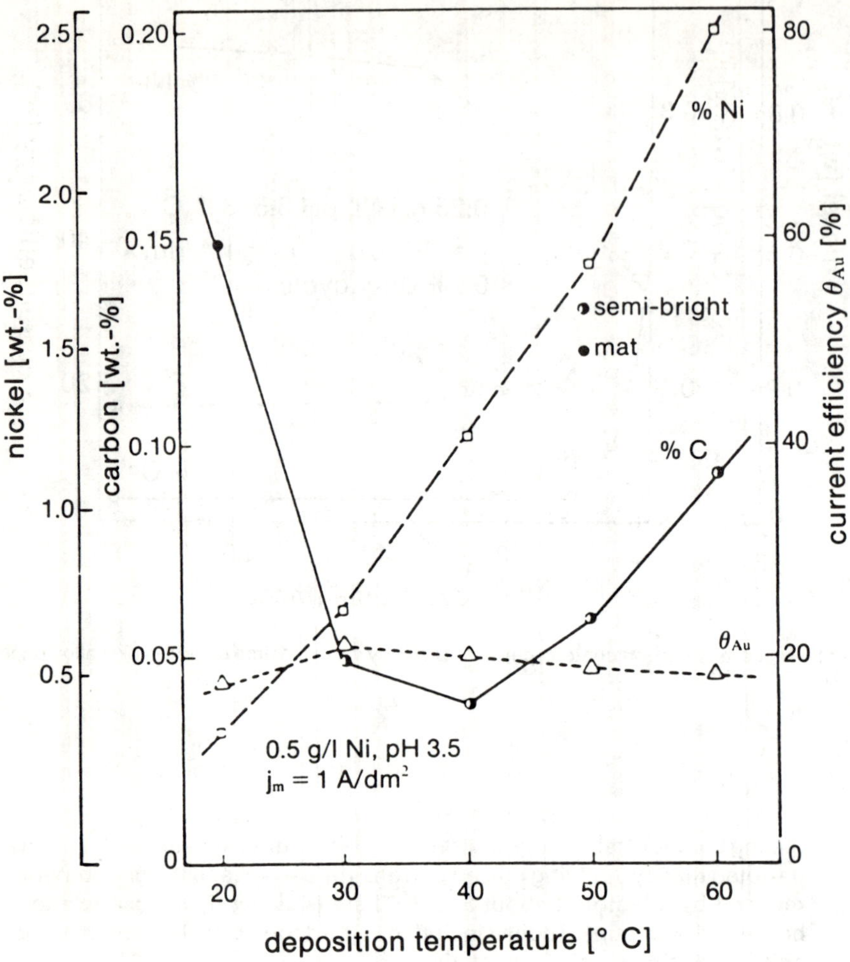

Fig. 9.12—Nickel and carbon content, current efficiency vs. deposition temperature of gold-nickel alloy deposits. On-time 2.5 msec, off-time 20 msec (23, 27).

In Table 9.3 the Vickers hardness numbers of DC and PC plated gold-cobalt deposits are tabulated. It will be noted that the hardness of PC plated deposits is somewhat higher than the hardness of DC plated layers, in spite of the lower level of nonmetallic inclusions. Obviously, the effect of the lower impurity level on hardness is offset by the higher cobalt content of PC plated deposits. It is also interesting to note that the hardness is practically independent of the cobalt concentration of the plating solution. Even very small cobalt concentrations are sufficient to produce high hardness in the alloy deposits.

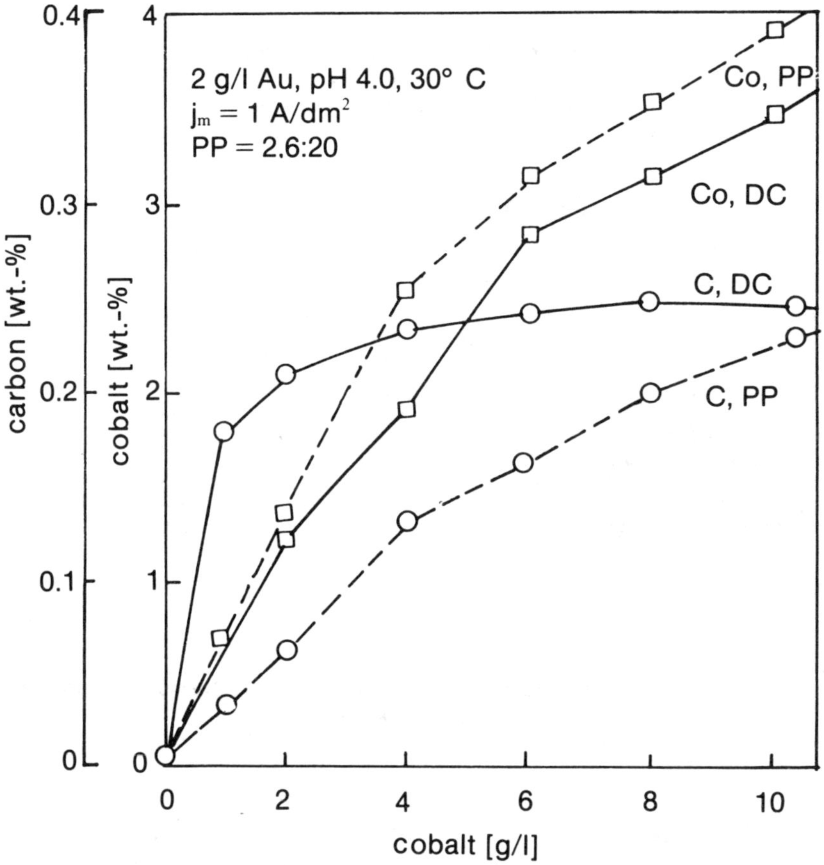

Fig. 9.13—Cobalt and carbon content of DC and PC plated gold-cobalt alloy deposits vs. Co concentration (higher level). On-time 2.6 msec, off-time 20 msec (23, 27).

Comparing the hardness values (Table 9.3) and the concentrations of nonmetallic elements in the deposits (Table 9.2), it becomes evident that the relatively high hardness of these deposits is not so much due to the alloying element (the concentration of which would be far too low), but rather to the codeposited nonmetallic substances. Such a view is also expressed by Okinaka et al. (16).

Summarizing the results by Knödler and coworkers, may be stated that the alloy metal content (cobalt, nickel) of the deposits from acidic cyanide solutions is increased by use of pulse plating, while the concentrations of nonmetallic elements (C, H, N, O) are reduced drastically. Composition and properties of gold alloy deposits can be varied widely by altering pulse plating parameters. Hardness of alloy deposits is increased slightly by pulse plating, but tensile stress is reduced. Brightness of gold-cobalt and, to a lesser extent,

TABLE 9.2
Carbon and Gas Contents of Gold-Cobalt Deposits
pH 3.5, 30° C, j_{DC} = 1 A/dm², j_m = 1 A/dm², j_p = 10 A/dm²
On-time 2.6 msec, off-time 20 msec

Co g/l	Current	Carbon and gas content ppm C	H	N	O	Atomic ratio (C = 1) C	: H	: N	: O
0	DC	221	46	48	90	1	2.48	0.19	0.31
	PC	38	7	29	69	1	2.19	0.65	1.36
0.05	DC	1610	435	1262	404	1	3.22	0.67	0.19
	PC	298	126	571	148	1	5.04	1.65	0.37
0.1	DC	2120	583	1819	464	1	3.28	0.74	1.45
	PC	385	153	650	152	1	4.74	1.45	0.30
0.3	DC	2166	582	1806	409	1	3.21	0.72	0.17
	PC	547	164	738	172	1	3.56	1.16	0.24
0.5	DC	2301	610	1831	529	1	3.16	0.68	0.17
	PC	521	149	667	136	1	3.41	1.10	0.20

TABLE 9.3
Vickers Hardness Number HV 0.01 of DC and PC Plated Gold-Cobalt Deposits
pH 3.5, 30° C, j_{DC} = 1 A/dm², j_m = 1 A/dm²,
j_p = 10 A/dm², on-time 2.5 msec, off-time 20 msec

Cobalt g/l	DC	PC
0.05	198	205
0.10	198	205
0.20	192	228
0.30	198	220
0.40	198	212
0.50	191	220

of gold-nickel deposits is diminished by pulse plating. The influence of pulse plating on the composition of alloy deposits may be explained in terms of an adsorption-desorption model. The decisive influence seems to be the restoration of the diffusion layer and the removal of electrolysis products during off-periods.

Further results on PC plating of gold alloys from acidic solutions were presented and discussed at the AES International Pulse Plating Symposium, held in Boston in 1979 (27).

In a paper given at this symposium, F.H. Leaman and B.A. Kownacki (28) compared PC and DC for continuously plating contacts for electrical connectors with nickel underplate and cobalt-hardened gold. It was found that factors that improve DC plated deposits in indexing and continuous cascade spray cells also

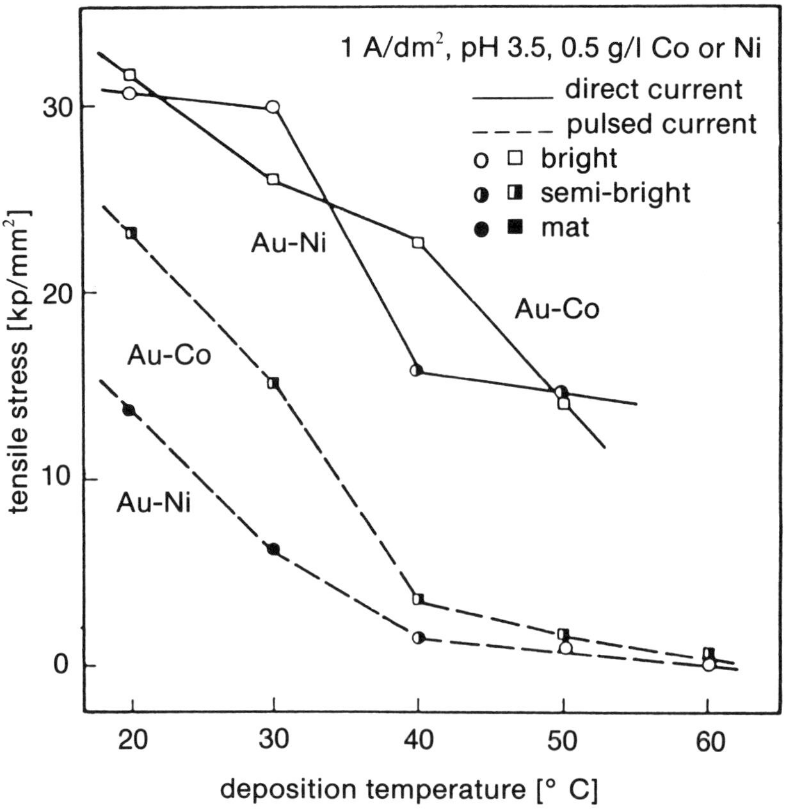

Fig. 9.14—Internal stress of DC and PC plated gold-cobalt and gold-nickel alloy deposits vs. deposition temperature (on-time 2.3 msec, off-time 20 msec) (23, 27).

improve PC plated deposits. All pulse plated deposits showed greater density values than DC deposits. Solder tests revealed slightly improved solderability for the PC plated samples, particularly at the lower duty cycles. Although the cobalt content of the deposits increased in both DC and PC plated deposits at higher current densities, it was somewhat lower with pulse plating at higher current densities and higher duty cycles. These data disagree with that of Knödler (23) and other authors who found the amount of cobalt in PC plated deposits to be higher. Porosity was greatly reduced by pulse plating at all of the conditions tested. Figure 9.15, taken from the work of Leaman and Kownacki, shows the effect of current density on the porosity of gold-cobalt deposits. DC plated gold deposits were found to have a high porosity, as noted by other authors. Deposit hardness of PC gold-cobalt coatings was found to be slightly lower than that of DC plated deposits at comparable current densities. The effect of a pulsed current nickel underplate on porosity of subsequent gold-cobalt alloy deposits was also studied. It was found that pulsed nickel plating reduced the porosity of

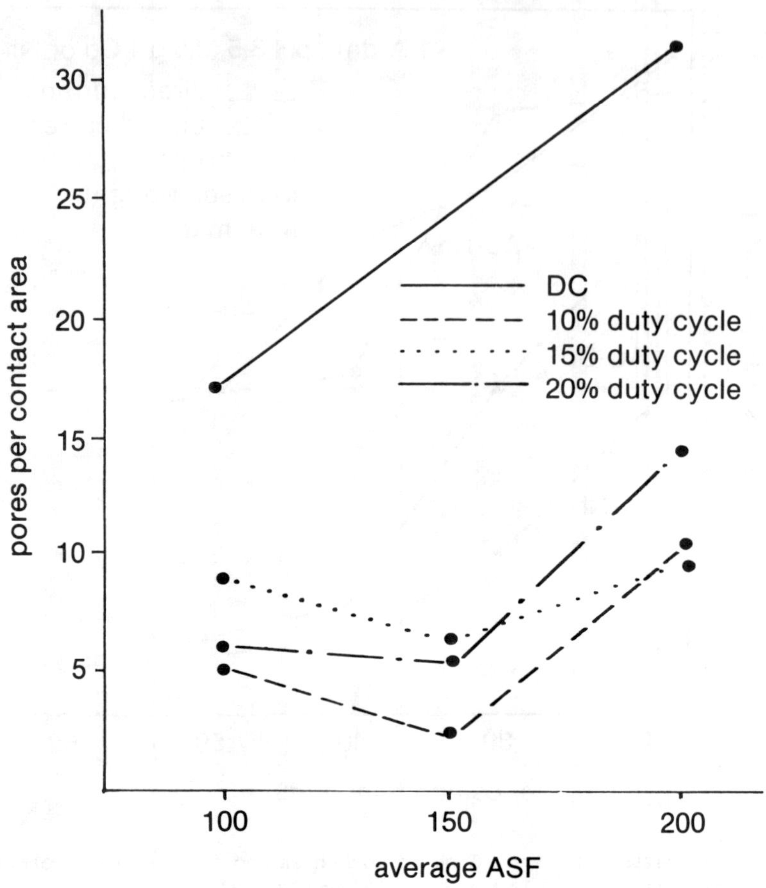

Fig. 9.15—Porosity vs. average current density of gold-cobalt deposits (1000 ppm Co, pH 4.2, 95° F) (28).

the gold deposit by about 40 percent. Both soft and hard gold deposits exhibited decreased resistivity with increasing current density for both DC and PC plating. Wear tests showed that wear resistance of cobalt-hardened gold is significantly increased by pulse plating, while the wear resistance of soft gold is not enhanced.

In a later paper, Knödler and coworkers (29) investigated the influence of higher pulse frequencies and the effect of periodic reverse plating on gold-cobalt and gold-nickel deposition from acidic cyanide solutions. Fig. 9.16 shows the effect of pulse frequency, i.e., the inverse of cycle time, on cobalt and carbon contents and also on the current efficiency of gold-cobalt deposition at three different duty cycles (10, 35, and 60 percent). It can be seen from the figure that cobalt contents remain nearly constant over the whole frequency range investigated. This is not surprising, as high pulse frequencies

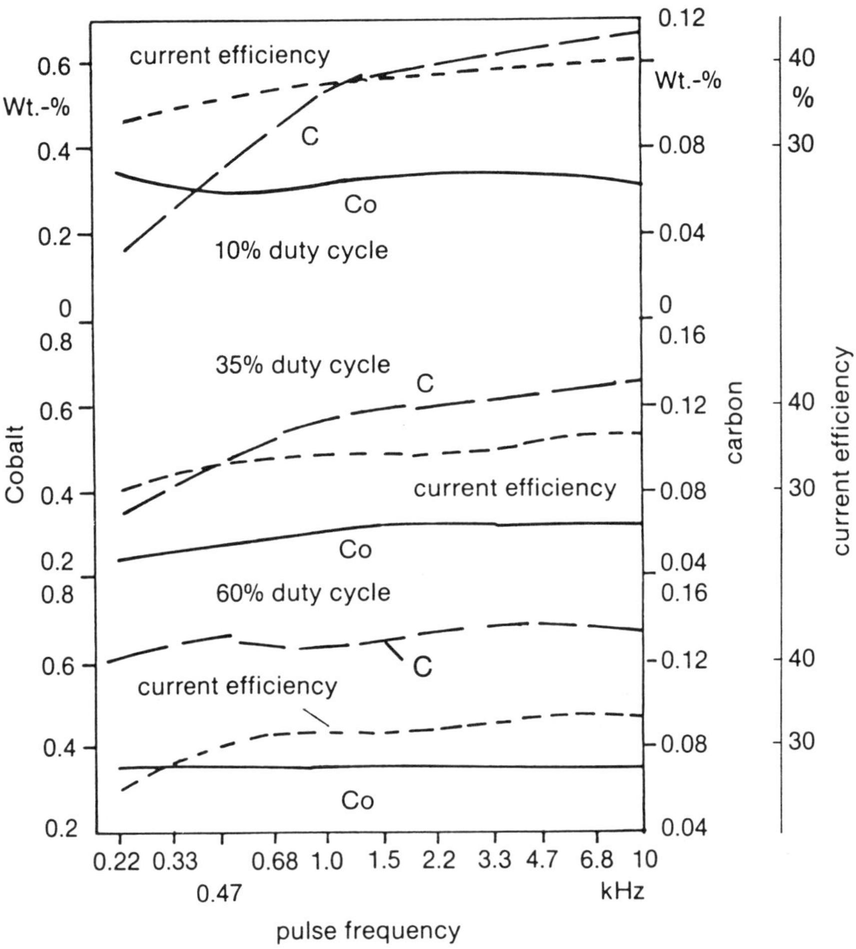

Fig. 9.16—Cobalt and carbon content, current efficiency vs. pulse frequency of PC plated gold-cobalt alloy deposits (Bath composition: 8 g/l Au, 0.25 g/l Co, pH 4.0, 30° C). Average current density 1 A/dm^2, duty cycle 10, 35 and 60% (29).

are equivalent to on- and off-times that are too short for diffusion processes in the diffusion layer to become noticeable. On the other hand, there is a distinct decrease of the carbon contents at low frequencies, that is, at longer on- and off-times. This is in agreement with the results shown in Fig. 9.10. The time for the carbon-containing reaction product to attain a stationary value in the cathodic diffusion layer is much lower than for the cobalt ions. The effect is most pronounced at low duty cycles, i.e., for longer off-periods.

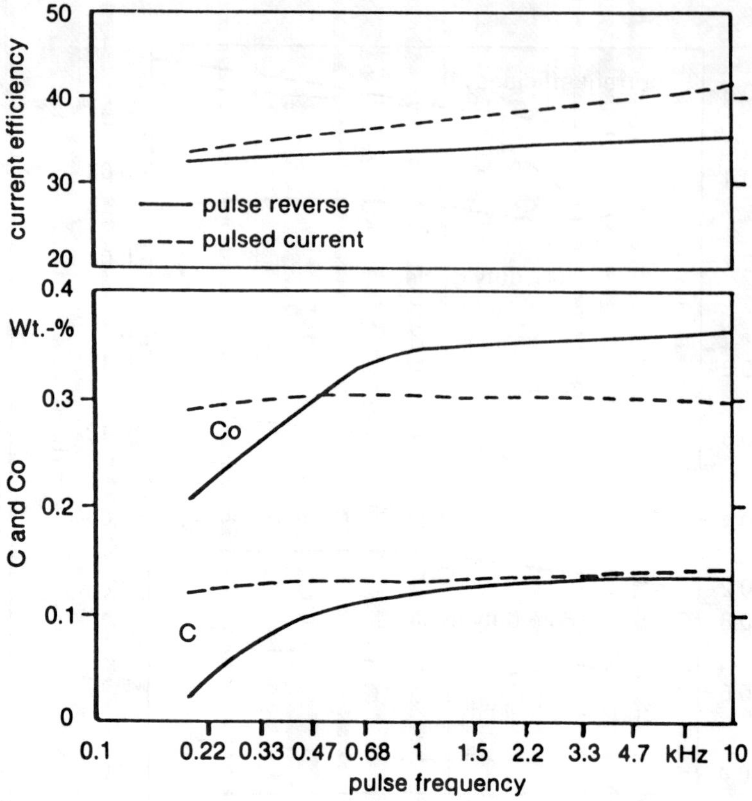

Fig. 9.17—Cobalt and carbon content, current efficiency of gold-cobalt alloy deposits with PC and PR from same bath as Fig. 9.16. Duty cycle 60%, average current density 1 A/dm^2, cathodic PC density 1.66 A/dm^2 for PC, 2.25 A/dm^2 for PR, anodic PC density 1 A/dm^2 with PR.

The effect of periodic reverse current on the electrodeposition of gold-cobalt alloys is shown in Fig. 9.17. At low pulse frequencies the carbon content of PR plated deposits is significantly lower than that of PC plated layers. The explanation is that during the anodic current pulse the carbon-containing species, which is formed during the cathodic pulse, is desorbed from the cathode, thus being prevented from incorporation into the deposit. At high pulse frequencies, on- and off-times are too short for diffusion processes to have any significant effect on the composition of the alloy deposits.

Comparable results were obtained in the gold-nickel system. The final conclusion is that pulse frequency, in order to have any effect on the composition of gold alloy deposits, must not be too high. A pulsed frequency of 1 kHz (corresponding to a cycle time of 1 msec) seems to be the upper limit. This is simply the result of the limited rate of the diffusion processes in the cathodic diffusion layer.

The experimental work of Ibl and coworkers in the field of PC plating of gold has opened up new perspectives as regards the effect of pulsed current on the

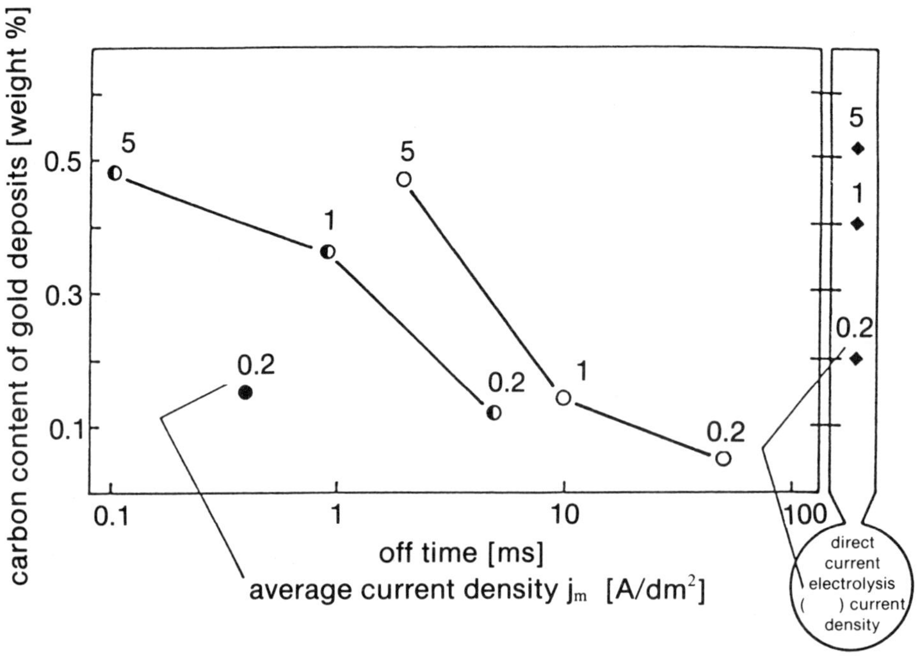

Fig. 9.18—Carbon content of gold-cobalt alloy deposits vs. off-time. Pulsed current density ● 1 A/dm^2, ◐ 10 A/dm^2, ○ 100 A/dm^2; on-time T_{on} = 0.1 msec (33).

concentration profile of the cathodic diffusion layer and mass transport and capacitance effects. According to Ibl, one may distinguish two diffusion layers, one near the cathode, which pulsates, and the other farther outside, which is essentially stationary after a number of pulses have been applied. By using very short pulses, Ibl succeeded in realizing extremely thin pulsating diffusion layers, which allow high instantaneous pulsed current densities. Ibl also pointed to the necessity of strictly controlling the hydrodynamic conditions during the experiments with pulse plating. Working with a well-defined flow system (30), Ibl, Puippe and Angerer made a systematic study of PC plating of various metals, among others of gold-cobalt deposits (31-33). As regards the effect of pulse parameters on the composition of the alloy deposits, fair agreement exists between these results and those of Knödler (23). Ibl extended the experimental range to higher pulsed current densities. Fig. 9.18, which is taken from Ibl's paper (33), shows the effect of the off-time on carbon content of gold-cobalt alloys for a pulse length of T_{on} = 0.1 msec and three different pulsed current densities (1, 10, and 100 A/dm^2). Average current densities are indicated in parentheses. On the right side of Fig. 9.18 the carbon content of DC plated deposits is plotted. It can be seen in the figure that the carbon content of PC plated alloy deposits is generally lower than that of DC plated deposits, and that

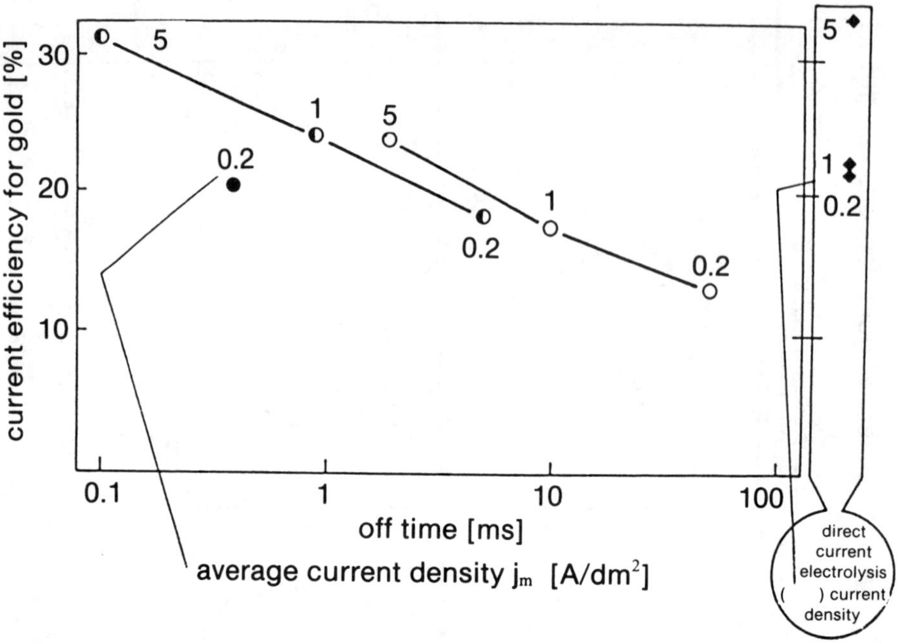

Fig. 9.19—Current efficiency vs. off-time of gold-cobalt alloy deposits at $T_{on} = 0.1$ msec; key as in Fig. 9.18 (33).

carbon contents decrease with increasing off-time. Corresponding results were obtained for a pulse length $T_{on} = 10$ msec, except for long off-times, where no further change in carbon content is observed.

The effect of pulsed current on the current efficiency of the alloy deposition as found by Ibl and coworkers is shown in Fig. 9.19 for a pulse length of 0.1 msec. On the right side is shown the current efficiency for DC plating at the same average current density. As the off-time is increased, carbon contents steadily decrease. These results are in good agreement with Knödler's results (23). Ibl and co-workers interpreted their results in terms of an adsorption-desorption model. Hydrogen is desorbed during off-periods. When a new pulse begins, the surface is less covered with hydrogen than during DC plating and hydrogen deposition is facilitated, the result being a lower current efficiency for gold deposition. Carbon is thought to be linked with adsorption of metal cyanide on the electrode surface. The observed influence of electrolysis conditions on the carbon incorporation can be explained by desorption of cyanide during the off-period.

Ibl and coworkers also studied the effect of pulsed current on the porosity of alloys gold-cobalt (31, 33, 34). Fig. 9.20 (34) shows the influence of the thickness of the coating on the porosity (expressed as number of pores on a 25 cm^2 area) of DC and PC plated gold and gold-cobalt deposits. Porosity of PC plated deposits

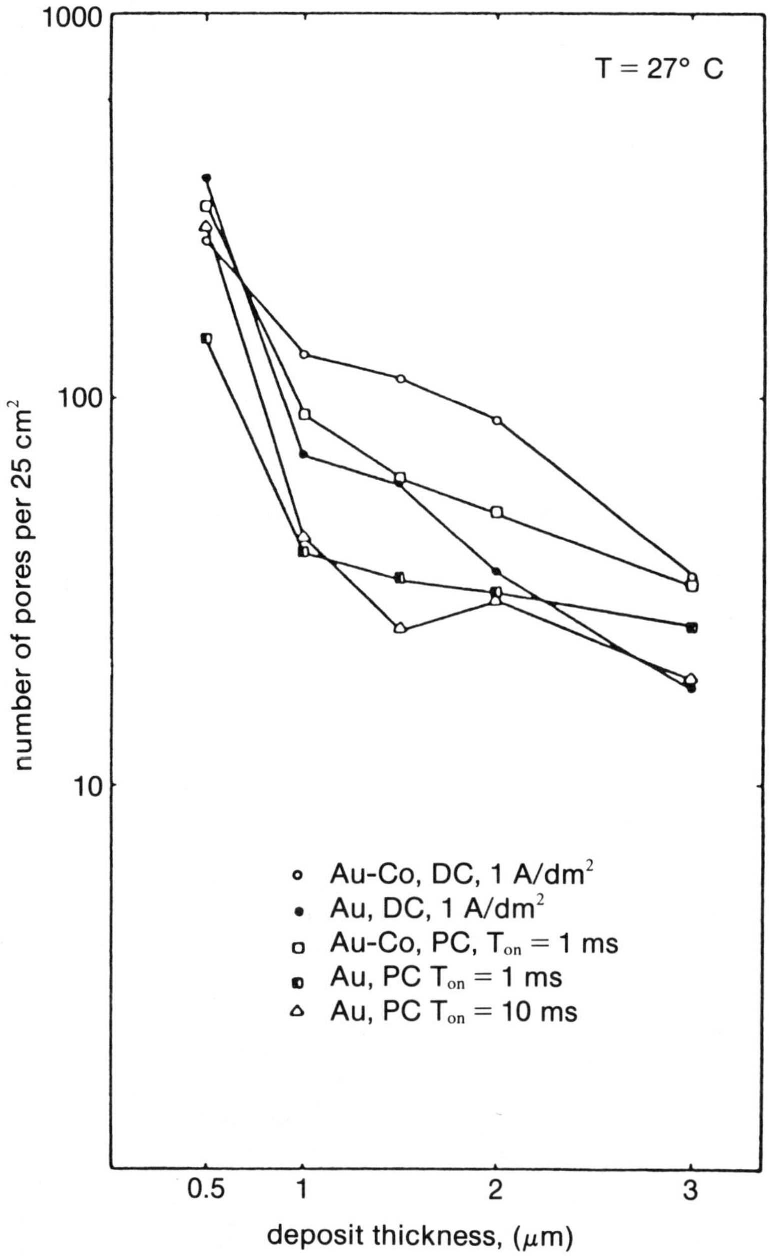

Fig. 9.20—Porosity vs. thickness of gold and gold-cobalt alloy deposits (34).

is lower for both gold and gold-cobalt alloys. This is in agreement with earlier results by Rehrig (5). Porosity of gold-cobalt deposits decreases slightly with increasing off-time. Deposits from cobalt-free gold electrolytes are less porous than gold-cobalt alloy deposits. The effect of pulse plating on the porosity is

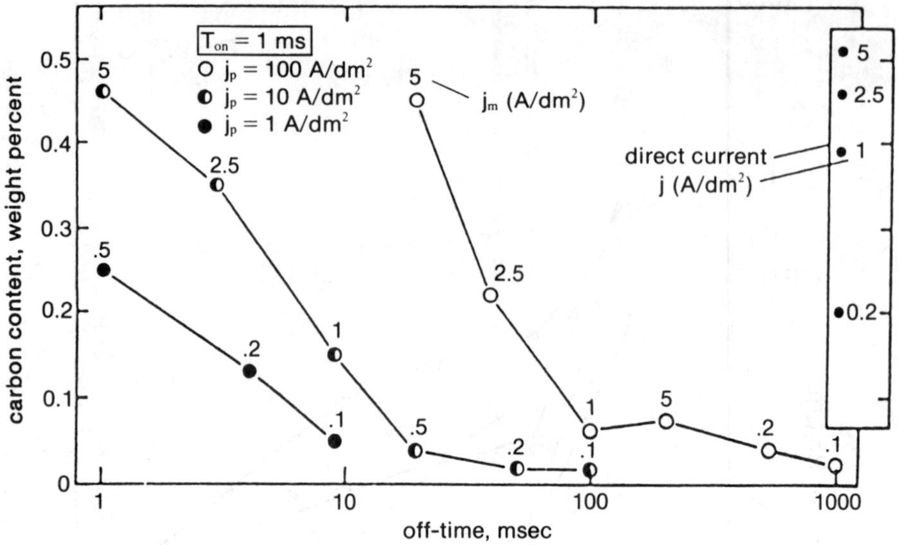

Fig. 9.21—Carbon content vs. off-time of gold-cobalt alloy deposits; $KAu(CN)_2$ 15 g/l, Co 0.1 g/l, pH 3.7; at T_{on} = 1 msec (35).

interpreted as being the result of structural changes in the deposit. Pulse reverse has also proved to reduce porosity of gold and gold alloy electrodeposits. For instance, the porosity of nickel-hardened gold was reduced by using low-frequency PR (about 1 Hz) instead of DC (83).

In further studies on the PC deposition of gold alloy, Ibl and coworkers extended the range of pulse plating conditions. As an example of the results of these studies, the influence of the off-time on carbon content of gold deposits from a citrate bath containing 0.1 g/l Co is shown in Fig. 9.21 (35, 36). At constant pulse length T_{on} = 1 msec and constant pulsed current density (j_p), the carbon contents decrease with increasing off-time T_{off}. The average current density (j_m) is indicated above each point in the figure. Results for DC plating, with current densities corresponding to the average PC densities, are shown for comparison on the right side of Fig. 9.21. Carbon contents of DC plated deposits are higher than those of PC plated deposits.

When the off-time (T_{off}) and the product of PC density and on-time ($j_p T_{on}$) are kept constant, the ratio of the amount of carbon or hydrogen desorbed during off-time to the amount of deposited metal is the same for different pulse conditions. Under such conditions the carbon content was nearly the same for each experiment, as can be seen from Table 9.4 (35).

The decrease of the carbon content with increasing off-time tends to level off at high values of off-time, indicating that desorption of the carbon-containing species on the cathode reaches a limiting value, as shown in Fig. 9.22 (37). The agreement with results of other authors is very good.

TABLE 9.4
Influence of Pulse Current Density (j_p) at Constant ($j_p \cdot T_{on}$) and T_{off} on Carbon Content of Gold Deposits

Off-time T_{off} (msec)	j_p (A/dm^2)	Pulse time, T_{on} (msec)	Carbon content, %
10	1	10	0.18
10	10	1	0.15
10	100	0.1	0.15
100	100	1	0.06
100	10	10	0.05
100	1	100	0.04
1000	100	1	0.02
1000	10	10	0.02
1000	1	100	0.04

The effect of PC on the porosity of gold deposits was measured by the SO_2 test. Some of the results of these studies are shown in Fig. 9.23 (35). Porosity decreases with increasing off-time. This is attributed to the desorption of hydrogen during off-time because porosity is related to the presence of bubbles on the cathode surface during deposition.

It is concluded from these studies that the large range of experimental conditions for pulse electrolysis leads to a greater variety of results than is possible with DC electrolysis. The optimum conditions for desired coating properties lie in a narrow range that cannot be predicted theoretically.

The effect of pulsed current on the morphology of gold-cobalt deposits was also studied by Puippe and Ibl (38). At constant pulse time (1 msec) and constant PC density (10 A/dm^2), the grain size of gold-cobalt alloy deposits increases with increasing off-time, as revealed by SEM micrographs. This is in contrast to results with cadmium deposition where a grain refinement was observed. The explanation is that when gold is deposited, an activation of growth centers or a recrystallization seems to take place. The bigger grains are more stable thermodynamically and if the system has enough time it will reach its more stable state.

Mäusli, Reid and Steinemann (39) examined the structure of DC and PC plated gold-cobalt deposits by transmission electron microscopy at extreme resolution. Pores with diameters ranging from 5 to 50 Å could be detected. Grain size of PC plated deposits with an on/off ratio of 10/20 msec/msec is about 3 times higher than in DC plated deposits. As off-time is decreased, pores become less numerous, disappearing totally at on/off ratios of 10/50 and 10/100 msec/msec, while grain size continues to increase. This effect is explained by desorption of hydrogen and/or cyanide-containing material during off-periods. The lower density of DC plated gold-cobalt layers and their apparent lack of ductility can also be explained by the presence of pores. The pores are filled with

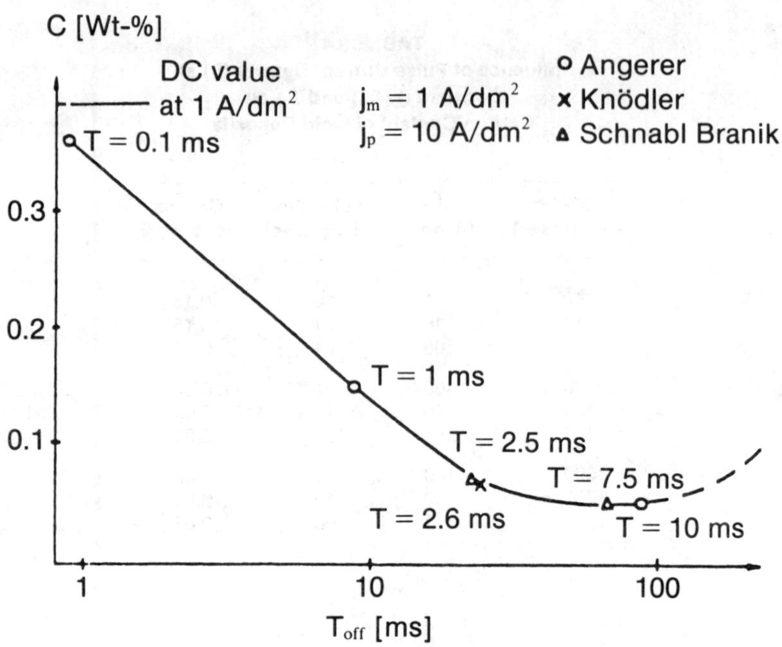

Fig. 9.22—Carbon content vs. off-time of gold-cobalt alloy deposits at constant average current density j_m (1 A/dm^2) and PC density j_p (10 A/dm^2). Same solution as in Fig. 9.21 (37).

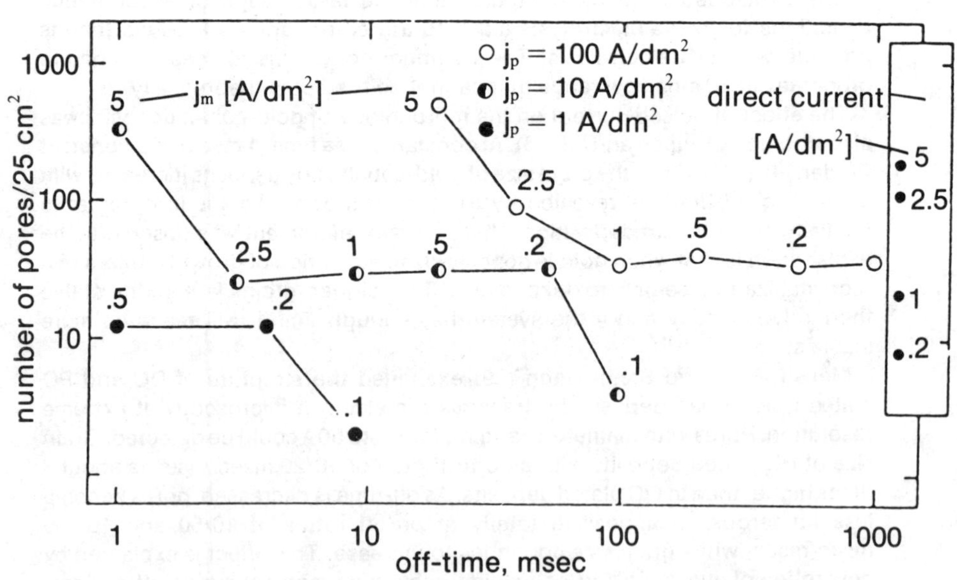

Fig. 9.23—Porosity vs. off-time of gold-cobalt alloy deposits at T_{on} = 1 msec. Same solutions as in Fig. 9.21 (35).

"polymer" (the true nature of the carbonaceous inclusions was not known at the time of this investigation). On heating at 400° C the polymer decomposes, as found by Harris and Darby (40, 41). Pore structure in DC plated gold-cobalt deposits is quite insensitive to thermal shock treatment. This is explained by the high pore density $10^{19}/cm^3$) and the small grain size which prevents coalescence of pores. On the other hand, lack of pores in PC plated gold-cobalt deposits proved detrimental for contact applications, where structural strength is all-important and high ductility is not an advantage.

The effect of pore structure on the properties of gold-cobalt deposits was investigated in more detail in a later paper (42). The deposits for this investigation were produced in a conventional citrate-buffered electrolyte containing 10 g/l gold and 0.07 g/l cobalt. The structure of a deposit with PC 10/100 msec/msec was completely different from that of a DC plated deposit, with a marked increase in grain size and complete absence of pores or inclusions. Bulk resistivity of a PC plated gold-cobalt deposit (10/100 msec/msec) is only about half of that for a DC coating. This is reflected in measurements of contact resistance, results of which are presented in Fig. 9.24. Values of contact resistance for the PC coatings are lower over the whole range of loads. The slope of the curves is close to 1/3, indicating that over this range of loading, deformation of the contact surface is elastic.

Figure 9.25 shows data for another series of measurements carried out over a wider range of loads. Values for PC and DC coatings are similar to those in Fig. 9.24, but show a change in the slope of the curves at a load of about 100 g. This may represent a transition from purely elastic to partial plastic deformation. The contact resistance of PC coatings was reduced by etching with dilute aqua regia. This may be due to the removal of surface contamination. The data for the DC deposits indicate the retention of a tenacious surface film after etching. There were also large differences in wear behavior between DC and PC plated deposits, as shown by wear measurements according to the crossed rods method. PC plated coatings showed characteristics of adhesive wear by seizure, while DC coatings subjected to the same condition showed only a slight polishing effect with negligible abrasive wear (Fig. 9.26 and 9.27).

Hardness stability, according to Flühmann et al. (42), is a highly important property of gold-cobalt deposits for contact applications. Hardness of a DC coating is maintained on short heating up to 400° C, at which temperature no basic change in structure is observed. With a PC coating similar heating results in recrystallization and softening. Such a condition could easily arise due to a local increase in temperature resulting from sliding contact.

The authors reach the conclusion that in DC plated gold-cobalt alloy coatings high wear resistance is achieved at the expense of other desirable properties such as ductility, low resistivity and contact resistance, and bondability. PC plating offers interesting possibilities of achieving an optimum compromise between various characteristics by suitable choice of the on/off cycle to regulate the incorporation of stabilizing inclusions. The authors also point to the possibility of using duplex coatings produced simply by switching the current mode in the same bath. For example, a DC layer could be used with a thin top coat of a high-purity PC layer for improved solderability or low contact resistance. On the other hand, an initial PC layer with low porosity could be followed by a DC layer for improved wear resistance.

Stimetz and Hren (43), in their study on pulse plating of hard gold deposits, also found higher cobalt contents in PC plated deposits, in agreement with most

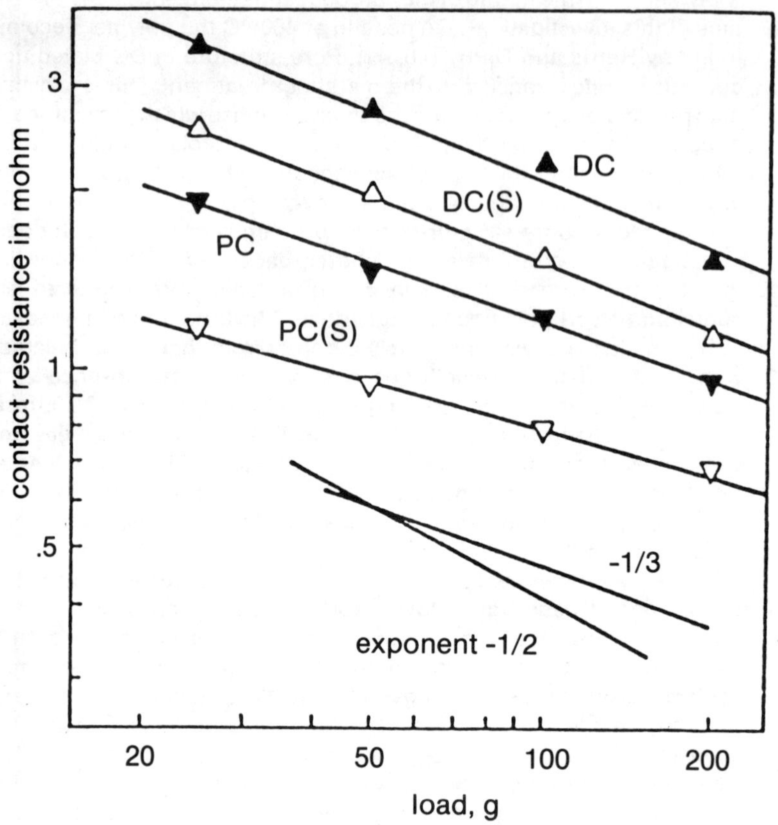

Fig. 9.24—Contact resistance of DC and PC 10/100 plated gold-cobalt coatings obtained with the crossed bar method for static contact and after sliding under load (42).

other authors. Using a current density of 0.3 A/dm^2 and a duty cycle of 10 percent, a compact, semi-bright deposit was obtained. At current densities of 0.8 and 1.2 A/dm^2 with DC plating and PC plating with a 50 percent duty cycle, the deposits were compact and semi-bright. A reduction of the duty cycle at these current densities resulted in a rounded modular surface topology and a dull appearance. The transmission electron and electron diffraction analyses indicated no significant difference in microstructure between DC and PC plated gold alloy deposits. The grain size was in the range of about 120 to 600 Å for all deposits analysed. Twin density in PC deposits with a 50 percent duty cycle was different from other deposits.

Inglot, Socha and Pezinski (44) also employed a weakly acidic citrate gold solution with small additions of either cobalt or nickel. Pulse plating with an on/off ratio of 0.33/3.3 msec/msec reduced the current efficiency of the alloy plating, and also the porosity of the deposits in the range of 0.5 - 2 μm. At higher

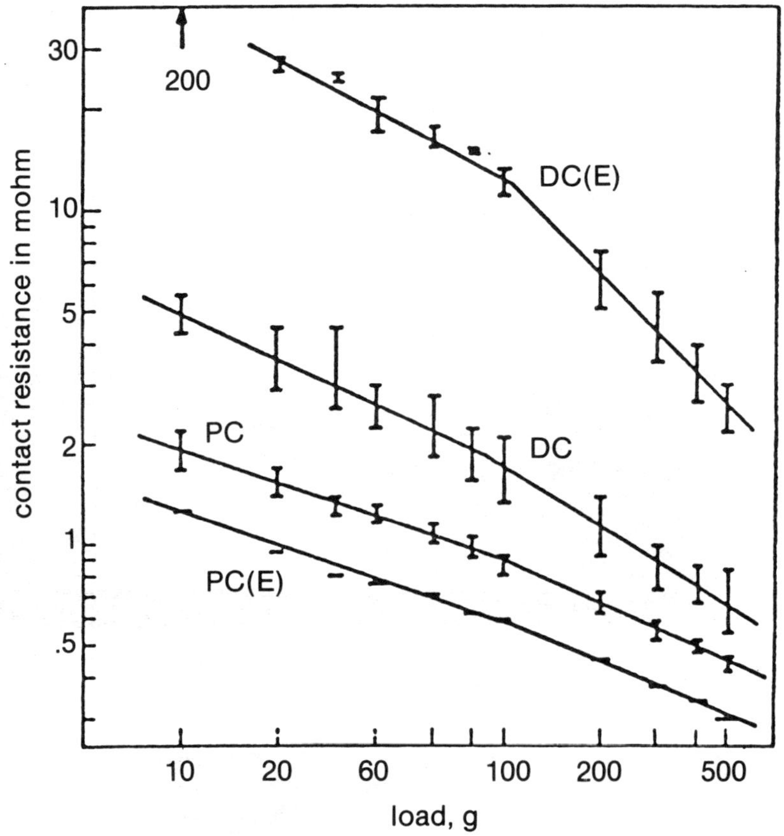

Fig. 9.25—Contact resistance of DC and PC plated gold-cobalt coatings obtained with the crossed bar method as plated and after aqua regia etch. (42).

thicknesses, no difference in porosity between DC and PC plated deposits was observed. Wear resistance of PC plated gold-cobalt deposits was higher than that of DC plated coatings, but with gold-nickel alloy deposits the opposite effect was observed. It is recommended that PC plated gold-cobalt or gold-nickel deposits be used as an underlayer for low-carat deposits.

Most of the work on pulse plating of gold alloys seems to have been done with gold-cobalt and gold-nickel from weakly acidic cyanide solutions. There are, however, some other gold systems which have received some attention. For example, gold-iron alloys can be deposited from such solutions analogous to gold-cobalt and gold-nickel alloys. In his studies on electrodeposition of gold-iron alloys Knödler (45) used a citrate-buffered gold cyanide solution to which small amounts of Fe(III) ions were added. The effect of increasing Fe(III) concentrations on the current efficiency and the composition of the alloy deposits is shown in Fig. 9.28. Bright and hard gold-iron alloys, similar to gold-cobalt and gold-nickel alloys, can be obtained by DC deposition. For comparison

Fig. 9.26—Scanning electron microscopy of wear track on PC 10/100 gold-cobalt coating. Adhesive (seizure) wear and ploughing is apparent (42).

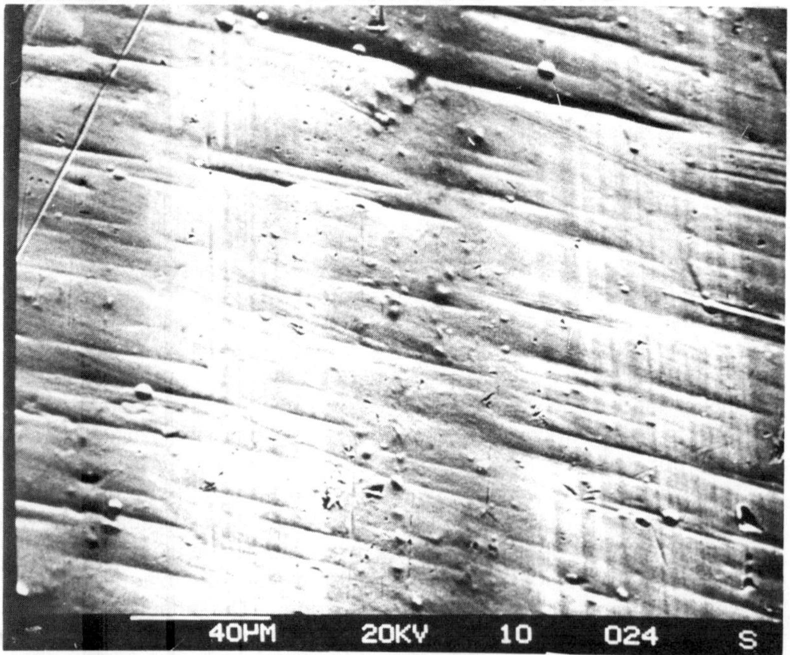

Fig. 9.27—Scanning electron microscopy of wear track on DC gold-cobalt coating. A light polishing indicates negligible wear (42).

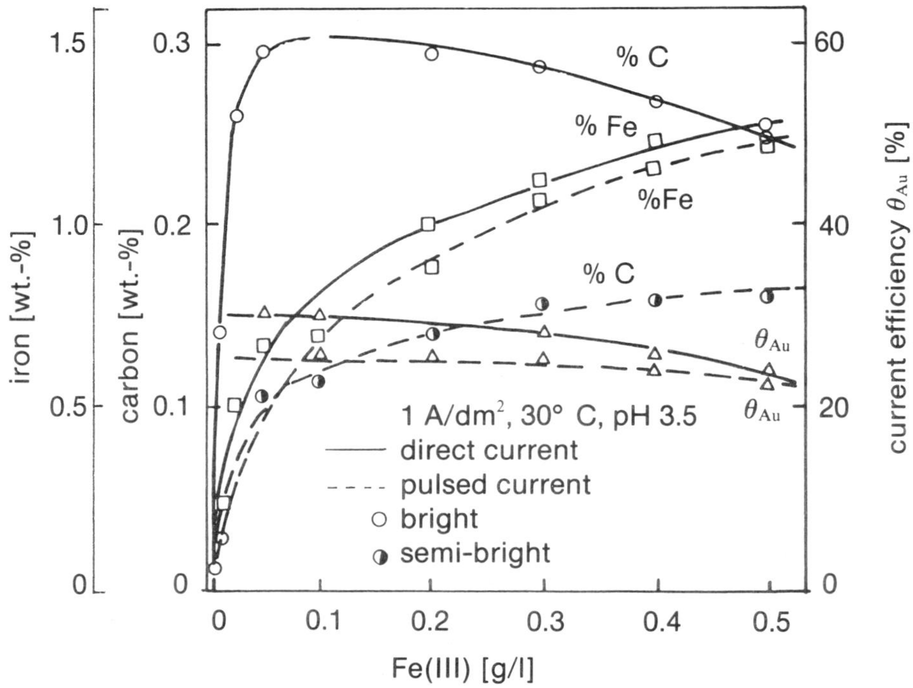

Fig. 9.28—Iron and carbon content, current efficiency vs. Fe(III) concentration of DC and PC plated gold-iron deposits. On-time 2.3 msec, off-time 20 msec (45).

purposes the results with PC plating are included in the figure. This figure can be compared to Fig. 9.6 and Fig. 9.7 for gold-cobalt and gold-nickel respectively. An increase in the Fe(III) concentration causes an increase in the iron content of the deposits. Iron contents are higher, by a factor of about 3, than nickel or cobalt contents in alloys deposited under comparable conditions. Carbon contents are higher by a factor of about 1.5 and show a maximum at 0.1 g/l Fe(III). Iron contents of PC plated deposits are slightly smaller than those of DC plated deposits. This is in contrast to the results in the gold-cobalt and gold-nickel systems, where pulse plating causes an increase in the alloy metal content. Carbon contents, on the other hand, are reduced drastically by use of pule plating. Current efficiency of PC deposition is slightly lower than that of DC deposition. Brightness of deposits is reduced by pulse plating.

The effect of increasing the average current density by increasing the on-time is shown in Fig. 9.29. Carbon and iron contents of the deposits increase slightly increase both in DC and PC plated gold-iron deposits.

Summarizing the results of gold-iron deposition, it may be said that pulse plating does not have such pronounced effects as in gold-cobalt or in gold-nickel deposition. Current efficiency and brightness of electrodeposits is reduced by pulse plating. Differences between DC and PC plated deposits are

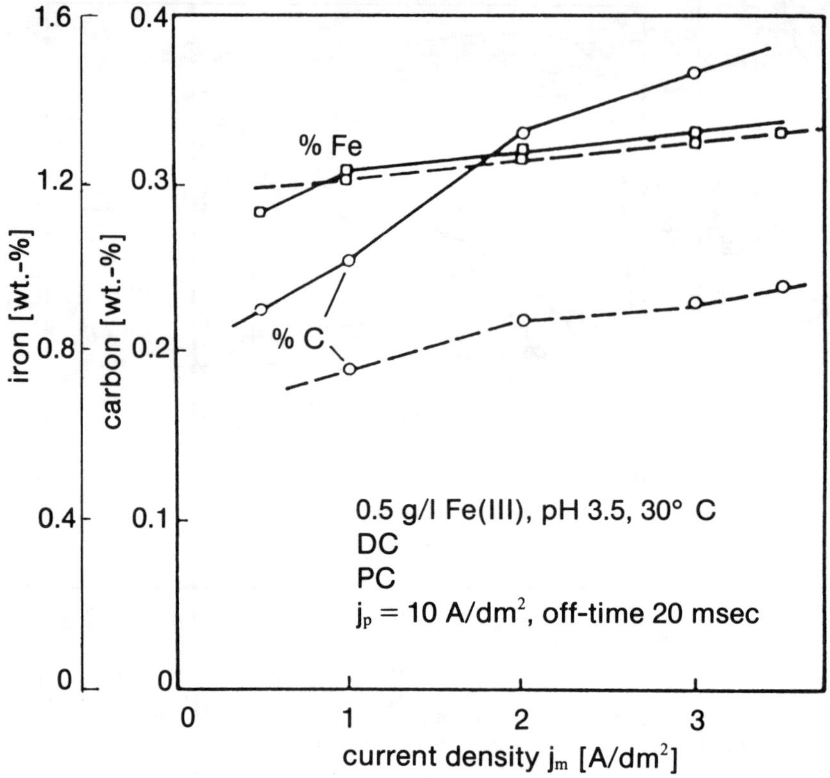

Fig. 9.29—Iron and carbon content vs. average current density of DC and PC plated gold-iron alloy deposits. Off-time 20 msec, on-time variable (45).

much smaller, compared to gold-cobalt or gold-nickel deposits. This may be explained partly by the fact that iron is present as Fe (III), which must be reduced to Fe(II) before it is incorporated in the deposit. No significant improvement could be observed by use of PC instead of DC for the electrodeposition of gold-iron alloys.

J.C. Catonne and coworkers (46) carried out an experimental study on the optimization of the brilliancy of gold-silver deposits by pulse plating. Considering the high number of variables in pulse plating, and in order to obtain a maximum of information from as few experiments as possible, the authors adopted the principle of experiment planning. In the case of a pure gold bath, this required planning in two dimensions, realized for four different values of the on/off ratio. The optimum zone of brilliancy corresponds very precisely to the minimum zone of porosity. The deposits obtained in this zone are always less porous than DC plated deposits. Microhardness of deposits increases with PC density. At comparable rates of deposition, metal distribution under pulse plating conditions is always poorer than under DC conditions.

By careful planning, it was possible to reduce the number of experiments in the study of gold-silver deposition to about twenty. DC plated deposits were all dull. As for the microhardness, results with PC were almost identical to those obtained with DC. Metal distribution of PC plated gold-silver alloys was inferior to that under DC plating conditions. It was concluded that by careful planning it is possible to reduce the number of experiments necessary to define the optimum range of PC plating parameters.

B. Sutter, from his studies on pulse plating of gold-nickel alloys, comes to the following conclusions (47):

(1) The maximum pulse length is inversely proportional to the square of the PC density. Higher pulse lengths cause the deposits to "burn".
(2) The interval between the current pulses must be sufficient for relaxation processes to occur in the diffusion layer.
(3) Relaxation time should be essentially proportional to the square root of the product of pulse length and PC density. With shorter relaxation times, the PC conditions approach those of DC plating.
(4) Maximum plating rate decreases with increase in peak current density.
(5) Using PC with an average current density higher than that permitted under DC plating conditions causes the deposits to burn.

In order to optimize the process of pulse plating of gold it is important to take into account the cathode decay time, which was determined after a series of systematic trials involving the pulsing of the cathode potential during electrolysis (48). Using gold plating baths in industrial or pilot production lines, Sutter and coworkers obtained the following results:

Porosity—Low frequency pulsed currents (62-100 Hz) can reduce the porosity of gold-cobalt and gold-nickel deposits. Under favorable conditions the thickness of gold can be reduced to one third, but the surface of the basis metal plays an important role.

Throwing power—When metal distribution difficulties result from lack of homogeneity in current flux, they can be partly solved by the use of pulse plating.

2.3. *Applications of pulse plating of gold*

According to Avila and Brown (49), industrial pulse plating was first attempted at the Western Electric Engineering Research Center in Princeton, N.J., in June 1967. In the following year, the need for a high conductivity gold electrodeposit for integrated circuitry initiated the development of a pulse plating unit which supplied an on-off DC pulse with on- and off-times in the millisecond range and operated with current peaks up to 8 A. Experiments performed with a phosphate gold electrolyte demonstrated that for equivalent resistivity, approximately one-half as much thickness was required for pulse plated gold as for conventional DC plated gold.

C.A. Burrus (50) used a simple capacitor discharge pulse generator for pulse electroplating of high-resistance materials, poorly contacted devices, and extremely small areas. The high starting voltage aids in overcoming problems caused by poor contacts, thin insulating layers or other defects. Examples of practical applications are the gold plating of n-type GaAs and surfaces in series with rectifying junctions. T.N. Bhar and G.M. Lamb (51), using a simple pulse plating unit, succeeded in gold plating thin film circuits to a thickness of 18 μm in a gold sulfite plating bath. Thin film chokes and antennae for laser radiation have also been successfully gold plated by this method.

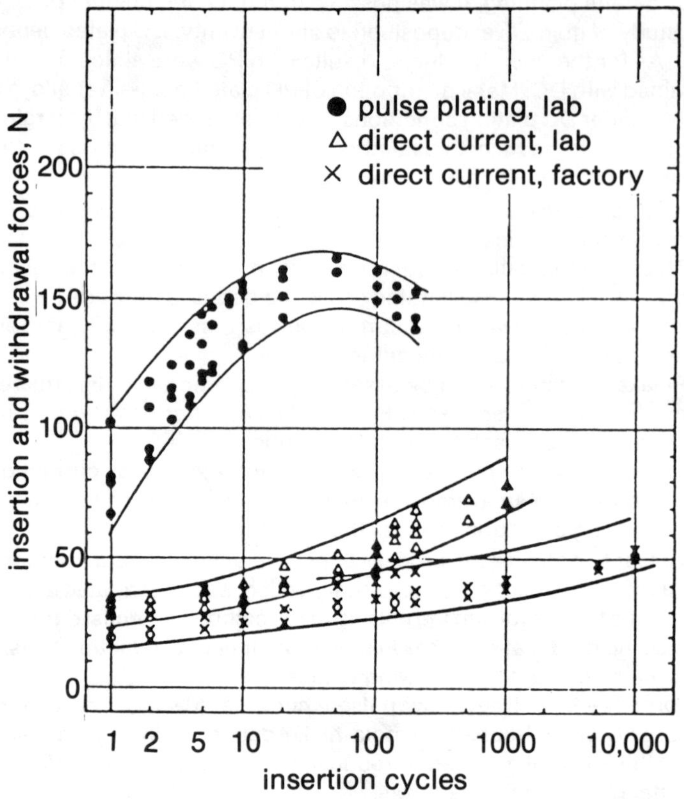

Fig. 9.30—Insertion and withdrawal forces of DC and PC gold-cobalt deposits on connectors vs. insertion cycles (52).

K.L. Schiff and R. Schnabl (52) examined the wear behavior of connectors with DC and PC plated gold-cobalt coatings by measuring the insertion and withdrawal forces. Fig. 9.30 shows that these forces were higher for PC plated deposits from the beginning and increased with the number of cycles. SEM investigations revealed that wear of PC plated coatings was higher than that of DC plated coatings, although hardness and ductility were nearly the same. It is concluded that the better wear behavior of DC plated gold-cobalt layers is due to the presence of carbonaceous impurities, which probably act as a lubricating agent. Examination of wear tracks by SEM revealed a higher wear for PC plated gold-cobalt deposits.

These studies were continued by M. Branik and R. Schnabl (53). Results of work on the effect of pulse plating on the composition of gold-cobalt alloys were in agreement with those of other authors. The suitability of the gold coatings for connector applications was examined by means of wear and corrosion tests, and ductility was measured by bending tests. Wear was estimated by measuring the insertion and withdrawal forces in connector assemblies. The wear resistance depended markedly on the current modes. Fig. 9.31 and Fig. 9.32 are taken from Branik and Schnabl's work. As pulsed current density (indicated by numbers on the right side, in A/dm^2) is increased, insertion and withdrawal forces also increase. Only for PC layers deposited with a pulsed current density of 2 A/dm^2 and on/off-times of 2.5/3.0 msec/msec were values of these forces similar to those obtained with DC plated layers. Wear tracks of the PC deposit (22) showed strong adhesive wear and prow formation. It is concluded from these results that the higher carbon content in DC plated deposits is responsible for the better wear behavior in contact applications. PC plated deposits, which have low carbon inclusions, have a higher ductility, but at the same time there is also a marked decrease of the wear resistance under sliding conditions. Thus, no advantages can be gained using this plating method, according to Branik and Schnabl.

The views expressed by Branik and Schnabl are in agreement with conclusions put forward by Souter (19), which have already been discussed. These views remain to be reconciled with the contradictory view arrived at by Antler (20). In this connection reference should also be made to the paper of Flühmann, Reid, Maüsli and Steinemann (42) reviewed above.

In a later study J. Ehrhardt and H. Nowacki applied both DC and PC to the high-speed deposition of gold alloy coatings on electrical connectors (54). The authors used a weakly acidic gold-cobalt solution with current densities up to 25 A/dm^2. Very smooth gold deposits were obtained by DC and by PC plating with an on/off ratio of 3/7 msec/msec under the following conditions: 24 g/l gold, temperature 70° C, current density 20 A/dm^2. Wear was estimated by measuring the insertion and withdrawal forces in connector assemblies and by the electrographic porosity test. Wear was found to be lowest for gold coatings PC plated with an on/off ratio of 3/7 msec/msec. DC plated deposits came next, while PC plated deposits with on/off ratios of 1/10 and 3/30 showed high wear. Also, porosity was lowest for PC plated deposits. The authors concluded that pulse plating can be used successfully for contact applications if a suitable on/off ratio is selected. Simultaneously, gold thickness can be reduced without deterioration of the quality of the gold coating. The conclusions reached by Ehrhardt and Nowacki are in disagreement with the results of Branik and Schnabl, and also with those of other studies. It must, however, be pointed out that there is a virtually infinite variety of pulse plating conditions, and optimum conditions may be restricted to a narrow field of plating parameters.

D.L. Rehrig (55) reported on the high-speed selective gold plating of lead frames for microelectronic packaging. An automated "spot plating" process at the Allentown Work of Western Electric reduced the gold needed for lead frames by 90 percent. The plating procedure includes (a) nickel plating the entire stamped copper frame with 2 - 10 μm of nickel, (b) selectively depositing a DC gold plate 0.4 - 0.65 μm on the external leads for solderability, and (c) selectively building the gold thickness to 2.5 μm in the bond area of the frame. An asymmetric AC current, as shown in Fig. 9.33, was used for step (c) after production experience had shown that DC deposition was unsuitable for producing bondable gold deposits at high current densities, i.e., greater than 30 A/dm^2. A proprietary high-speed gold solution was used. The basic standards for bondability were pull strength requirements. It could be concluded that a critical threshold in anodic current density of about 20 A/dm^2 must be reached to ensure bondable gold. As cathodic current density was increased from 100 to 300 A/dm^2, grain size increased, and deposits appeared less lustrous and more

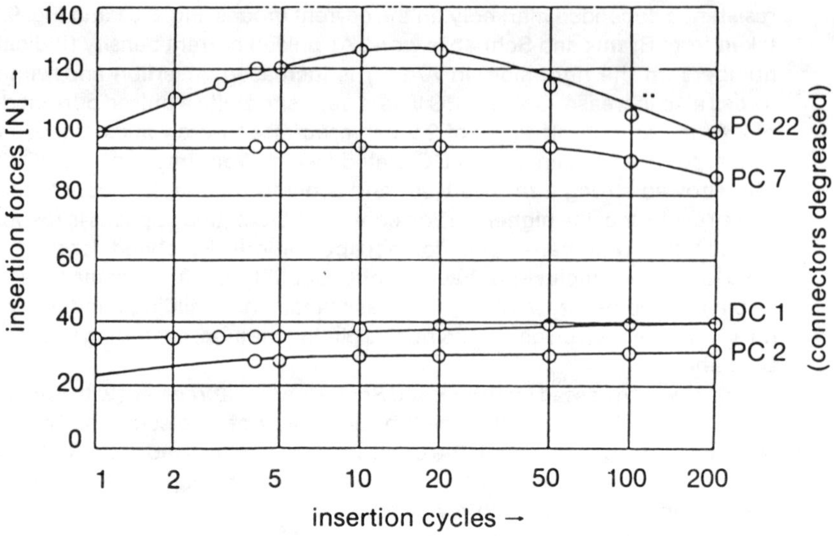

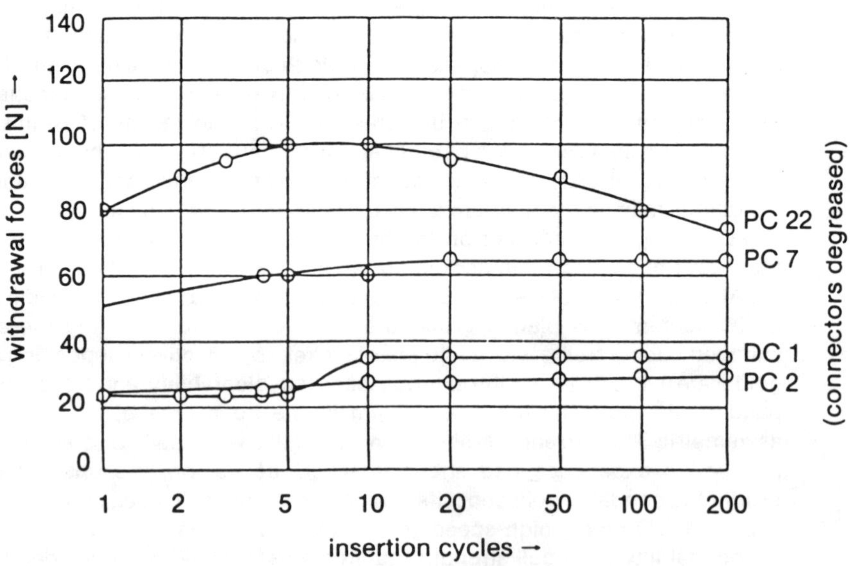

Fig. 9.31, 9.32—Insertion and withdrawal forces of connectors plated with DC and PC gold-cobalt coatings over nickel vs. insertion cycles. Numbers on right side indicate pulsed current density in A/dm^2 (53).

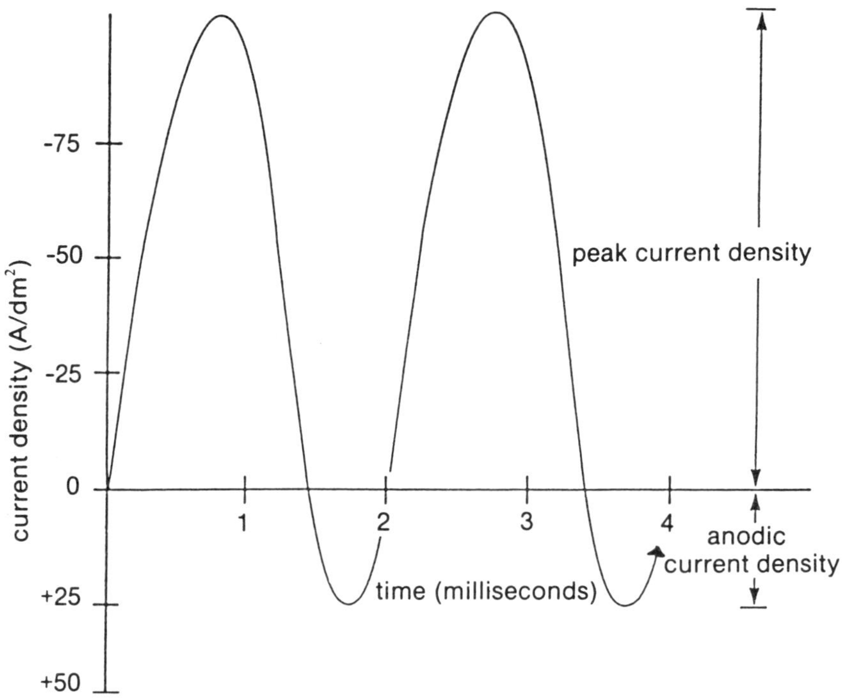

Fig. 9.33—Asymmetric AC current used in the spot plating process at Allentown by Rehrig (55).

mat, although bondability was satisfactory in each case. As the frequency of the AC was increased from 60 to 250 Hz, surface projections on a microscopic scale increased slightly, as did the bondability. There is a direct relationship between thermocompression bondability and electrodeposit microstructure. Projection of grains from the deposit surface is critical for acceptable bonding results.

In a second paper, co-authored by R.V. Dafter Jr., R. Haynes and D.L. Rehrig (56), results were reported of thermocompression bonding of high-speed selectively plated pure gold deposits using both DC and asymmetric AC for 6 different proprietary solutions. Asymmetric AC waveform yielded more acceptable bonding adhesion over a wider peak current density range than DC plating. The current densities for minimizing the number of bonding failures depended on the specific solution used. Bonding strength also depended on the gold metal concentration and deteriorated in the presence of copper and nickel in two of the solutions investigated.

D.W. Endicott and R.K. Asher (57) described a method for selectively electroplating electronic multi-lead headers using an automated endless belt apparatus. Gold deposits proved best for bonding purposes. Scanning electron microscopy and Auger electron microscopy were used to study the effect of waveform on the topology and function of the deposit. A pulse rectifier that can superimpose a standard DC (base current) onto a squarewave DC pulse was used. The base current, peak pulsed current and pulse duty cycle are important parameters that must be controlled to ensure deposits that will bond despite temperature variation during the bonding. A duty cycle of 16.7 percent with on-and off-times of 1 and 5 msec respectively met the bonding requirements with the minimum amount of diffusion of the basis metal (Kovar).

In a paper on the optimization of gold usage, J.S. Stevenson (58) came to the conclusion that improvements in the quality of gold deposits on printed circuits and connectors by use of pulse plating are only marginal. Stevenson did not expect practical applications on a larger scale for printed circuits.

A more optimistic view was held by A. Blair (59). Based on the results of experimental work, Blair claimed the following advantages of pulse plating: (a) reduction of porosity, (b) reduction of minimum thickness, (c) improvement in throwing power, (d) higher current density range, (e) improvement in bondability.

In a paper read at the AES Second International Pulse Plating Symposium (1981), J. Lochet (60) described the use of pulse plating to produce a special functional gold deposit on semiconductor devices shown in Fig. 9.34. A minimum thickness of 3 μm of gold on the "hockey stick" area was necessary for the formation of a Si-Au eutectic. Gold deposits had to be of uniform lemon color. Using the standard power supply it was impossible to produce the desired color in the thickness range required. The problem could be solved by using pulse plating with an on/off ratio of 0.1/0.3 msec/msec. This cycle also gave (a) a higher gold thickness, (b) a slightly better metal distribution, (c) a more consistent and reproducible gold structure, (d) a slightly lower average voltage. The initial "flash" part (0.4 μm) of the process was found to be most critical. A solution at a slightly acidic pH and a gold concentration of at least 24 g/l gave satisfactory results. For the remaining 2.8 μm an alkaline gold plating solution was used, which produced very pure deposits. By use of pulse plating, this minimum thickness could be further reduced to 2.25 μm.

In a study by D. Towne (61) a comparison was made of DC and PC in barrel plating of connectors. Porosity, resistivity, and wear resistance were also tested. A statistical analysis indicated that gold usage during barrel plating could be reduced by pulse plating for some sample parts, depending on their shape.

Pulse plating has also been applied to microwave integrated circuit (MIC) fabrication (62). Substrates for MICs are first metallized by RF sputtering with thin films of chromium and gold, then electroplated in a gold citrate solution using pulse plating with a 10 percent duty cycle, i.e., 0.1 msec on, 1 msec off. Dense, fine-grained deposits resulted from the use of higher peak voltage and higher limiting current density. Thus it was possible to eliminate the striking step prior to the pulse plating process. The fine-grained structure and low porosity of the PC plated gold not only contributed to the higher conductivity and low loss circuit performance, but also eliminated the need for a barrier layer between the layers of the thin film metallization system.

Use is also made of a special kind of modulated pulse periodic reverse system (Polar) for electrodeposition of coatings (63). In this system, a series of cathodic current pulses is followed by a series of anodic pulses, thus combining the effect of pulse plating with that of periodic reverse current. It is pointed out that gold plating bath composition has to be modified in some cases in order to get optimum results from the use of this system.

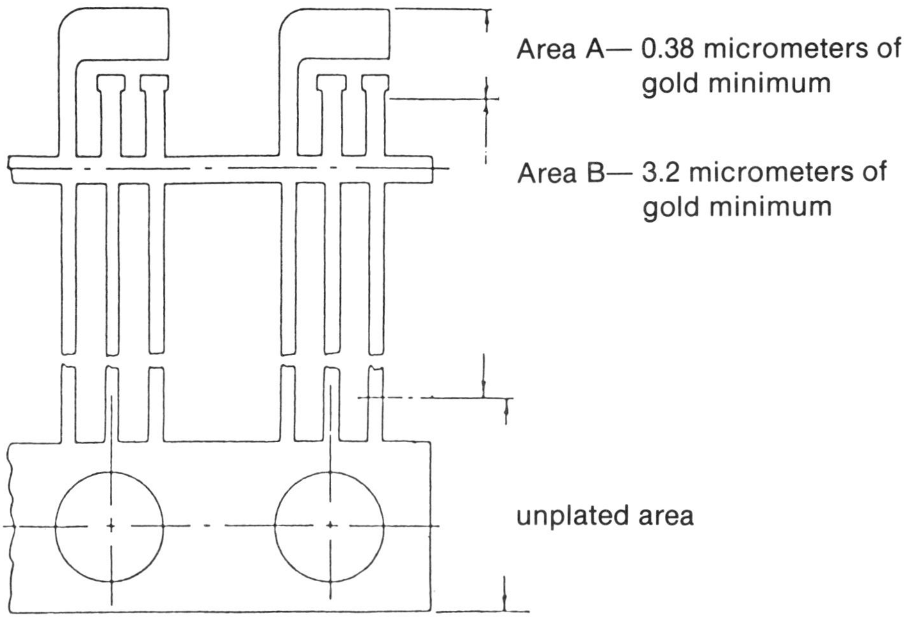

Fig. 9.34—Semiconductor device for reel selective gold plating (60).

The use of pulse plating for electroforming of gold for decorative and industrial applications is mentioned by D. Withey (64). Thicknesses of several hundreds micrometers can be obtained now that the initial difficulties in finding suitable electrolytes are overcome. This should ensure that many more applications for heavy gold electrodeposits will be found in the future.

3. Pulse plating of precious metals other than gold

3.1. *Palladium*

Since the drastic increase of the price of gold in the late seventies, palladium electrodeposits have received much attention as a substitute for gold electrodeposits. Studies have shown that equivalent thicknesses of palladium and gold have the same degree of porosity and corrosion resistance. Since the density of gold is more than one and a half times that of palladium, equivalent thicknesses of palladium are 60 percent lighter than gold. Palladium is much harder than gold, but has the disadvantage of polymer film formation in certain atmospheres. This can be overcome by a thin gold coating. Palladium has a much higher contact resistance than gold, and this can likewise be overcome by a gold flash or by increasing the contact loading.

Palladium is readily deposited from a number of aqueous electrolytes of which only three are of commercial importance: (a) electrolytes based on diammino-dinitrito-palladium ("P" salt), (b) electrolytes based on tetrammino-palladium, (c) highly acid electrolytes.

A detailed study of the morphological, crystallographic, electrical and electrochemical characteristics of palladium electrodeposits from a mildly acid phosphate electrolyte was carried out by R.J. Morrissey (65). The deposits from this bath are bright and fine-grained but subject to microcracking, the extent of which depends on bath temperature, current density, and deposit thickness. At bath temperature above 55° C deposits are substantially microcrack-free at the expense of a concurrent decrease of brightness and throwing power. Microcracking is regarded as being caused by a volume increase due to recombination of hydrogen atoms diffusing out of the palladium deposit.

In a paper read at the AES Second Pulse Plating Symposium (1981), Morrissey (66) described efforts which have been made, by pulse plating and by chemical modification of the bath, to eliminate microcracking and to improve brightness and plating range. Substitution of a citrate-phosphate electrolyte for the all-phosphate electrolyte produced a grain morphology totally different from that obtained in the all-phosphate system. Microcracking was virtually eliminated. By use of pulse plating it was possible to reduce grain size and improve brightness. Optimum pulse parameters, as established by Hull cell experiments, were 1 msec on- and 9 msec off-time.

A search for improved brightening agents led to the discovery of a class of compounds which produce further reduction of grain size and very high levelling power. Pulse deposition from baths containing this class of brightener effects only a slight improvement in deposit brightness and plating range and increases the degree of microcracking. From visual appearance of Hull cell specimens, optimum pulse parameters are 10 microseconds on- and 40 microseconds off-time. All brightener systems investigated so far produced strongly [110]-oriented deposits. Pulse deposition does not improve brightness or plating range at ordinary current densities. At long on-times both of these parameters are degraded.

In an investigation on the morphology of pulse plated deposits, J.Cl. Puippe and N. Ibl (38) examined the effect of PC on palladium electrodeposition from a tetrammine bath. In these experiments the pulse time was less than the transition time and the average current was well below the limiting value for DC plating under the same hydrodynamic conditions. Thus the depletion in the diffusion layer was small and the observed effects, therefore, were not due to the influence of mass transport phenomena. When pulsed current with a high PC density (5000 A/dm^2, pulse time 1 msec) is used, a decrease in grain size is observed compared to palladium deposited with DC of the same average current density (5 A/dm^2). This is explained by an increase of the nucleation rate with current density and seem to be a general feature of pulse plating (see Chapter 3).

It is known that palladium has a high affinity for hydrogen. Thermodynamic studies of the system Pd-H have shown that two phases can coexist at room temperature, of which the stable α-phase has an atomic ratio H/Pd of 0-3 percent. With more absorbed hydrogen, at atomic ratios between 58 and 69 percent, the β-phase is obtained. Both phases coexist when H/Pd lies between 3 and 58 at.-%. During electrodeposition of palladium, codeposition of occurs. The behavior of the codeposited hydrogen and its influence on the properties of DC plated deposits has been the subject of several studies (67, 68). It is concluded from these studies that the hydrogen content of palladium deposits should be as low as possible to avoid high tensile stresses and microcracking. An electrochemical method for estimating the hydrogen content of palladium deposits has been described by Hedrich (67).

The effect of pulse plating on the hydrogen content and the properties of palladium deposits from a "P" salt solution was studied by Locarnini and Ibl (69). Preliminary experiments with DC had shown that current efficiency is reduced by raising the current density. This may be interpreted as being due to the fact that at high pH and high current densities the potential for palladium deposition is shifted towards more negative values. The H/Pd ratio was highest at pH 9 and a current density of > 15 A/dm^2. It was found that current efficiency is lower with pulse plating than with DC plating. The current efficiency decreases with increasing pulsed current density, increasing pH, and increasing pulse length (Fig. 9.35). This may be explained by an increase of the interfacial pH and NH_3 concentration with increasing pulsed current density, increasing on-time and increasing pH of the solution.

Figure 9.36 shows the hydrogen content of palladium deposits as a function of on-time for various pulse plating conditions. There is a striking difference between DC and PC plating. Whereas in DC plating a decrease in current efficiency causes an increase in the ratio H/Pd, it is the contrary in PC plating. This suggests that there is an additional effect influencing the ratio H/Pd. It is assumed that hydrogen which is adsorbed during the cathodic pulse, is more or less desorbed during off-times. Hardness of PC plated deposits is higher than that of DC plated deposits. There is, however, no correlation between the hydrogen content and the hardness. In palladium deposits cracking sometimes occurs after the electrolysis has stopped. This has been ascribed to the desorption of hydrogen. However, in palladium deposits PC plated at long on-times (where hydrogen content is small) cracks form more readily than in DC plated deposits with higher current density (where H/Pd is larger). The tendency for crack formation is lower for smaller thicknesses. Crack-free deposits of 1 μm thickness were obtained at pulsed current densities of 100-500 A/dm^2. At the same time the porosity was much lower for PC plated deposits than for DC plated ones. This may be a relevant advantage of PC plating of palladium. For higher thicknesses (> 2.5 μm), pulse plating brings no improvement with respect to crack formation.

PC deposition of palladium was also studied by K. Hosokawa and coworkers (70) with special reference to the effect of codeposition of hydrogen on the properties of the deposits. A neutral phosphate solution containing benzoic acid in low concentration was used for this investigation. Fig. 9.37 shows the effect of PC density on current efficiency for two baths with a palladium concentration of $2.1 \cdot 10^{-2}$ M and $4.2 \cdot 10^{-2}$ M respectively. Current efficiency remains high up to high current densities. The results suggest the following mechanism: During the off-period the hydrogen atoms adsorbed on the electrode surface exchange their charge with the palladium ions from the bulk of the solution. Reduction of palladium ions by adsorbed hydrogen during the off-period would lead to the electrodeposition of bright palladium layers. According to this mechanism the off-time must be one of the most important factors in pulse plating.

The influence of other parameters, such as T_{on}, T_{on}/θ and average current density was also studied. The results show that the average current density is more important than T_{on} with respect to the properties of the deposits. Under the conditions given, crack-free and pore-free palladium deposits were obtained. Thus, the application of pulse plating to the electrodeposition of palladium is considered to be most useful.

Fukumoto and coworkers (71) investigated the electrodeposition of palladium from baths of dinitrodiammine, dibromodiammine and dichlorodiammine by

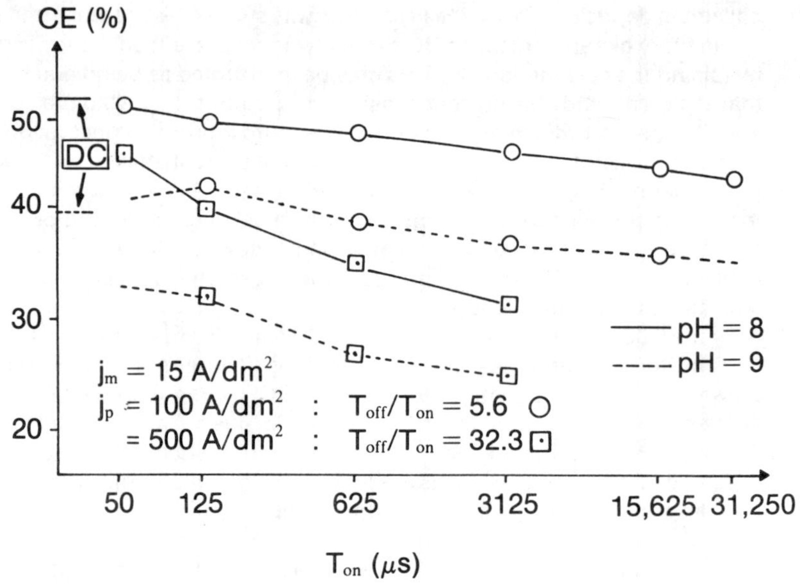

Fig. 9.35—Current efficiency of palladium deposition from a "P" salt solution as a function of on-time in pulse electrolysis (69).

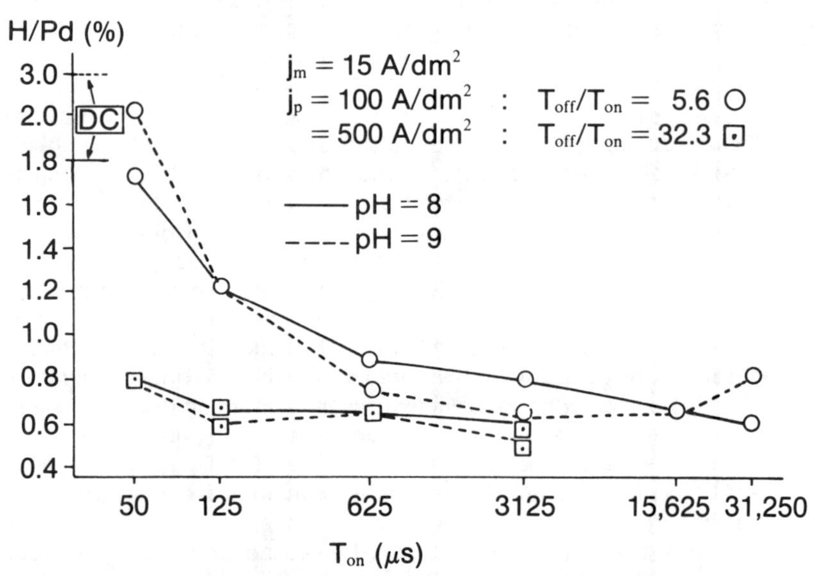

Fig. 9.36—Hydrogen content of palladium deposits from a "P" salt solution as a function of on-time in pulse electrolysis (69).

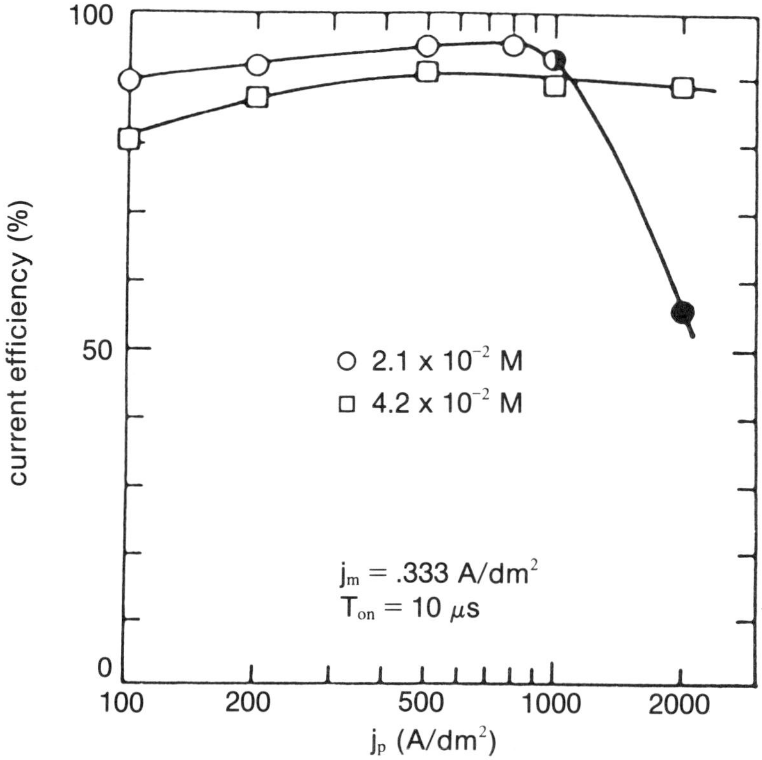

Fig. 9.37—Influence of pulsed current density (j_p) on current efficiency at $j_m = 0.333$ A/dm^2 and $T_{on} =$ 10 microseconds for palladium deposition: •, black; ◑, semi-bright; and o, bright deposit (70).

using pulsed current electrolysis. Surface morphology, grain size, internal stress and texture, deposition potential and hydrogen content of the deposits were examined as a function of pulse parameters. A close correlation was found between crystallographic features of the deposits and the mode of pulsed current. In agreement with earlier work by Puippe and Ibl (38), a decrease in grain size was observed in PC electrolysis using higher current density, longer on- and shorter off-time, which led to higher cathodic overpotentials. Although the desorption of hydrogen from the palladium deposits was expected to occur during off-time, this did not always occur. The effect of PC electrolysis was most remarkable in the dinitrodiammine bath, which yielded a palladium deposit containing less adsorbed hydrogen than that obtained by DC electrolsis.

A. Knödler, Ch.J. Raub and E. Raub (72), in their study on PC electrolysis of palladium and palladium alloys, used a different kind of plating solution, composed of

5 g/l	Pd as K_2PdCl_4
14 g/l	$NaNO_2$
40 g/l	NaCl
25 g/l	H_3BO_3
pH	4.7

Palladium is present in this electrolyte as a nitrito-complex. Soluble palladium anodes were used, thus avoiding the complications due to anodic oxidation of bath constituents. One of the aims of this study was to produce palladium deposits low in hydrogen, as hydrogen is known to affect the properties of the deposits adversely. Therefore, PC and pulse reverse (PR) were applied. It was expected that hydrogen would diffuse out during the off-period or be oxidized during the anodic current pulse of the cycle. The effect of cycle time on the hydrogen content of palladium deposited from the electrolyte described above is shown in Fig. 9.38. Average current density was 1 A/dm^2 and duty cycle 30 percent. As expected, the hydrogen content of PC plated deposits decreases, first slowly, then more rapidly with increase in cycle time. Obviously, a minimum off-time of about 4 msec is required for hydrogen desorption to become noticeable. An even more rapid decrease in hydrogen content is observed when pulse reverse is used because hydrogen is eliminated from the deposit not only by diffusion, but also by anodic oxidation. At low cycle times, hydrogen contents of PR plated deposits are higher than those of PC plated ones. This is because cathodic pulsed current density is higher with PR than with PC if the same average current density (1 A/dm^2) is used. Hydrogen contents of palladium deposits from this type of electrolyte are significantly higher than of deposits from, e.g., tetrammine type solutions.

The above solution for palladium deposition was used subsequently for the electrodeposition of alloys of palladium with cobalt and nickel, simply by adding either cobalt or nickel chloride. Alloy deposits from this type solution contained 0.5 to 1 wt.-% of either cobalt or nickel. By lowering the palladium concentration it was possible to increase the nickel content up to 3 percent. No noticeable difference was observed between DC and PC plated alloy deposits. An interesting feature of alloy plating from this type solution is that hydrogen contents are drastically reduced by codeposition of either cobalt or nickel.

Pulse plating of Pd-Ni has been investigated by Y. Fukumoto, Y. Kawashima and T. Hayashi (71a). The electrolyte consisted in 0.1 M $Pd(NH_3)_2Cl_2$, 0.2 M $NiSO_4$, 0.38 M $(NH_4)_2SO_4$ and NH_4OH and had a pH of 7.5. An average current density of 1 A/dm^2 and a total current of 30 to 100 mC/dm^2/pulse were identified as optimum pulse parameters, producing smooth and bright deposits. An increase in pulsed current density, as well as an increase in off-time, led to an increase in the Pd content of the electrodeposited alloy. Preferred crystal orientations were also investigated as a function of pulse parameters.

3.2. *Silver*

Measurements in a Haring cell with a 1 M silver nitrate solution under both DC and a variety of PC conditions indicate that macrothrowing power in the deposition of silver is unaffected by application of pulsed current and is also independent of current density (3). This is in agreement with the results of other investigators regarding the high reversibility of the electrode reaction.

Fukumoto and coworkers (73) applied PC electrolysis to the electrodeposition of silver from a silver-ammonia-complex bath and also from a conventional silver cyanide bath. The surface morphology and crystal structure of the silver deposits were found to be closely related to the cathodic overpotentials established under PC electrolysis. The morphology of PC deposits from the silver-ammonia-complex bath does not seem to differ significantly from that of DC deposits. Changes in the preferred orientation of the silver deposits, however, were closely related to the pulsed current density and pulse on-time.

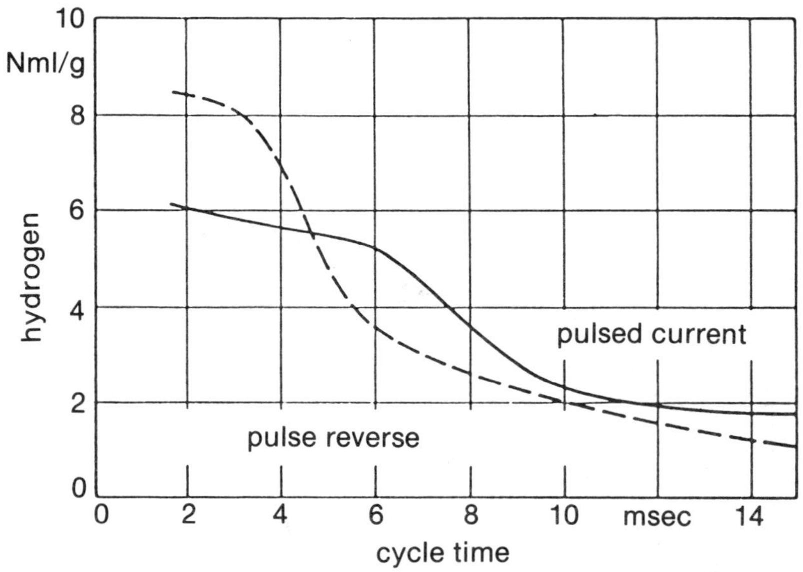

Fig. 9.38—Hydrogen content of PC and PR plated palladium deposits from an acidic plating bath. Average current density $j_m = 1\ A/dm^2$, duty cycle 30%, cathodic PC density 3.33 A/dm^2 with PC, 4.3 A/dm^2 for PR, anodic current density 0.4 A/dm^2 with PR. Bath temperature 50° C (72).

Thus, the development of the fiber structure of silver deposits from these baths depended on the mode of PC electrolysis. A smooth and fine-grained deposit was obtained from the silver cyanide bath by PC electrolysis with on-time in the range of 0.1 to 10 msec and at an average current density of 1 A/dm^2. The adsorption of cyanide ions seems to play an important role in the deposition of silver under PC conditions.

Further results were reported by Fukumoto and coworkers (74) at the AES Second Pulse Plating Symposium (1981). In the silver-ammonia bath, increase of on-time and of PC density resulted in the refinement of the silver deposits. PC deposits from a cyanide bath with a duty cycle of 1/15 were smooth and fine-grained. When the duty cycle was lowered to 1/150, deposits became rough and granular. This can be explained by the fact that much more pronounced influence can be expected in PC electrolysis, the deposition reaction being controlled mainly by changes in the adion concentration. Changes in surface appearance and grain size of deposits may result from the effect of overpotential associated with PC electrolysis on adsorption of the complexant on the electrode surface and also the reactivation of the growth centers or the re-arrangement of the adatoms, which influence the specific crystal structure and grain size. The overpotential is therefore the most important factor in the preferred orientation of the silver deposits. Influence of the on-time on the overpotential of the silver electrode in PC electrolysis is shown in Fig. 9.39. PC plated silver deposits from the silver-ammonia bath were found to have finer grains in the lower range of on-times and to become granular with increase of

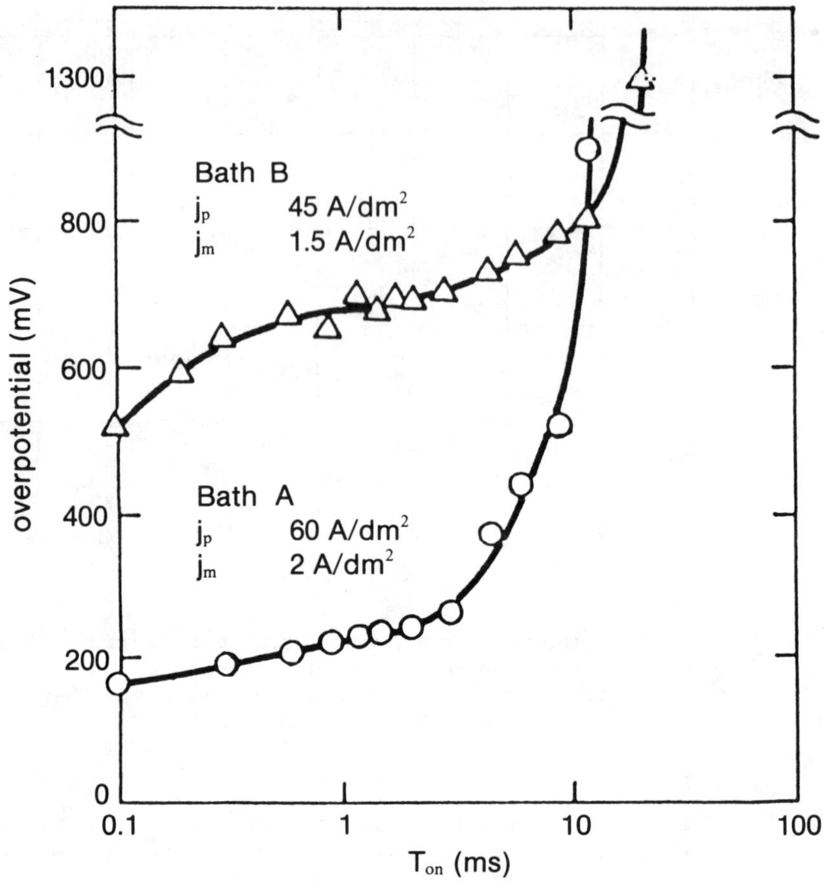

Fig. 9.39—Effect of on-time on the overpotential of silver deposition from a silver-ammonia bath (Bath A) and a silver-cyanide bath (Bath B) under PC conditions (74).

on-time. Silver deposits from the cyanide bath always showed the same [111]-orientation and the overpotential increased with increase of the on-time. Increases in PC density and on-time resulted in an increase of the overpotential leading to higher adatom concentrations on the electrode surface. This in turn leads to an increase in the rate of nucleation and to a decrease of the critical radius of the stable crystallite (75).

Besides palladium, Hosokawa and coworkers (70) also investigated the electrodeposition of silver by PC electrolysis from cyanide solutions. Typical relations between current efficiency and the pulse length T_{on} at a constant cycle time are shown in Fig. 9.40. For each peak current density there is obviously an optimum pulse length with respect to current efficiency. Grain size of the silver deposits becomes smaller as on-time increases in the range from 10 to 50 microseconds. When the on-time exceeds the optimum value, the surface of the

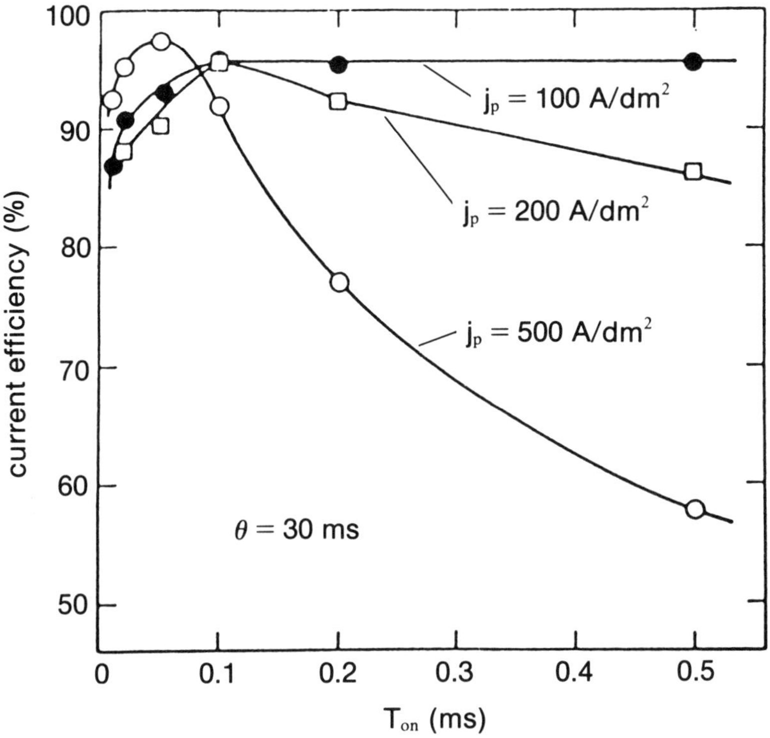

Fig. 9.40—Influence of pulse length (T_{on}) on current efficiency of silver deposition from a silver-cyanide solution at $\theta = 30$ msec (70).

silver deposit becomes rough. The influence of PC density is shown in Fig. 9.41. Current efficiency reaches a maximum abruptly and then decreases markedly with increasing PC density. The decrease is due to the fact that PC current density exceeds the limiting current density of silver deposition. Smooth deposits can be obtained in the PC density range below 2000 A/dm^2.

The current efficiency changes with the cycle time θ at constant pulse length $T_{on} = 10$ microseconds and various PC densities as revealed by Fig. 9.42. With increasing θ, the current efficiency first increases and reaches a constant value after a maximum at $\theta = 3$ msec. The surface of the deposits becomes smooth with an increase in θ up to 10 msec and then becomes rough again.

Influence of on-time on current efficiency is shown in Fig. 9.43. It is evident that in the range of $j_p > 500$ A/dm^2 the current efficiency goes through a maximum which coincides with the transition time τ. The smoothest surface is obtained at an on-time $T_{on} = 50$ microseconds. Results suggest that the morphology of the silver deposits is controlled mainly by θ (which corresponds approximately to T_{off} when $T_{on} << T_{off}$) for $T_{on} < \tau$ and mainly by T_{on} when $T_{on} > \tau$.

It should be noted that the duty cycle in Hosokawa's investigations is extremely low.

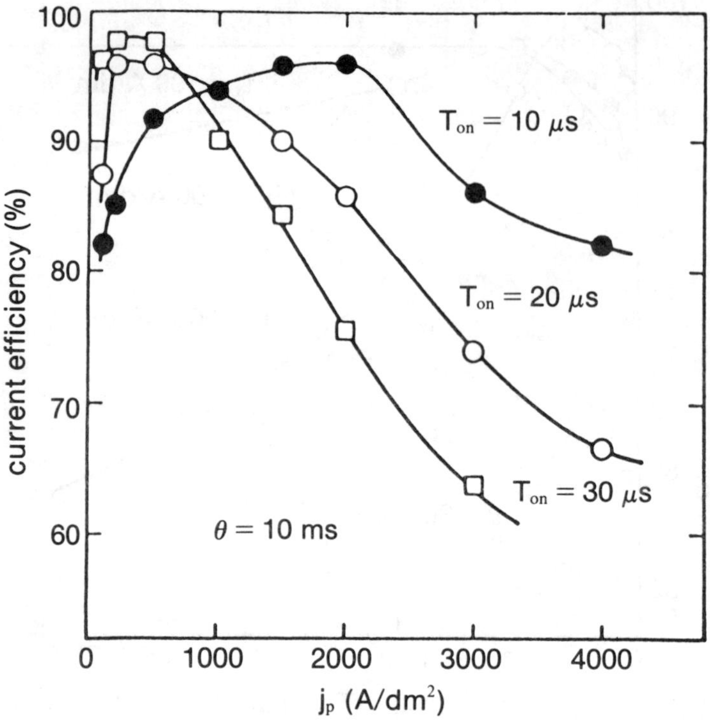

Fig. 9.41—Influence of PC density (j_p) on current efficiency of silver deposition from a silver-cyanide solution at $\theta = 10$ msec (70).

Hosokawa and coworkers extended their studies to the electrodeposition of silver from ethylenediamine, ammonia, and thiocyanate baths (76). Smooth deposits from these baths could be obtained only by the use of pulse plating. The morphologies of PC plated deposits from the cyanide bath was similar to those obtained with DC plating. With increase of the off-time, grain size of the deposits from the ammonia and the thiocyanate bath increased. The influence of pulse parameters on the deposits from the ethylenediamine bath was quite different. The surface of the deposits changed only slightly when the cycle time θ was increased. On the other hand, large grains appeared as the pulse length T_{on} increased at constant PC density and constant duty cycle. From the results the following conclusions are drawn:

- High overpotentials are one of the important characteristics of pulse plating. Pulse electrolysis might, therefore, be more effective in electrodeposition processes which exhibit relatively low polarization in DC electrolysis.
- The off-time is the most important parameter in pulse plating. Three phenomena are effective on the morphology of the deposits: (a) recrystallization, (b) adsorption of ions or addition agents, and (c) corrosion. Recrystallization would involve two steps, namely surface diffusion of adatoms and local cell mechanism. The rate of the latter process depends on

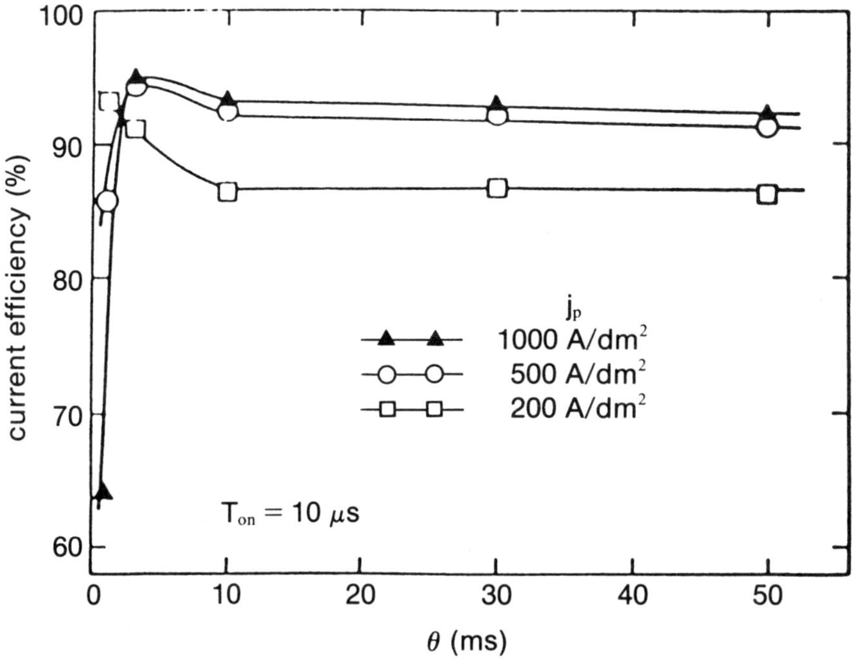

Fig. 9.42—Influence of pulse period (θ) on current efficiency of silver deposition from a silver-cyanide solution at T_{on} = 10 microseconds (70).

the exchange current density. If the effects of the off-time are mainly due to a local cell mechanism, the exchange current density should be low for the pulse plating to become effective. If the exchange current density is high, large-grained deposits result because of high recrystallization rate. Very stable complex ions such as in thiourea and cyanide baths are not suitable for pulse plating. The morphology of PC plated deposits from the cyanide bath was similar to that obtained by DC plating.

- The adsorption of ions or additives on the active sites is also important for the levelling of the surface. Results obtained with the ethylenediamine bath suggest that adsorption of ethylenediamine plays an important role in forming smooth and fine-grained deposits.

3.3. *Silver alloys*

One of the earlier studies on pulse plating of silver alloys is that by Leidheiser Jr. and Ghuman (77) described in Chapter 11.

O.V. Moiseeva et al. (78) studied the simultaneous discharge of silver and copper from a thiocyanate solution, using a pulsed current of rectangular waves. The purpose of this investigation was to form high-quality binary alloys containing 5-70 wt.-% silver. Diffusion limitations could be overcome using high currrent densities with short on-times and long-off-times.

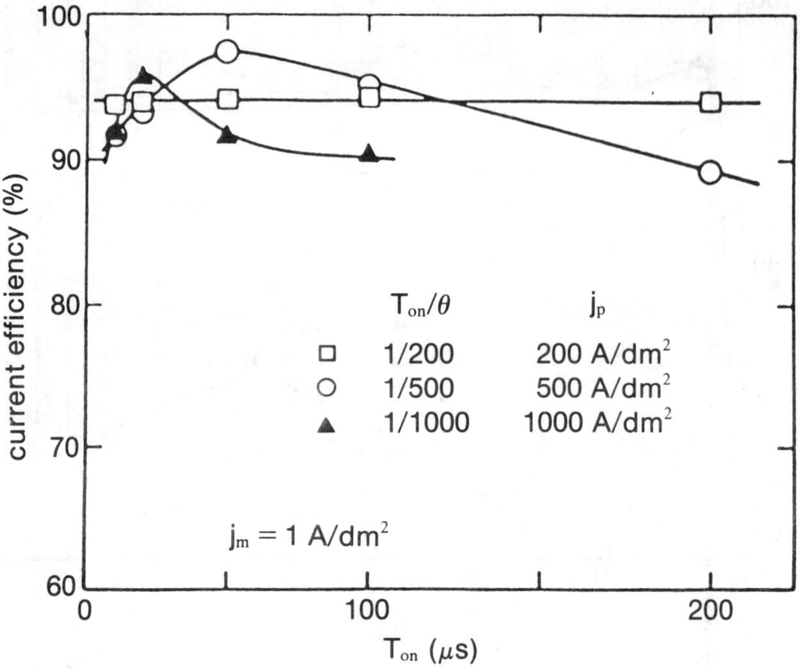

Fig. 9.43—Influence of on-time (T_{on}) and off-time (T_{off}) on current efficiency of silver deposition from a silver-cyanide solution at j_m = 1 A/dm^2 (70).

Pulse electrodeposition of silver alloys from cyanide solutions has also been studied by A. Knödler, Ch.J. Raub and E. Raub (79). Codeposition of cadmium and zinc respectively by DC plating occurs only after the limiting current density of silver deposition is exceeded. The effect of PC plating on alloy deposition is mainly that of raising the limiting current density of silver deposition. The composition of the resulting silver-zinc and silver-cadmium alloys is predominantly governed by the hydrodynamic conditions of the plating bath. No significant differences in composition between DC and PC plated alloy deposits were found to exist.

Antimony is used as a brightener in silver plating solutions. For the PC electrodepositon of silver-antimony alloys a silver cyanide bath was employed to which Sb(III) ions were added in increasing amounts. Sorbitol proved to be an efficient complexing agent for Sb(III). The effect of the duty cycle at constant average current density (1 and 2 A/dm^2 respectively) on the composition of silver-antimony alloys is shown in Fig. 9.44. This implies a decrease of the pulsed current density as can be realized in the lower part of Fig. 9.44. As the duty cycle exceeds 40 percent, a sudden decrease of the antimony content of the alloy deposits may be noted. This decrease may be interpreted as being due to the fact that for higher duty cycles the PC density is lower than the limiting current density for silver deposition. Substantial co-deposition of antimony takes place only when this limiting current density is exceeded, which in turn is influenced by the pulse parameters and the hydrodynamic conditions of the plating bath.

The effect of pulse frequency on the composition of PC plated silver-antimony alloys is shown in Fig. 9.45. As the antimony concentration in the plating solution is raised, antimony contents of the alloy deposits are increased, as expected. No significant change in alloy composition is observed over the frequency range investigated. This confirms the conclusion reached and discussed above, that high frequencies of the pulsed current exert only an insignificant influence on alloy composition, which in turn is the result of the limited rate of diffusion in the cathodic diffusion layer. Changes in alloy composition may be expected only at pulse frequencies below 1 kHz, other pulse parameters being held constant.

3.4. *Other platinum metals*

There is a short notice on pulse plating of decorative rhodium coatings (80). Better mechanical properties and higher throwing power are claimed as a result of pulse plating. More recent investigations on pulse plating of rhodium (81) show that the best pulsing cycle consists of an on-time of 0.1 msec and an off-time of 0.9 msec. Under these conditions and with an average current density of 8 A/dm^2, crack free deposits were obtained with a thickness of 22 μm.

Ruthenium deposits usually lack ductility and produce stress cracks at thicknesses above 1 μm. Codeposition of nickel with ruthenium can improve ductility slightly. A paper by A. Blair and A. Ferrara (82) reported some preliminary observations of the effects of pulse plating in this alloy system. Four electrolytes, each containing ruthenium as a nitrilochloro complex, were studied. Three of the solutions contained a suitable nickel complex, each at a different concentration. Each solution was investigated using DC and PC of various duty clcles. It became soon obvious that a 50 percent duty cycle with less than 5 msec on-time gave good results. Nickel concentration in the solution appears to have no effect on grain size of the deposits. However, pulse duration has a marked effect on this deposit feature, short pulses giving the smallest grain size. Stress cracking was apparent at 1.0-1.25 μm at all pulsing conditions with the pure ruthenium electrolyte. In the presence of nickel, this feature was reduced even with DC. The best conditions appeared to be at nickel level B (no concentration given) and a pulse duration of 1 msec, 50 percent duty cycle, at which deposits of 2.5 μm were produced crack-free. Average current densities ranged from 1.1 A/dm^2 at 5 g/l Ru to more than 10.8 A/dm^2 at 12 g/l Ru with increased agitation. Cathode efficiency remained high throughout at approximately 11 mg/A min. Contact resistance showed little dependence on pulse conditions. Most deposits showed significant porosity, mostly due to stress cracking, but the authors anticipate that low nickel concentrations and longer pulse times may give improvements in this respect. It is possible that future research will produce ruthenium deposits under high current density conditions ($>$ 10 A/dm^2) that will possess good corrosion resistance, low contact resistance and reasonable ductility, which may open up a new field of applications for this relatively low-priced precious metal.

4. Concluding remarks

In recent years the practical application of pulse plating to the electrodeposition of precious metals has been greatly hampered by the lack of knowledge of its effect on the properties of the electrodeposits. Progress in the theoretical foundations of pulse

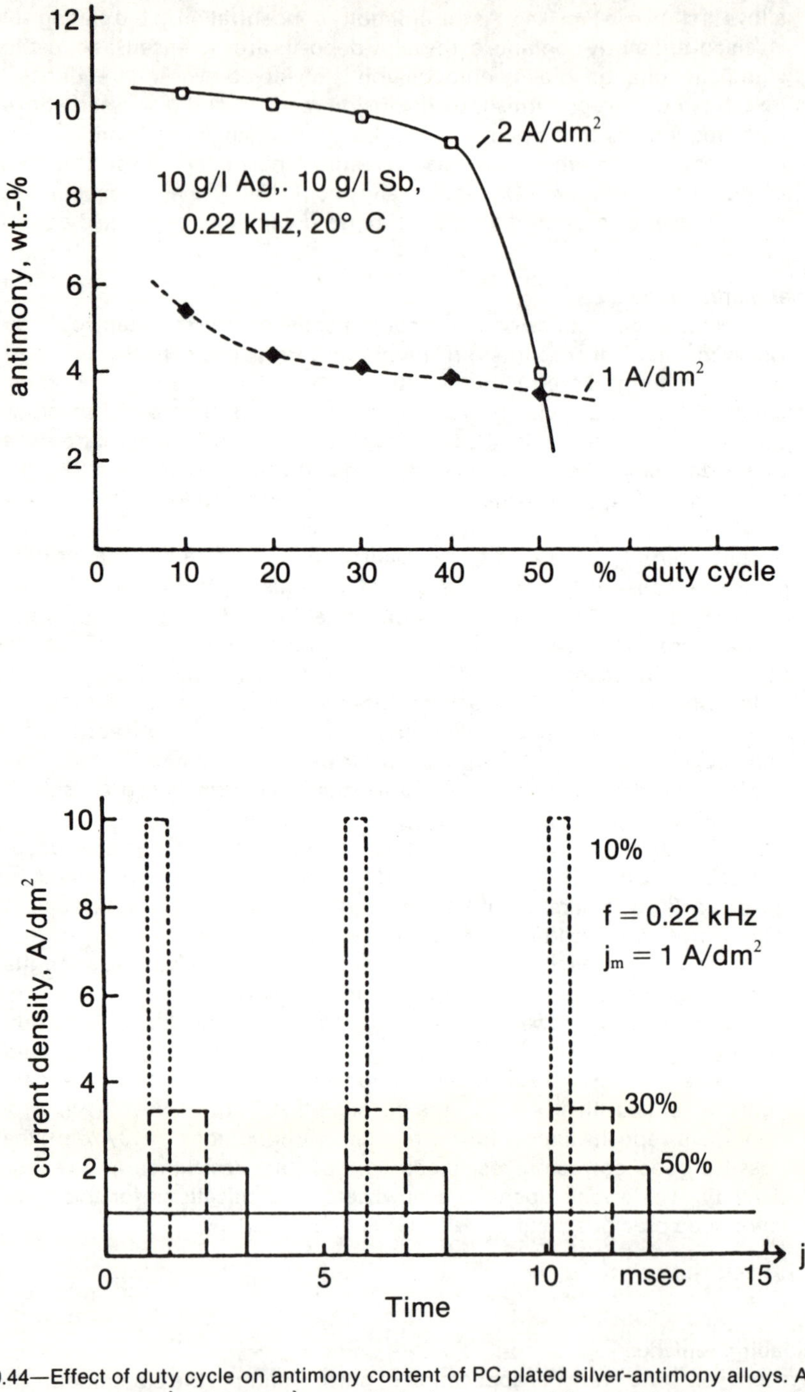

Fig. 9.44—Effect of duty cycle on antimony content of PC plated silver-antimony alloys. Average current density 1 A/dm^2 and 2 A/dm^2 respectively. Lower part of figure: Scheme of pulsed current used for deposition of silver-antimony alloys (79).

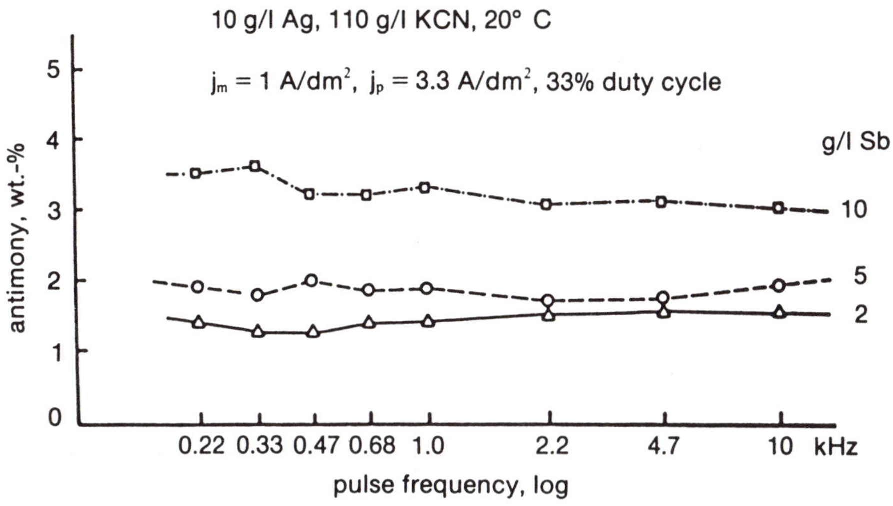

Fig. 9.45—Effect of pulse frequency on antimony content of silver-antimony alloys for three different antimony concentrations of the plating bath (79).

plating and the experimental evidence accumulated so far now make possible the production of precious metal electrodeposits of superior quality and of a wider range in properties as compared to those deposited by conventional methods. A systematic study may be necessary in order to obtain a desired property, since the latter may occur only in a more or less restricted range of conditions. An individual property may sometimes be achieved only at the expense of other characteristics that may be desirable in certain other applications. Pulse plating offers a large variety of interesting possibilities of achieving an optimum compromise between various characteristics by suitable choice of pulse plating parameters. There is also the possibility of using duplex coatings, produced simply by switching the current mode in the same bath. Although the theory of pulse plating has advanced very far, the effect of pulse plating on properties of precious metals deposits cannot be predicted in each case and depends not only on the plated metal itself but also on the bath type used. It may even become necessary to adapt the bath composition and/or addition agents to the specific needs of pulse plating.

One principle that must never be lost sight of is the limited rate of diffusion in the cathodic diffusion layer. This is also why the over-all plating rate of PC deposition cannot exceed that of DC deposition.

As the number of variables' combinations in pulse plating is nearly infinite, possible improvements may come about only by painstaking research work or, if one is lucky, by chance. This is still more true for alloy plating since another variable, i.e., the concentration of the alloy metal, is involved.

In any application the possibilities offered by pulse plating must be weighed against certain disadvantages such as reduced brightness and lower current efficiency.

Pulse plating seems to be predestined for precious metal electrodeposition where the higher equipment costs may be outweighed by substantial savings in metal consumption and higher quality deposits.

It would be desirable if companies or individuals who already use pulse plating successfully for electrodeposition of precious metals and alloys would be less hesitant in disclosing their experiences with pulse plating. Thus, the technique of pulse plating may well lead to further improvements in the highly promising field of precious metal plating and in electrochemical technology in general.

References

1. H.Y. Cheh; *Journal Electrochemical Society,* 118 : 551 (1971)
2. H.Y.Cheh; *Journal Electrochemical Society,* 118 : 1132 (1971)
3. H.Y. Cheh, P.C. Andricacos and H.B. Linford; *Plating and Surface Finishing,* **64**, (7) : 42 (1977)
4. P.C. Andricacos, H.Y. Cheh and H.B. Linford; *Plating and Surface Finishing,* **64,** (9) : 42 (1977)
5. D.L. Rehrig; *Plating,* **61** : 43 (1974)
6. D.L. Rehrig, H. Leidheiser, Jr. and M.R. Notis; *Plating and Surface Finishing,* **64,** (12) : 40 (1977)
7. R.J. Morrissey and A.M. Weisberg; *Transactions Institute Metal Finishing,* **58** : 97 (1980)
8. J.W. Dini and H.R. Johnson; *Gold Bulletin,* **13** : 31 (1980)
9. J.W. Dini and H.R. Johnson; *Plating and Surface Finishing,* **67,** (1) : 53 (1980)
10. Y.Y. Wang, C.S. Tung and C.C. Wan; *Metal Finishing,* **73**, (9) : 21 (1980)
11. Ch.J. Raub and J. Lendvay; *Galvanotechnik,* **67** : 95 (1976); Metalloberfläche, **29** : 165 (1975)
12. J. Horkans and L.T. Romankiw; *Journal Electrochemical Society,* **124** : 1499 (1977)
13. M. Visvanathan and Ch. J. Raub; *Surface Technology,* **4** : 339 (1976)
14. G.B. Munier; *Plating,* **56** : 1151 (1969)
15. E.T. Eisenmann; *Journal Electrochemical Society,* **124** : 1957 (1977)
16. Y. Okinaka, F.B. Koch, C. Wolowodiuk and D.R. Blessington; *Journal Electrochemical Society,* **125** : 1745 (1978)
17. H. Leidheiser Jr., A. Verles, M.L. Varsanyi and I. Czako-Nagy; *Journal Electrochemical Society,* **126** : 391 (1979)
18. F.H. Reid; *Metalloberflache,* **30** : 453 (1976)
19. U.W. Souter; *Conference Proceedings 19th Annual Holm Seminar,* Chicago (1973)
20. M. Antler; *Conference Proceedings 19th Annual Holm Seminar,* Chicago (1973)
21. M. Antler; *Plating,* **60** : 468 (1973)
22. M. Antler; *IEEE Transactions: Parts, Hybrids, Packaging,* **Vol. PHP,** (1) (March 1974)
23. A. Knödler; *Galvanotechnik,* **68** : 383 (1977)
24. Ch.J. Raub and A. Knödler; *Gold Bulletin,* **10** : 38 (1977)
25. Ch.J. Raub and A. Knödler; *Plating and Surface Finishing,* **65** (9) : 32 (1978)
26. A. Knödler; *Metalloberfläche,* **28** : 465 (1974)
27. A. Knödler; *Conference Proceedings AES 1st International Pulse Plating Symposium, Boston, MA (April 19-20, 1979)*
28. F.H. Leaman and B.A. Kownacki; *Conference Proceedings AES 1st International Pulse Plating Symposium,* (1979)
29. A. Knödler, Ch.J. Raub and E. Raub; *Metalloberfläche,* **38** : 379 (1984)
30. M. Braun; *Dissertation No 5015,* ETH, Zurich (1973)
31. H.F. Angerer; *Dissertation No 6213,* ETH, Zurich (1978)
32. H. Angerer and N. Ibl; *Journal of Applied Electrochemistry,* **9** : 219 (1979)
33. N. Ibl, J.Cl. Puippe and H. Angerer; *Surface Technolotgy,* **6** : 278 (1978)
34. J. Bielinski, N. Ibl and H. Angerer; *Oberflache-Surface,* **20** : 2 (1979)
35. K. Hosokawa, H. Angerer, J.Cl. Puippe and N. Ibl; *Plating and Surface Finishing* **67** (10) : 52 (1980)
36. J.Cl. Puippe, N.Ibl, H. Angerer and H.J. Schenk; *Oberfläche-Surface,* **20** : 77 (1979)
37. N. Ibl; *Conference Proceedings Interfinish '80*
38. J.Cl. Puippe and N. Ibl; *Plating and Surface Finishing,* **67** (6) : 68 (1980)
39. P.A. Mäusli, F.H. Reid and S. Steinemann; *Metalloberfläche,* **32** : 369 (1978)
40. S.J. Harris, E.C. Darby, K. Bridger and A.E. Mason; *Transactions Institute of Metal Finishing,* **54** : 115 (1976)

41. S.J. Harris, E.C. Darby, K. Bridger and A.E. Mason; *Gold Bulletin,* **9** : 81 (1976)
42. W. Flühmann, F.H. Reid, P.A. Mäusli and S. Steinemann; *Plating and Surface Finishing,* **67,** (6) : 62 (1980)
43. C.J. Stimetz and J.J. Hren; *Conference Proceedings AES 1st International Pulse Plating Symposium* (1979)
44. B. Inglot, J. Socha and A. Pezinski; *Powl. ochr. (Poland)* **9,** (3) : 31 (1981)
45. A. Knödler; *Metalloberfläche,* **38** : 539 (1984)
46. J.C. Catonne, J.D. Jehanno and J. Royon; *Conference Proceedings, Interfinish '84,* 191
47. B. Sutter; *Oberfläche-Surface,* **215** : 16 (1984)
48. B. Sutter, F. Mortier, M.A. Laroche, J. Reby and G. Vasseur; *Conference Proceedings, Interfinish '84,* 201
49. A.J. Avila and M.J. Brown; *Plating,* **57** : 1105 (1970)
50. C.A. Burrus; *Journal Electrochemical Society,* **118** : 833 (1971)
51. T.N. Bhar and G.M. Lamb; *Journal Electrochemical Society,* **126** : 1514 (1979)
52. K.L. Schiff and R. Schnabl; *Metalloberfläche,* **32** : 423 (1978)
53. M. Branik and R. Schnabl; *Metalloberfläche,* **33** : 350 (1979)
54. J. Ehrhardt and H. Nowacki; *Metalloberfläche,* **36** : 405 (1982)
55. D.L. Rehrig; *Conference Proceedings, AES 1st International Pulse Plating Symposium* (1979)
56. R.V. Dafter Jr., R. Haynes and D.L. Rehrig; *Conference Proceedings, AES 1st International Pulse Plating Symposium* (1979)
57. D.W. Endicott and R.K. Asher; *Plating and Surface Finishing,* **68**, (1) : 46 (1981)
58. J.S. Stevenson; *Galvanotechnik,* **71** : 21 (1980)
59. A. Blair; *Metalloberfläche,* **34** : 201 (1980)
60. J. Lochet; *Conference Proceedings, AES 2nd International Pulse Plating Symposium* (1981)
61. D. Towne; *Conference Proceedings AES 2nd International Pulse Plating Symposium* (1981)
62. W.H. Chang; *Conference Proceedings AES 2nd International Pulse Plating Symposium* (1981)
63. A. Holtzman; *Oberfläche-Surface,* **22** : 16 (1981)
64. D. Withey; *Gold Bulletin,* **16** : 70 (1983)
65. R.J. Morrisey; *Plating and Surface Finishing,* **67**, (12) : 44 (1980)
66. R.J. Morrissey; *Conference Proceedings AES 2nd International Pulse Plating Symposium* (1981)
67. H.D. Hedrich and Ch.J. Raub; *Metalloberfläche,* **31** : 512 (1977)
68. H.D. Hedrich and Ch.J. Raub; *Surface Technology,* **8** : 347 (1979)
69. J.M. Locarnini and N. Ibl; *Conference Proceedings AES 2nd International Pulse Plating Symposium* (1981)
70. K. Hosokawa, Y. Tsuru, M. Matsunaga and T. Kuwamura; *Conference Proceedings AES 2nd International Pulse Plating Symposium* (1981)
71. Y. Fukumoto, Y. Kawashima, K. Handa and T. Hayashi; *Journal of the Metal Finishing Society of Japan,* **34** : 576 (1983)

71a. Y. Fukumoto, Y. Kawashima and T. Hayashi; *Plating and Surface Finishing,* **73,** (3) : 62-66 (1986).

72. A. Knödler, Ch.J. Raub and E. Raub; *Metalloberflache,* **38** : 379 (1984)
73. Y. Fukumoto, H. Kitanishi and T. Hayashi; *Journal of the Metal Finishing Society of Japan,* **32** : 302 (1981)
74. T. Hayashi, H. Kitanishi and Y. Fukumoto; *Conference Proceedings AES 2nd International Pulse Plating Symposium* (1981)
75. Y. Fukumoto, H. Kitanishi and T. Hayashi; *Journal of the Metal Finishing Society of Japan,* **33** : 24 (1982)
76. K. Hosokawa, Y. Tsuru, K. Tanimura and M. Matsunaga; *Conference Proceedings, Interfinish '84,* 197
77. H. Leidheiser Jr. and A.R.P. Ghuman; *Journal Electrochemical Society,* **120** : 484 (1973)
78. O.V. Moiseeva et al.; *Elektrokhimiya,* **17** : 464 (1981)
79. A. Knödler, Ch.J. Raub and E. Raub; *Metalloberfläche,* **38** : 439 (1984)
80. R. Olson; *Plating and Surface Finishing* **68** (2) : 38 (1981)
81. A. Blair and A. Ferrara; *Conference Proceedings, AES 2nd International Pulse Plating Symposium* (1981)
81. W.S. Miu and Y.S. Fung; *Plating and Surface Finishing,* **73** (3) : 58-61 (1986)
82. A. Blair and A. Ferrara; *Conference Proceedings, AES 2nd International Pulse Plating Symposium,* Chicago (1981).
83. L. Sjögren, B. Asthner, L.-G. Liljestrand and L.B. Rëvay; *Plating and Surface Finishing,* **73** (3) : 70-73 (1986)

Chapter 10

Pulsed Electrodeposition of Rhenium

J.Cl. Puippe

1. Introduction

The physical properties of rhenium, which include a high melting point, high work function, and electrical conductivity of the oxide, make this metal attractive from the applicational viewpoint, with special reference to possible uses in space environments, erosion-resistant electrical contacts, diffusion barriers, heat shields, and emitters. In the present state of the art, however, the production of sound and stable coatings by electrodeposition from aqueous solutions has proved extremely difficult, despite a fair amount of investigational work on a variety of electrolytes, which include, among others, sulfate, phosphate and oxalate solutions, as proposed by Fink and Deren as early as 1934 (1).

The baths investigated up to the present time may conveniently be divided into two groups on the basis of current efficiency for metal deposition and the quality of the coatings. With regard to current efficiency, the electrolyte based on sulfate was superior to those containing complexing agents such as phosphate or oxalate, but even so, the efficiency for the former bath was only 5-7 percent in room temperature deposition at a pH of 0.8 and a current density of 10 A/dm^2. Furthermore, the deposits obtained under these conditions were not stable in the presence of air or moisture. Deposit stability was improved in a plating process developed by Camp (2), based on a slightly acid complexed bath, but again current efficiency was extremely low, in the order of 1-3 percent. In addition, both types of process mentioned suffered a common disadvantage in terms of a porous coating structure and development of severe cracking on heat treatment at around 1,000° C. It was noted, however, that the degree of cracking of deposits after heating decreased with increasing current density. In general, the current density that may be applied in electrodeposition processes is limited by mass transport considerations, related to diffusion effects at the cathode and in particular to the thickness of the diffusion layer. In pulse plating, as shown by Ibl (3, 4) (see also Chapter 5 of this book), two diffusion layers are formed, comprising a pulsating layer adjacent to the cathode and a stationary layer extending to the bulk of the electrolyte. Using very short pulses the pulsating layer remains very thin; hence the concentration gradient is very high, allowing the application of very high current densities during the pulse without exceeding the mass transport limitation. Since increase of current density appears to favor improved properties of electrodeposited rhenium, pulse plating thus offers interesting possibilities in this context.

This approach was adopted by Hosokawa et al. (5, 6) as a means of improving current efficiency for rhenium deposition and producing deposits on tungsten wire

TABLE 10.1

Experimental Conditions and Range of Pulse Parameters

Electrolyte	$KReO_4$	15 g/l
	H_2SO_4 (d 1.84)	5 ml/l
pH	0.8 ± 0.1	
Temperature	20 - 90° C	
j_p	100 - 20,000 A/dm^2	
T_{on}	1 μs - 10 ms	
T_{off}	10 μs - 100 ms	
j_m	1 - 20 A/dm^2	

which would remain crack-free after heating. The remainder of the present chapter is devoted primarily to the results reported by these authors.

The basic electrolyte composition and the range of deposition conditions studied are summarized in Table 10.1; the pH value of 0.8 ± 0.1 was selected as most suitable on the basis of preliminary DC plating tests. The cathode was a tungsten wire of 0.8 mm diameter, and the anode a spiral of platinum wire of 1 mm diameter. Cleaning procedure for the cathode included a dip for 1 minute in a solution of 290 g/l $K_2Fe(CN)_6$, 9.7 g/l $Na_2C_2O_4$ and 91 ml 30 percent NaOH to remove oxide films, the cleaned cathode then being stored in a desiccator prior to plating.

A commercial pulse generator was used to supply square wave current pulses. Instantaneous (peak) pulsed current density (j_p), pulse duration (T_{on}) and off-time (T_{off}) were measured from the time-current curve on an oscilloscope through a current transformer and a current probe amplifier, average current density (j_m) being calculated from the relationship

$$j_m = j_p \quad T_{on}/(T_{on} + T_{off}) \qquad [1]$$

The frequency of the pulse cycles was also measured by a frequency counter. The number of coulombs in a single pulse was derived from the time-current curve. Current efficiency for rhenium deposition was determined from the gain in weight of the cathode and the total coulombs passed (i.e., j_m x total plating time), taking the number of participating electrons as seven. No electrolyte stirring was deemed necessary in view of the vigorous evolution of hydrogen which occurred at the cathode.

Selection of appropriate values for T_{on} and j_p was based on calculation of transition times according to the Sand equation [2]:

$$\tau = \frac{\pi D\, c_0^{\,2}\, (nF)^2}{4\, j_p^{\,2}} \qquad [2]$$

TABLE 10.2

Transition Times at Various Single Pulses

j_p (A/dm^2)	100	700	5000	10,000	20,000
τ (μs)	13000	260	5.0	1.3	0.31

where D = diffusion coefficient for perrhenate ion,
1.3×10^{-5} cm^2/sec (5)
c_0 = bulk concentration of perrhenate ion,
5.2×10^{-5} mol/cm^3
n = 7

From the values of pulsed current densities and corresponding τ shown in Table 10.2, on the assumption that the current efficiency for rhenium deposition is 100 percent, it is evident that very high current densities can be employed with sufficiently short pulse lengths.

2. Factors affecting current efficiency

2.1. *Pulse conditions*

Fig. 10.1 shows the influence of pulsed current density and pulse duration on current efficiency at a constant average current density of 10 A/dm^2. In all cases a sharp maximum is observed, the current efficiency near the maximum value increasing with increase in the pulsed current density. The effect of varying the average current density was studied in two ways:

(a) by variation of pulsed current density with a constant pulse duration of 10 μs, and θ/T_{on} of 500;
(b) by variation of total cycle time (θ), with constant pulsed current density (5000 A/dm^2) and pulse length (10 μs).

The results are shown in Figures 10.2 and 10.3 respectively. From Fig. 10.2 it is seen that the current efficiency for rhenium deposition goes through a maximum of 50 percent in the j_m range from 5 to 8 A/dm^2, while from Fig. 10.3 it is evident that increasing the pulse period θ and consequently the off-time (T_{off}) exerts a marked increase on the efficiency in allowing restoration of the perrhenate ion concentration in the cathodic diffusion layer, presumably up to the value of the bulk concentration, since the current efficiency reaches a maximum of 58 percent at 20° C when the off-time exceeds 12 ms.

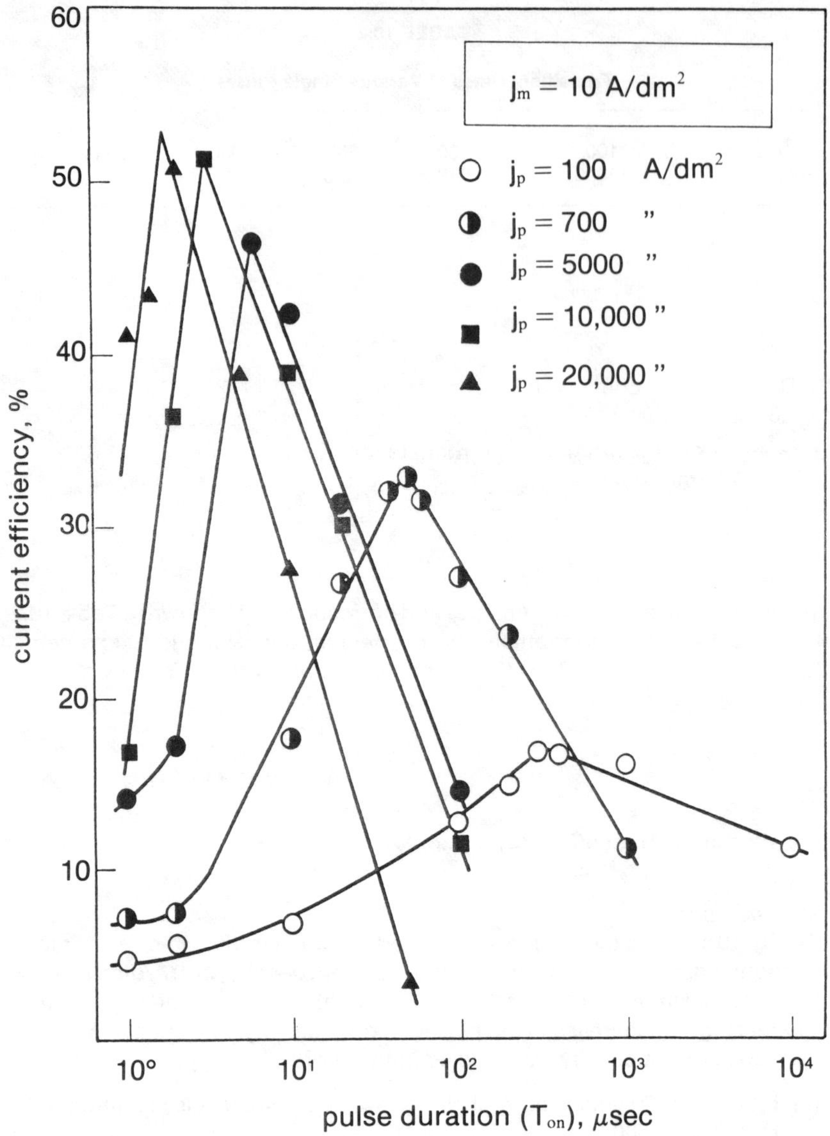

Fig. 10.1—Influence of pulse duration (T_{on}) on current efficiency at a constant average current density (j_m) of 10 A/dm².

2.2. *pH*

As shown in Fig. 10.4, pH variation had relatively small effect on current efficiency in the range 0-1.0. In the range from 1.0-2.0 only grey deposits were obtained, and at higher values little deposition occurred.

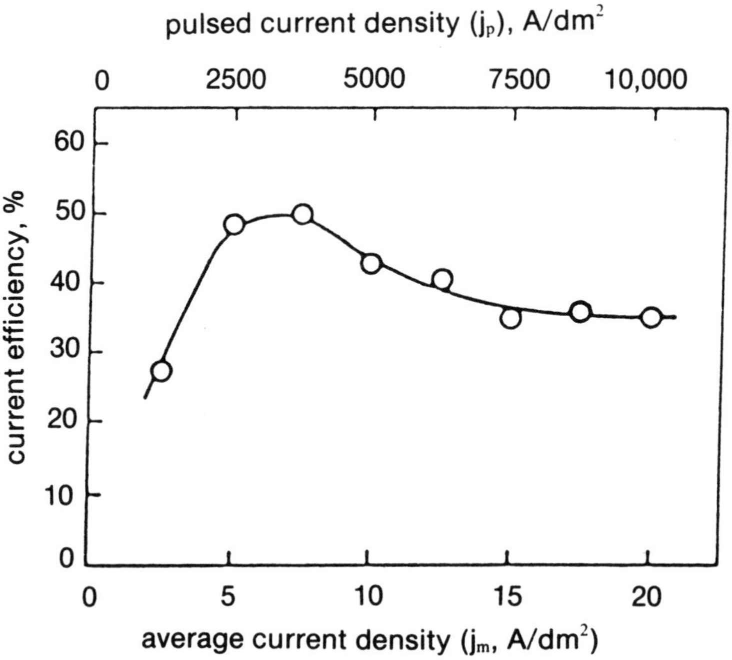

Fig. 10.2—Influences of average current density (j_m) and pulsed current density (j_p) on current efficiency at T_{on} = 10 μs and θ = 5000 μs.

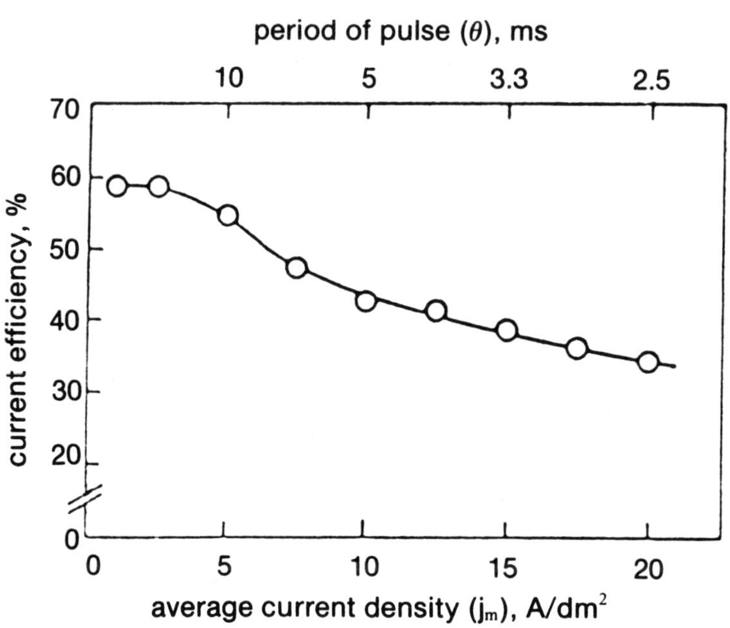

Fig. 10.3—Influences of average current density (j_m) and period of pulse (θ) on current efficiency at j_p = 5000 A/dm^2 and T_{on} = 10 μs.

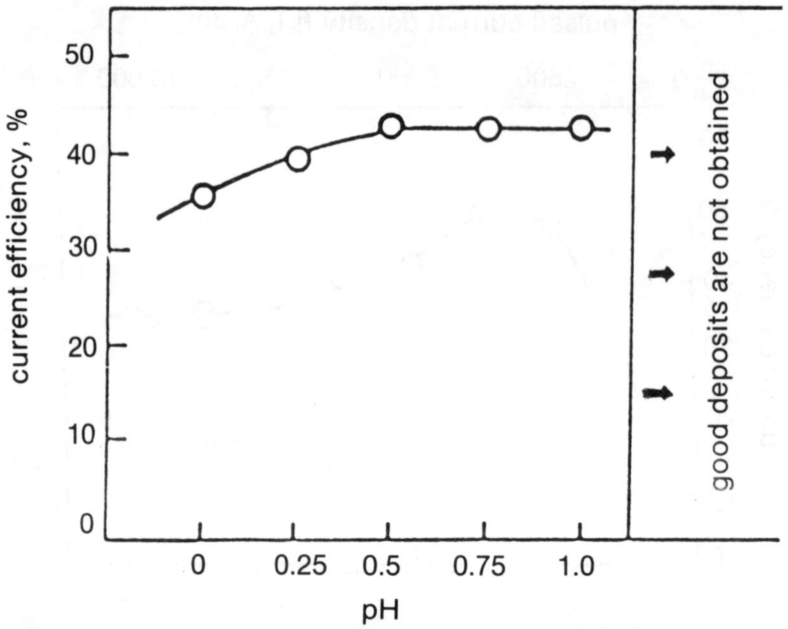

Fig. 10.4—Influence of pH on current efficiency at $j_m = 10$ A/dm^2, $j_p = 5000$ A/dm^2, $T_{on} = 10$ μs and $\theta = 5000$ μs.

2.3. *Temperature*

As shown in Fig. 10.5, current efficiency increased with temperature to over 60 percent at 80° C.

3. Brightness and tarnish resistance of deposits

Deposits of approximately 2 μm thickness produced under conditions corresponding to the right hand side of the maxima in Fig. 10.1 were lustrous as plated, with brightness and tarnish resistance increasing with increasing pulse length, T_{on}. However, the current efficiency decreased. For example, deposits obtained with pulsed current densities above 5000 A/dm^2 and pulse lengths near to the maxima were somewhat dull in appearance and turned black within a few weeks in normal atmospheric exposure, while coatings produced with pulsed current density of 10 KA/dm^2 and pulse length of 100 μs with current efficiency of 12 percent, were mirror-bright and remained unchanged in appearance after three months. Under the latter conditions, bright coatings up to 10 μm in thickness, and possibly higher, could be produced. Brightness decreased to a considerable extent with increase in deposition temperature above 50° C, but the effect of changes in bath pH in the region from 0 to 1 was not significant. The authors, however, recommend a pH of 0.7 to 0.9 for optimum tarnish resistance and current efficiency.

Under conditions represented by the left hand side of the maxima, where current efficiency increases with pulse duration, the colors of deposits obtained with a constant pulse length of 1 μs varied from lustrous steel grey, through dull white, to mat black with increasing pulsed current density from 0.7 to 10 kA/dm^2. An increase in

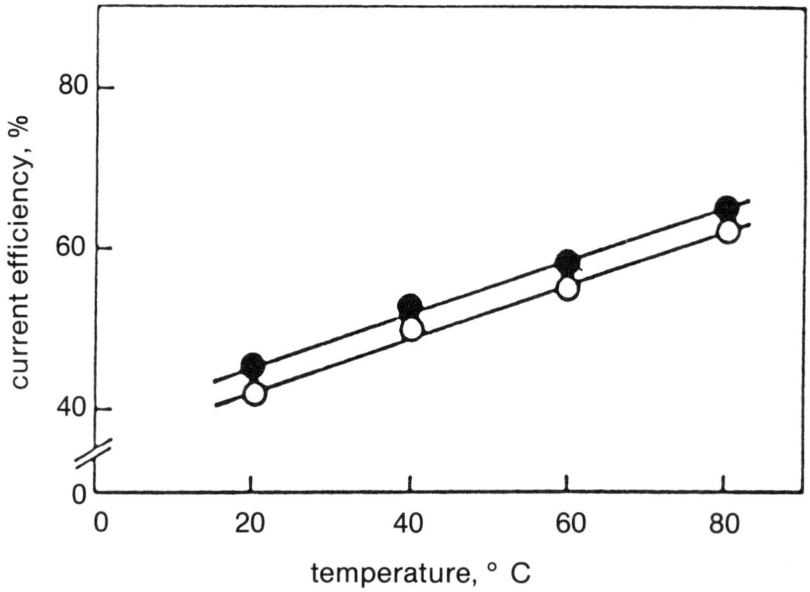

Fig. 10.5—Influence of temperature on current efficiency at $j_m = 10$ A/dm^2, $T_{on} = 10$ μs. The j_p and θ:○, 5000 A/dm^2, 5000 μs;●, 10 kA/dm^2, 10,000 μs.

pulse length resulted in a change of color from steel grey to lustrous silver-white at the lower current density, and from mat black to dull white at the higher value.

4. Crack formation

The effect of heating on crack development was studied by scanning electron microscopy on surfaces and sections of deposits, representative results of which are shown in Fig. 10.6. A coating produced by direct current plating (Fig. 10.6e) developed severe cracking after heating in hydrogen, with crack widths reaching about 1 μm. As expected, however, the high current density achievable by pulse plating resulted in a marked decrease in the size and number of cracks as well as an increase in coating cohesion and the adhesion of the deposit to the tungsten substrate. With a pulsed current density of 10 kA/dm^2 and pulse length of 10 μs, coatings were obtained which remained almost crack-free and firmly adherent to the substrate after heating (Fig. 10.6d). It may be concluded therefore that the increase in current efficiency by using pulsed currents, i.e., decrease in the overall yield of hydrogen, effectively prevents the formation of cracks. Cross sections of deposits corresponding to SEM pictures of Fig. 10.6a, b, c, d are shown in Fig. 10.7a, b, c, d respectively. Absence of cracking and good adhesion are here again evidenced for the deposit made with pulse plating ($T_{on} = 10$ μs; $j_p = 10$ kA/dm^2; $j_m = 10$ A/dm^2).

5. Discussion

The current efficiency, plotted in Fig. 10.1 as a function of T_{on} at constant j_p and j_m, goes through a sharp maximum. This result can be interpreted on the basis of changes

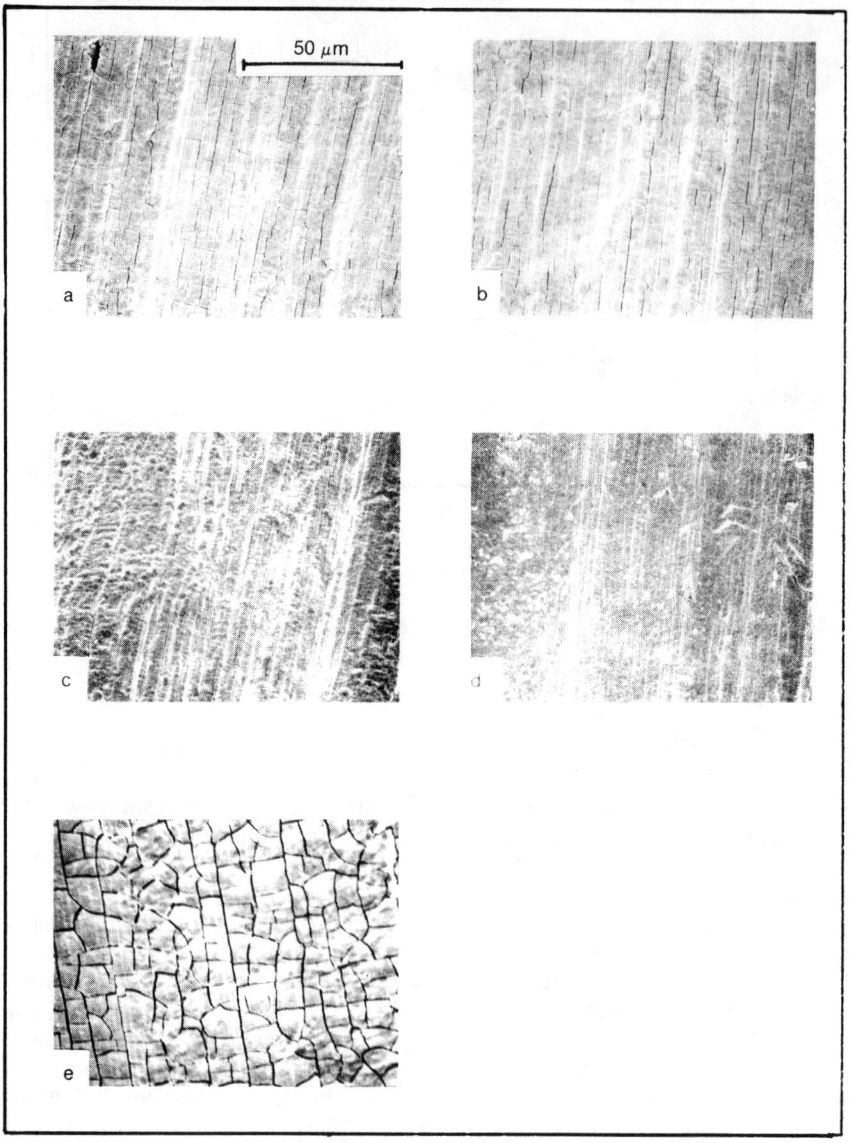

Fig. 10.6—Scanning electron micrographs of 2 μm thick rhenium deposits showing influence of PC density (j_p) on crack formation (bath Table 10.1). Average current density (j_m) for all deposits was 10 A/dm^2: (a) T_{on} = 1000 μs, j_p = 100 A/dm^2; (b) T_{on} = 10 μs, j_p = 700 A/dm^2; (c) T_{on} = 10 μs, j_p = 5000 A/dm^2; (d) T_{on} = 10 μs, j_p = 10 kA/dm^2; (e) DC plated, j = 10 A/dm^2.

in the partial currents involved respectively in charging of the double layer by hydrogen ions, deposition of rhenium, and hydrogen evolution. The first partial current is predominant at the beginning of a pulse and decreases thereafter, while the partial current for rhenium deposition increases during the pulse, which explains the ascending part of the curves before the maximum. The charging times t_c, calculated by

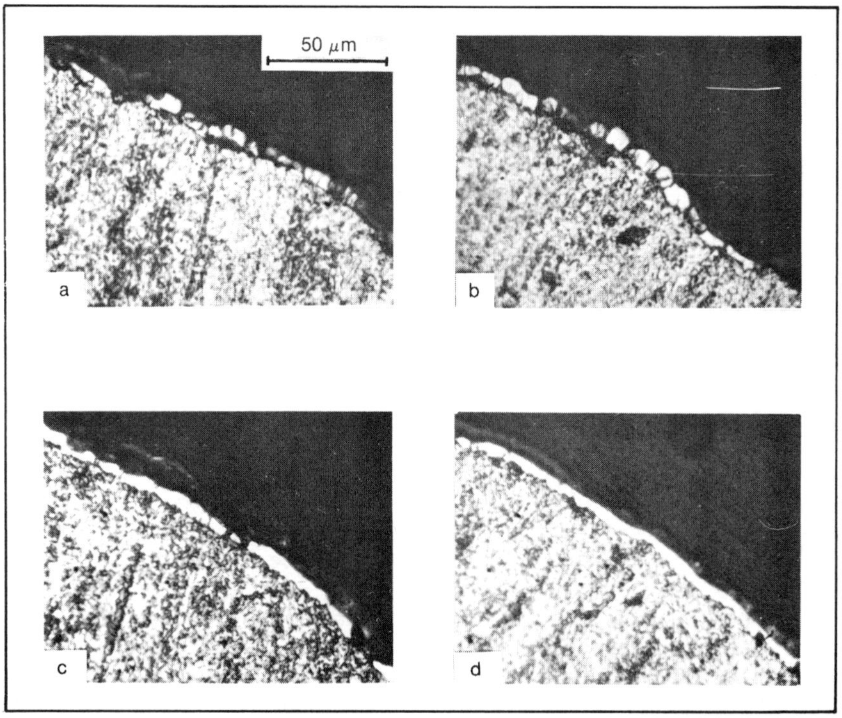

Fig. 10.7—Cross section of rhenium deposits showing influence of pulsed current density on crack formation (bath Table 10.1). Key as (a), (b), (c) and (d) in Fig. 10.6.

the method presented in Ref. 7, are in good agreement with this view (Table 10.3). After the maximum, hydrogen evolution becomes predominant over rhenium deposition, the maximum rate of which is diffusion-limited, the net result being a decrease in current efficiency with increasing pulse duration. This explanation is supported by comparison of the calculated transition times τ (Table 10.2) with the on-times $(T_{on})_m$ (Table 10.3), corresponding to maxima of the current efficiency curves. Taking into account the uncertainty of the diffusion coefficient (D) for rhenium, possible error in the assumption of 100 percent current efficiency in the computation of τ, and the fact that at least a part of the charging time of the double layer has to be added to the transition time (since the current is essentially capacitive at the beginning of the pulse), the values of τ and $(T_{on})_m$ are in fairly good agreement.

In considering possible reasons for the decrease in cathode efficiency on the right hand side of the maxima of Fig. 10.1, it is significant that, as shown in Table 10.3, the total quantity of charge supplied during a pulse, $j_p\ (T_{on})_m$, is virtually constant at the maxima on all the curves, with a value lying between 3.0 and $3.5 \cdot 10^{-4}$ As. This may be explained if it is assumed that an adsorbed monolayer of reducible ions is established on the cathode surface during off-periods, and that these adsorbed ions are preferentially discharged at the beginning of a pulse. Thereafter, for pulse periods exceeding $(T_{on})_m$, further ions must be supplied from the solution, and since these require a higher activation energy, i.e., overpotential, for discharge, preferential

TABLE 10.3

Various Values at the Maxima in Fig. 10.1 and Calculated Charging Times (t_c)

j_p (A/dm^2)	$j_p (T_{on})_m$ (As)	$(T_{on})_m$ (μs)	t_c (μs)	C.E. (%)
100	3.3×10^{-4}	330	199	17
700	3.5×10^{-4}	50	36	33
5000	3.0×10^{-4}	6.0	5.0	46
10,000	3.3×10^{-4}	3.3	2.4	51
20,000	3.4×10^{-4}	1.7	1.2	53

discharge of hydrogen takes place. On this view, the maximum in current efficiency will occur when the total charge supplied during a pulse corresponds to that required for discharge of the adsorbed species, i.e., $(T_{on})_m\ j_p = Q_{adsorbed}$.

From the reaction mechanisms for rhenium deposition described in the literature (8,9), the adsorbed species in the above scheme, dischargeable at a lower potential than hydrogen ions, could be Re^{7+}, ReO_4^-, or ReO_3^-. When the small amount of adsorbed material is consumed, discharge of hydrogen ions could involve reduction of ReO_4^- or ReO_3^-, or evolution as gas, in either case being lost for metal deposition.

6. Conclusions

The above investigations of rhenium electrodeposition on a tungsten substrate by square galvanostatic pulses using a simple perrhenate sulfuric acid solution offer several points of interest:

At room temperature a high current efficiency of 50 percent can be attained as compared to 6 percent in the case of direct current deposition. The formation of cracks, which is an undesirable feature of direct current deposits after heat treatment at temperatures in the order of 1,000° C, can be completely avoided by depositing the metal with pulsed current densities j_p higher than 10 kA/dm^2, pulse durations T_{on} smaller than 10 μs and average current density j_m 10 A/dm^2.

The current efficiency plotted as a function of T_{on} at constant j_p and j_m goes through a sharp maximum. This result is interpreted on the basis of the changes in the partial currents for charging of the double layer by hydrogen ions, rhenium deposition and hydrogen evolution. The ascending part of the curves at the left hand side of the maximum is due to capacity effects, and the descending part of the curves after the maximum is due either to mass transport limitations or to kinetic limitations associated with the formation of an adsorbed layer of rhenium ions which is preferentially reduced as compared to hydrogen ions.

Consequently, the current efficiency and the properties of the deposits are significantly changed by varying the pulse parameters j_p, T_{on} and T_{off}. The pulsed electrodeposition of rhenium presents an example where the optimum in terms of a given deposit property can be achieved by matching the pulse parameters with kinetic parameters such as adsorption time of a reacting species, and also underlines the fact that optimum conditions may sometimes lie in a narrow range of pulse parameters.

Finally, from the practical viewpoint, it should be noted that due to the very high current density required for deposition of rhenium under optimum pulse plating conditions, the process is applicable to only limited substrate areas because of apparatus limitations.

References

1. C.G. Fink and P. Deren; *Transactions Electrochemical Society,* **66** : 471 (1934)
2. E.K. Camp; *Plating,* **52** : 413 (1965); **53** : 318 (1966)
3. N. Ibl; *Metalloberfläche,* **33** : 51 (1979)
4. N. Ibl; *Surface Technology,* **10** : 81-104 (1980)
5. K. Hosokawa, J.Cl. Puippe and N. Ibl; *Proceedings of Interfinish 80,* Kyoto (1980)
6. K. Hosokawa, H. Angerer, J.Cl. Puippe and N. Ibl; *Plating and Surface Finishing,* **67,** (10) : 52-56 (1980)
7. J.Cl. Puippe; Chapter IV of this book
8. A.R. Meyer; *Oberfläche-Surface,* **11,** (3) (1970), **11** (4) (1970), **11,** (6) (1970) and **11**, (8) (1970)
9. A.R. Meyer; *Transactions Institute of Metal Finishing,* **46** : 209-212 (1968)

Chapter 11

Pulsed Electrodeposition of Alloys

A.J. Avila

1. Introduction

The difficulties in obtaining a uniform composition when electroplating alloy deposits have plagued the industry since the first alloy was deposited. Years of experience along with trial and error have permitted a reasonable control of the deposits. Even with all of these controls, the variation in composition from part to part on a plating rack can be excessive. In the solder plating of printed circuit boards, variations in compositions can sometimes be emphasized during reflow due to differences in deposit melting temperatures.

One of the approaches to alloy plating by direct current has been to arrange the bath composition in accordance with the rate of deposition of the ions concerned and the amount required in the deposit. For example, in a typical plating solution for a deposit of 30 percent iron and 70 percent nickel, the solution would contain approximately 16 percent iron and 84 percent nickel. The exact composition would vary with current density, with more iron in the deposit at a lower current density and less at a higher current density. Sometimes the rate of deposition of one species is strictly diffusion-controlled, as, for example, in palladium-silver alloy deposition from an ammoniacal bath (1,2). In this situation the alloy composition can be predicted from process parameters, including pulse parameters (2,3).

With pulsed current it is possible to maintain the composition of the deposit closer to that of the solution. In critical processes such as alloy gold deposits, where color is a factor, a range of color from pink to green can be achieved just by changing the on-time. Unfortunately, the state-of-the-art is such that finite numbers cannot be designated for frequency or concentration of the metal in the solution. However, the more concentrated the metals in the solution, the shorter can be the on-time.

It is difficult within the scope of this text to cover all possible alloy compositions that can be electroplated. An attempt will be made to review several alloy deposits in order to indicate the kind and type of metals that have been deposited by pulse plating. In all cases except for gold, only a limited amount of work has been reported. Pulsed solder plating, however, is gaining in use, particularly for printed circuit boards. Alloys discussed in other chapter of this book will not be covered here. Gold alloys and zinc alloys are presented in chapters 9 and 7 respectively.

2. History

One of the first references to pulsed current alloy plating occurred in 1934 with the granting of a patent to J. Winkler Jr. (4). Figure 11.1 shows a schematic of the Winkler

J. Winkler, Jr.

March 20, 1934 Electrodeposition of Metal Alloys 1,951,893

Filed May 4, 1932

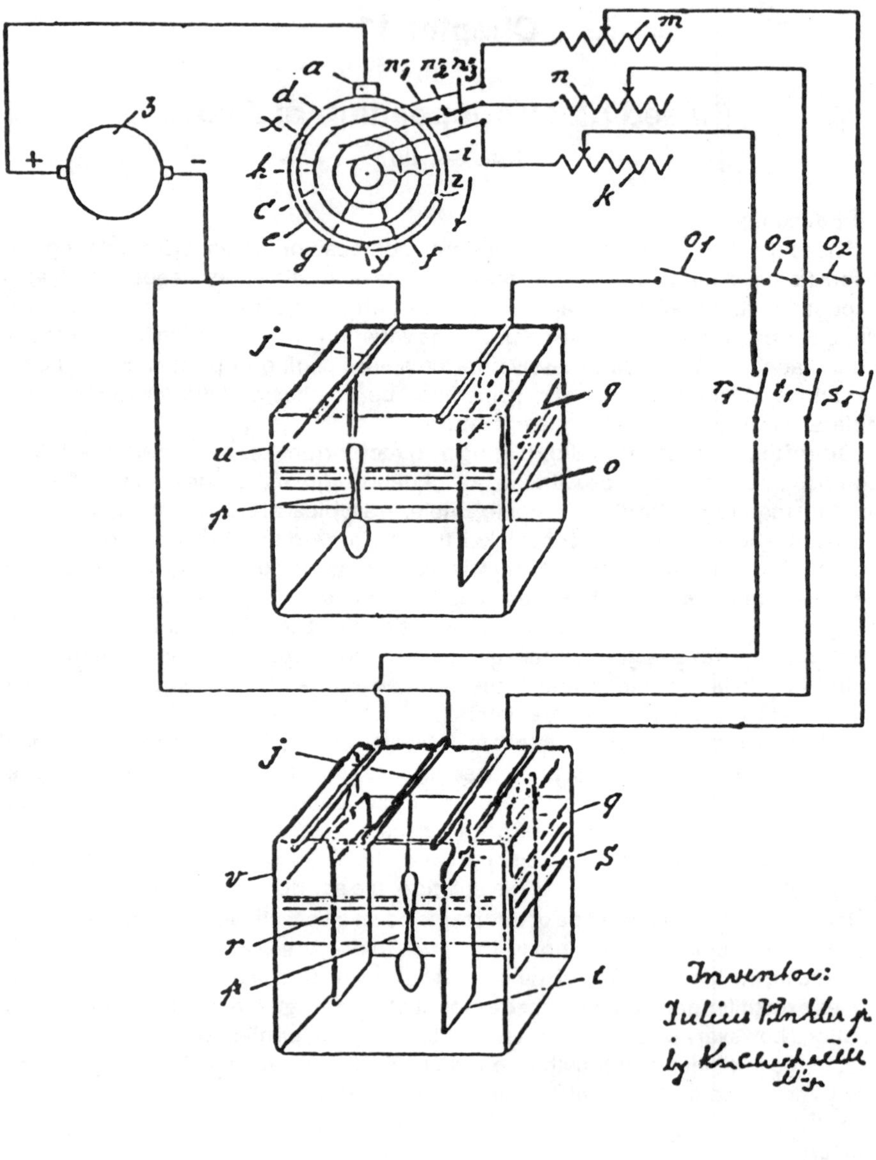

Fig. 11.1—Schematic representation of equipment for pulse plating according to J. Winkler Jr. patent.

proposal. The basis of this patent was a means to accurately control the deposition of an alloy with the use of a variable voltage pulse. Winkler believed that a variable voltage of a continuously changing high and low value could overcome the difficulties of alloy deposition. When the voltage is at its higher value, the cathodic current density and the cathodic drop of potential are comparatively high and the deposition of the more electronegative metals, such as nickel, is promoted. During the period of lower voltage the deposition of the more electropositive metals, such as gold or copper, is promoted. By selecting and carefully calculating the time and magnitude of the pulses, Winkler found that alloy composition could be controlled independently of the composition of the plating solution. Examination of Figure 11.1 shows a distributor-type control having three segmental metal strips, which are insulated from each other and connected to three slip or collector rings. These rings in turn are connected by sliding contacts with three separate anodes of the elements desired in the alloy. It is interesting to note that this is one of the first references to pulse plating as well as to alloy pulse plating.

G.E. Gardam and N.E. Tidswell described in 1954 an entirely different approach to electrodeposition by use of a pulsed current (5). Their method employed a solution in which the deposition potentials of the two metals are slightly different, a continuous current tending to deposit only the metal with the more positive deposition potential, to produce an alloy high in this metal. The current density is set sufficiently high so that the film of liquid immediately adjacent to the cathode is completely exhausted of metal in a short period of time. In other words, the pulsed current is set close to the DC limiting value, under which condition a continuous current would result in "burning" of the deposit. As the process of exhaustion nears completion, the cathode potential becomes more negative, and the other less easily deposited metal, begins to deposit. Before burning occurs, the current is cut off for a predetermined period to allow the solution to return to an equilibrium state. This cycle is repeated continuously. Figure 11.2 shows the influence of copper content and of plating cycle on the color of gold deposits.

3. Brass

Problems with the deposition of brass have been frustrating electroplaters for many years. While a good number of platers seem to have mastered the process, problems continually arise that require much effort to resolve. Pulsed current is not a panacea for perfect deposition of brass but it can offer an opportunity for better reproducibility. Referring again to the work of Gardam and Tidswell (5), Table 11.1 shows some results obtained by the application of their procedure to the pulse plating of brass, indicating the flexibility of the process.

In this particular context, it is important to remember that the state-of-the-art in 1954 did not provide millisecond pulsed units. As in the case of gold, both on- and off-times are important in affecting the composition and/or color of the alloy deposits.

From experience it has been established that a wide range of parameters will produce equally good results. In alloy plating, a 33 percent duty cycle of 10 msec on, 20 msec off, will be identical to 20 msec on, 40 msec off, and similar results can be anticipated. Table 11.1 also includes current density as a variable. The Hull cell is usually helpful in relating current density to a particular cycle. The reason for presenting this table is to emphasize the factors involved in alloy deposition. Understanding the mechanisms involved will help to ensure success with pulse plated alloy deposits.

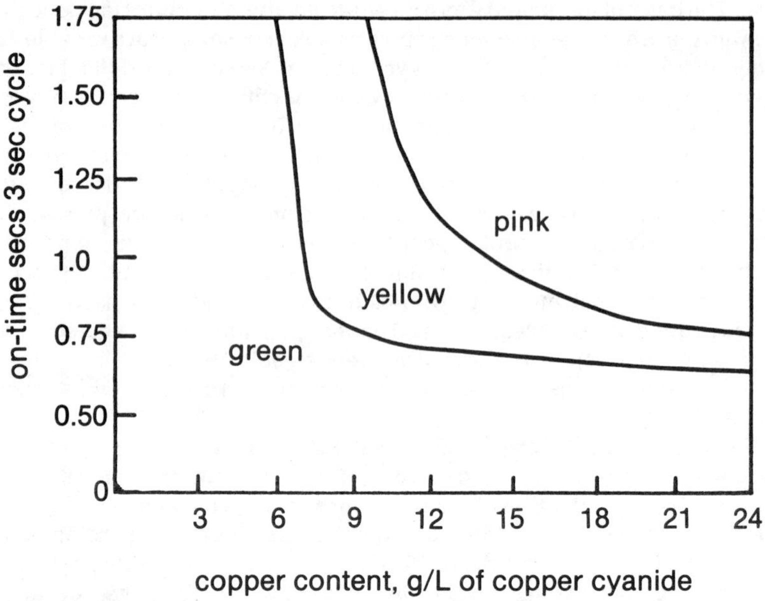

Fig. 11.2—Influence of copper content and of plating cycle on color of deposit.

4. Tin-lead

Solder plating follows the same kind of mechanism as that for gold and brass. A major application of solder plating is on printed circuit boards, where problems may stem from poor distribution and alloy composition. Poor distribution causes overplating in high current density areas that results in clogged holes during reflow. Wide variation in composition leads to uneven reflow, which is a source of rejects. Material savings where great quantities of solder are used can frequently justify the expenditure for pulse plating equipment. It is important to evaluate the cost of rejects or repairs to substantiate the advantage of the use of pulse plating. In an actual cost study, it was found that each 1 percent of plating defects increased the cost of a printed circuit board by 5 percent. The influence of pulse parameters on the alloy composition and on the current efficiency of Sn-Pb electrodeposits is discussed (16) in conjunction with a fluoborate type of bath.

5. Nickel-Iron

Perhaps the most critical need for alloy composition control is with magnetic alloy deposition. Age-resistant properties along with improved magnetic properties are the two most desirable qualities in magnetic films. It is practically impossible to deposit a nonmagnetostrictive film with direct current. The normal procedure with any direct current electroplated magnetic film, therefore, includes a stabilizing heat treatment at 200° C for at least two hours. This can be eliminated by the use of pulse plating.

In 1965, R. Gerard (6) published results of his work with pulsed current and thin magnetic films. The interesting aspect of the data (Figure 11.3) showed that as theoretically expected, pulsed current produced a deposit close to the solution composition. In this example the solution contains approximately 17 percent iron, the

TABLE 11.1
Deposition of Brass

Solution and Conditions		
Sodium Cyanide	49 g/l	
Copper Cyanide	30 g/l	
Zinc Cyanide	7 g/l	
Sodium Carbonate	4.7 g/l	
on-time seconds	**Current Density 0.32 A/dm^2**	**Current Density 0.64 A/dm^2**
	Color of Deposit	
½	orange	red
¾	orange-yellow	orange-yellow
1	orange-yellow	yellow
1¼	yellow	yellow
1¾	green	orange-green

iron content in the pulsed current deposit being also approximately 17 percent, while that in the direct current deposit is approximately 28 percent. Of course, frequency, on- and off-time, temperature, etc., were also factors, but the parameters were the same as those normally used for direct current. In conjunction with a development project on magnetic plated wire, the author was able to duplicate Gerard's results. Before the introduction of pulsed current, it was necessary to stabilize the magnetic film. The pulsed current deposits appeared to be completely stable.

Aging-resistant permalloy films with improved magnetic properties were plated by S.I. Tan, A.F. Mayadas and L.T. Romankiw (7) with the following composition:

$NiSO_4$ $6H_2O$	15 g/l
$FeSO_4$ $7H_2O$	2.25 g/l
Na-K tartrate	7.5 g/l
Sulfamic acid	1.0 g/l
Triton X-100	0.6 g/l
pH	3 - 4

An equal molar quantity of Fe^{3+} could be substituted for Fe^{2+} in the bath. They used a bath contained in a rectangular cell with the cathode having the same geometrical dimensions as the anode. Plating was performed by the pulse plating technique using square pulses of 0.56 A/dm^2 amplitude and a width of 15 seconds. Each current pulse was followed by 10 seconds of stirring the solution and a 15 to 30 second rest period. The current density could vary in the range of 0.4-0.6 A/dm^2. The films thus produced were stabilized at 200° C in an easy axis, 40 Oe field in air for at least 2 hours. Upon

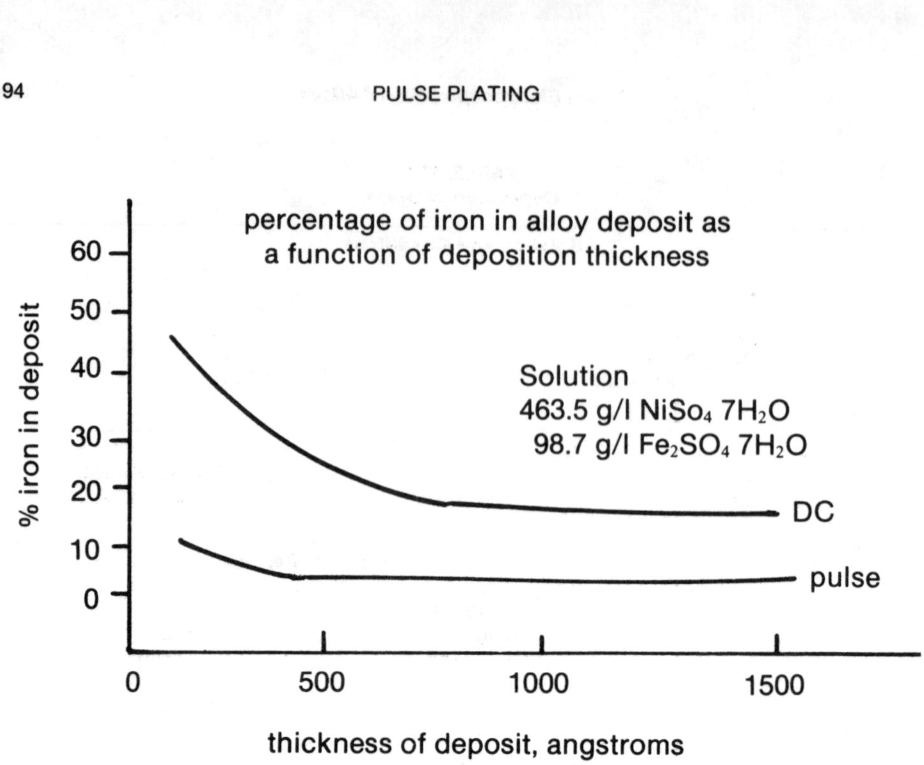

Fig. 11.3—Comparison of alloy composition deposited for pulsed current vs. direct current.

subsequent aging at 100° C in a 20 Oe field for 67 hours in air at 45 degrees to the easy axis direction, the films exhibited a high degree of resistance to aging, i.e., unusually small changes in static magnetic properties.

The effect of superimposing a sinusoidal alternating current on a direct current during the electrodeposition of Ni-Fe alloys was reported by Z. Kovac (8). Her work showed that by proper choice of conditions, electrodeposited Ni-Fe alloys could be prepared with a uniform composition throughout their thickness. From the same solution, the composition of the alloy was varied from 8 to 60 percent Fe merely by altering the frequency. It was found that using an electrolyte in which the bulk concentration of nickel was four times the bulk concentration of iron, at 100 Hz, Ni and Fe deposited at the same rate. Above 100 Hz, iron deposited at a higher rate than nickel.

Alternating current has a significant influence on many electrode processes and is widely used in such electrochemical investigations as the study of electrical double layers, the kinetics of the formation and dissolution of oxide films, fast electrode reactions, and in the electrodeposition and dissolution of metals. By superimposing AC and DC, the deposition kinetics of alloys is affected in a very practical manner. At low frequency the concentration wave is able to follow the slowly varying current, and the penetration depth is approximately the same as the DC reaction layer thickness. At higher frequencies, the formation and decomposition of metal complexes becomes less and less important, since they cannot follow fast changes of current. In addition, the penetration depth of the concentration wave becomes smaller. Thus, at high frequencies the DC behavior dominates.

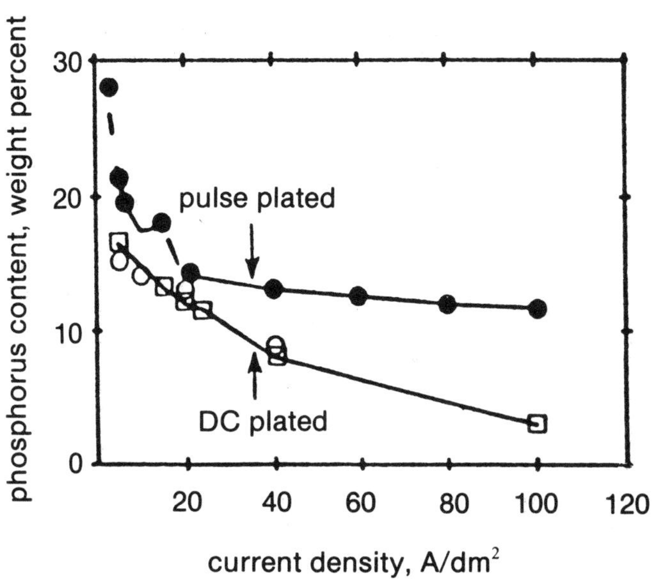

Fig. 11.4—Phosphorus content as a function of current density for PC and DC plated alloys. The symbol (o) designates data reported by Brenner (10).

6. Nickel-phosphorus

While nickel-phosphorus deposits are commonly produced by electroless plating based on the chemical reduction of metal salts in the presence of a catalyst, work has also been directed to the electrolytic deposition of these alloys.

In an extensive study by D.S. Lashmore and J. Weinroth-Aronson (9), it was shown that electrodeposited nickel-phosphorus alloys produced by pulse plating from a non-electroless solution are not only metallic glasses, but exhibit at least two amorphous structures. These structures or configurations are dependent upon the deposition parameters, since the deposition parameters influence the deposit composition to a large extent. Figure 11.4 shows how the phosphorus content in the deposit decreases with increasing current density for pulse and DC plated alloys.

It is possible to produce nickel-phosphorus alloys having a much greater phosphorus content by pulse plating. Optical micrographs demonstrate that pulse plated nickel-phosphorus alloy deposits yield a more uniform microstructure.

A most interesting characteristic of these metallic glasses was their wear behavior, which was comparable with that of hard chrome in dry sliding. The pulse plated deposits were consistently harder than the direct current deposits, with the highest hardness occurring at the highest current density.

Varying the on- and off-times from 0.2 to 100 msec, it was found that the condition of 2.0 msec on-time and 0.2 msec off-time yielded the highest microhardness values. Using these times as fixed value, the current density was varied. Table 11.2 shows the composition of the solution and the pertinent parameters that proved most successful under these conditions.

Heat treatment at 400° C transforms the usually amorphous nickel-phosphorus alloys, electroless or electroplated, to a crystalline state. The resulting matrix is nickel

TABLE 11.2
Composition and Operating Conditions of the Nickel-Phosphorus Electrolyte (9)

Nickel sulfate $NiSO_4$ $6H_2O$	150 g/l
Nickel chloride $NiCl_2$ $6H_2O$	45 g/l
Phosphoric acid H_3PO_4	50 g/l
Phosphorous acid H_3PO_3	40 g/l
pH	0.8
Temperature	78° C
Agitation	magnetic stirring bar

containing particles that apparently are nickel phosphide. These alloys are popular because they exhibit low friction coefficients and high hardness.

7. Chromium-molybdenum

Titanium is an almost ideal metal for space age applications. Its weight and strength properties meet all the specifications required for gas turbine and airframe components. Its weaknesses for these applications reside in limited wear, erosion and corrosion resistance. As a result it has been necessary to resort to high strength, corrosion-resistant alloys like Hastelloy X or the Rene series. One approach to the use of titanium has been to coat it with a material that will permit it to meet all requisite specifications. Eckler et al. (11) established that a chromium-molybdenum alloy deposit could be suitable for this purpose. Chromium-molybdenum alloy deposits produced by DC plating showed a threefold improvement in frictional characteristics as compared with conventional chromium plate, and performance in this respect can be doubled by the use of pulse plating. The bath used was self-regulating and operated at ambient temperature; it contained 300 grams per liter of chromic acid and 75 grams per liter of ammonium molybdate. Deposits were subjected to a short diffusional heat treatment at 650 - 900° C. On-times and off-times were varied from 0.1 to 99 msec with plating a time of 30 minutes at an average current density of 46.5 A/dm^2. The results are summarized in Figures 11.5 and 11.6, showing the effect of pulse plating, with a range of on-and off-times ($T_{on} = T_{off}$), on the surface morphology, cathode efficiency, and molybdenum content of deposits. It was found that the hardest alloy composition occurred when the molybdenum content was a maximum. Thus 3 msec on and 3 msec off represented the most suitable pulse plating conditions that produced alloys with 2 percent molybdenum, while with DC, 0.65 percent was the maximum obtainable molybdenum content in the alloy.

Figure 11.7 shows SEM photographs that reveal the influence of pulse length on the surface morphology for the Cr-Mo pulse plated deposits, with a constant off-time of 3 msec, corresponding to duty cycles approximating 17, 23, 50 and 87 percent. The lowest duty cycle examined yielded a very uniform fine grained deposit. The results obtained in these studies showed that pulse plating can be used to produce crack-free deposits with more than three times the molybdenum content of direct current plated coatings.

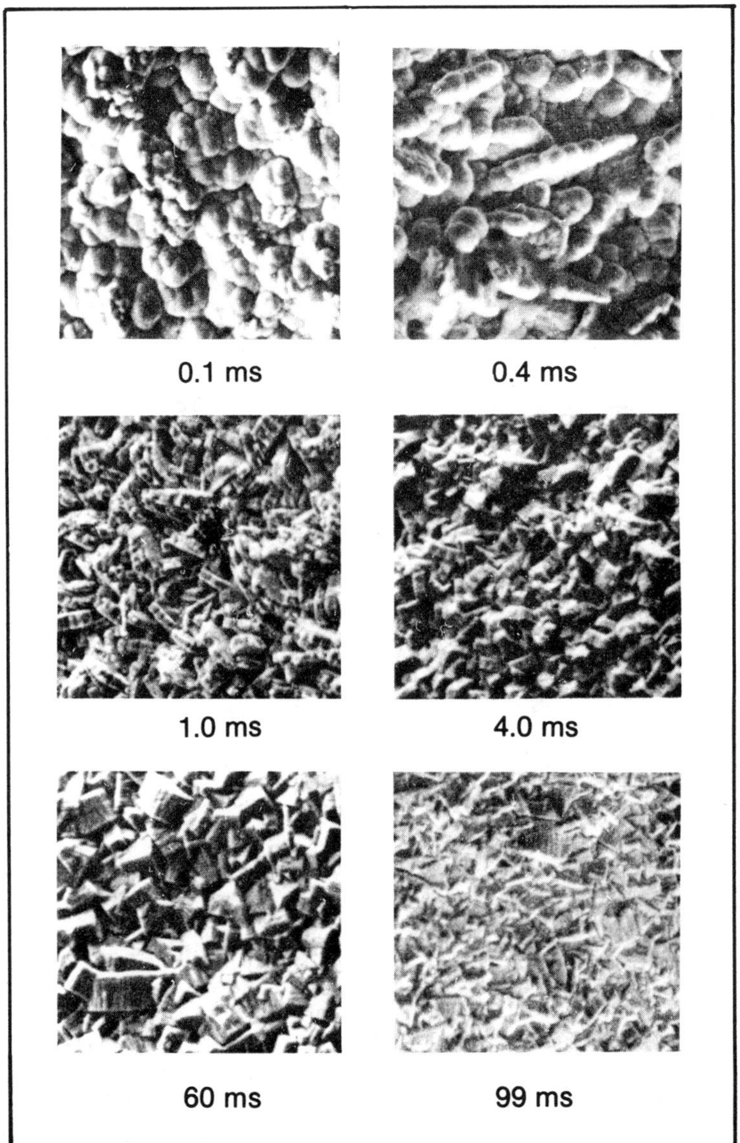

Fig. 11.5—Effect of pulse duration on surface morphology of Cr-Mo alloys with equal on- and off-times (1000 x).

8. Chromium-cobalt

Deposits of chromium-cobalt alloys with a wide range of composition have been prepared by pulsed potentiostatic electrodeposition from a bath containing chromic acid and cobalt chloride. M.H. Gelchinski, L. Gal-Or and J. Yahalom (12) reported that by changing the parameters of off-time and amplitude, alloys of various cobalt content were obtained at ambient temperature from the following electrolyte: chromic acid, 250 g/l; ammonium sulfate, 100 g/l; and cobalt chloride (hexahydrate), 100 g/l.

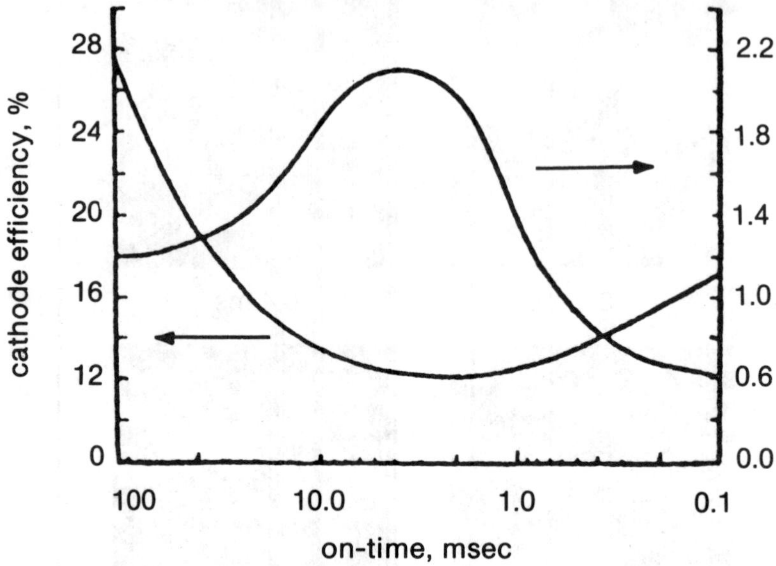

Fig. 11.6—Cathode efficiency and molybdenum content of Cr-Mo alloy as a function of on-time (in msec) with equal on- and off-times.

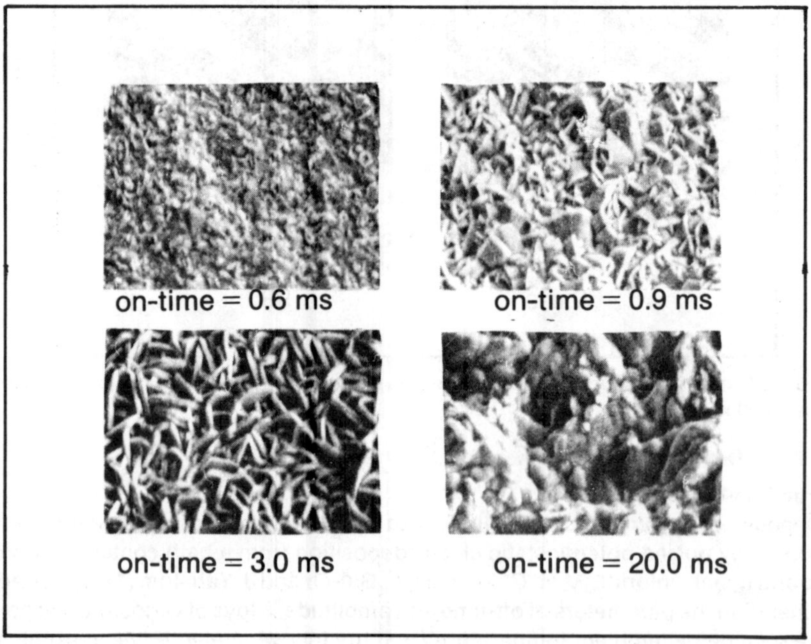

Fig. 11.7—Effect of pulse duration on surface morphology with an off-time of 3 msec (1000 x).

The range of experimental conditions is summarized in Table 11.3. The variation of the cobalt in the alloy as a function of amplitude and off-time is shown in Figure 11.8. An increase in the pulse amplitude causes a corresponding decrease in the cobalt content. On the other hand, an increase in the off-time leads to an increase in the cobalt content.

X-ray diffraction measurements revealed that deposits high in chromium content consisted of the A-15 phase along with the α-Cr (bcc) phase. Measurements indicated that the bcc lattice parameter decreased continuously with increase of cobalt content of the alloy. Deposits with more than 33 weight-percent cobalt were usually powdery.

Examination of the surface morphology by SEM showed that deposits obtained with short off-times were characterized by isolated islands or a three-dimensional growth which leads to a nodular morphology. Mirror-bright deposits were produced with an off-time of 3 msec. Increasing the off-time interval to 6 msec yielded a powdery deposit. The most plausible explanation of this behavior is that with long off-time, adsorption on the cathode takes place, which hinders lateral or two-dimensional growth of the deposit and encourages formation of new nuclei leading to three-dimensional growth, the deposits becoming powdery. Nodular deposits are observed at short off-times because the average current under these conditions approaches the DC limiting current.

9. Silver-tin

The preparation of Ag_3Sn coatings by pulsed electrodeposition was reported by Henry Leidheiser Jr. and A.R.P. Ghuman (13). Using direct current, deposition of silver occurred near the limiting current density, and a dendritic deposit structure resulted. This problem was overcome by utilizing pulsed current to obtain a uniform deposit composition over the thickness of the deposit. Pulsed current deposition was found to be a particularly effective technique for the preparation of high quality Ag_3Sn coatings with minimal amounts of elemental silver and tin.

All deposits were made from a solution containing AgCN, 0.020 M; $K_2SnO_3 \cdot 3H_2O$, 0.374 M; KOH, 1.25 M; and KCN, 1.40 M. Figure 11.9 represents the effect of on-time on the relative amount (expressed as X-ray intensity in arbitrary units) of Ag_3Sn, elemental silver and elemental tin in the alloy coatings. At a fixed current density, the concentration of elemental silver in the deposit was found to decrease and the concentration of elemental tin in the deposit to increase as the on-time was increased. For fixed on-off cycles, the amount of elemental silver in the deposit decreased and the concentration of elemental tin in the deposit increased as the current density was increased. Longer off-times produced a greater weight percent of silver in the deposit. Two reasons for this result are proposed. First, the longer off-time allows greater replenishment of silver complex ions in the cathode film. Second, the longer off-time allows greater reaction of the metallic tin in the surface of the deposit with ionic silver in the adjacent liquid.

The cross sections of many of the deposits were examined by electron microprobe and by scanning microscopy. In the case of compositions approximating Ag_3Sn, the composition was uniform throughout the thickness, and the distribution of the metals was homogeneous. The electrodeposited alloys contained minimal amounts of elemental silver and tin, were extremely ductile and could be easily polished to a high luster.

Pulse plated silver-tin alloys, particularly those with a composition approximating Ag_3Sn, show promise as decorative coatings since the tarnish resistance is greater than that of pure silver. These deposits have a pleasing appearance.

TABLE 11.3
Conditions for Various Pulsed Potentiostatic Electrodepositions

pulse width (msec)	off-time (msec)	pulse amplitude (mV SCE)	off-time amplitude (mV SCE)	approx. pulse current density (A/dm^2)	approx. average current density (A/dm^2)
3	1	-2000	-1000	53.8	40.4
	2				32.3
	3				26.9
	5				20.2
	6				18.0
3	2	-2200	-1000	77.2	46.3
	3				38.6
	4				33.0
	6				25.7
	8				21.0
3	1	-2500	-1000	83.9	62.9
	4				36.0
	6				28.0

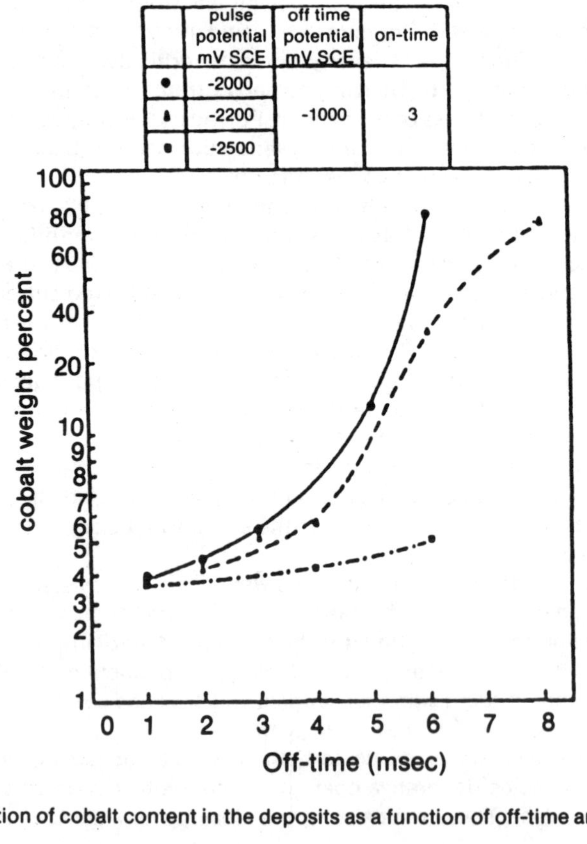

Fig. 11.8—Variation of cobalt content in the deposits as a function of off-time and pulse amplitude.

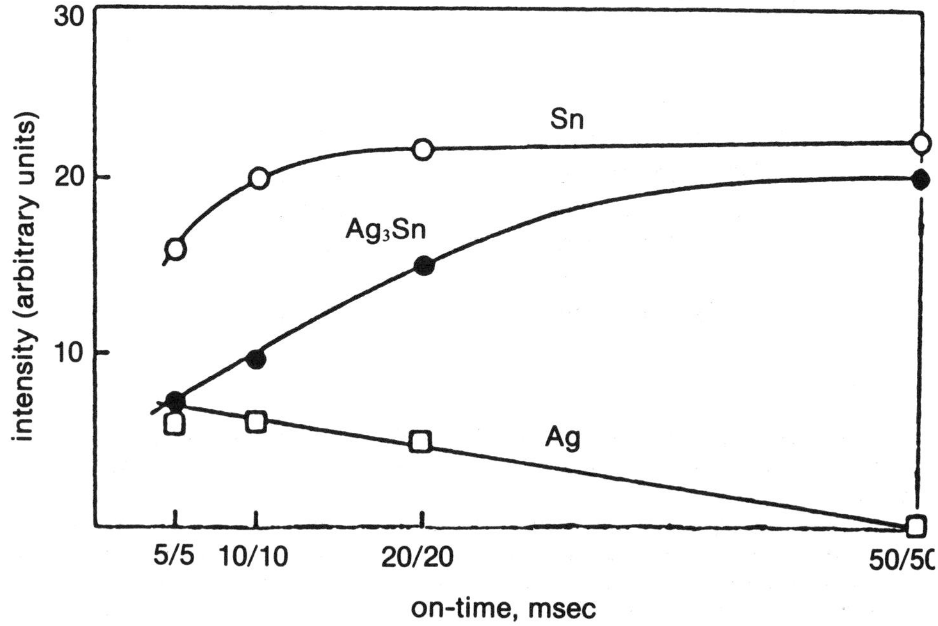

Fig. 11.9—Changes in Ag_3Sn, elemental silver, and elemental tin content in the alloy deposit as a function of pulse cycle with equal on- and off-times. Current density 2.26 A/dm^2. The ordinate represents the intensity of the strongest line in the X-ray diffraction pattern.

10. Cobalt-tungsten

In 1968, V.S. Rachinas (14) superimposed alternating current over direct current in the deposition of a cobalt-tungsten magnetic alloy. The plating solution composition was as follows:

Cobalt sulfate	125 g/l
Potassium tungstate	10 g/l
Magnesium sulfate	100 g/l
Boric acid	30 g/l

1.5 micrometer deposit in 5 to 10 minutes

Rachinas showed that with increase in the AC/DC ratio the coercivity first increased to a maximum and then decreased. The magnetic resonance did not appear to be affected. Quality of the deposit was substantially improved over direct current deposits in respect to improved brightness and decrease in the rate of corrosion by a factor of 4 to 6.

A cobalt-tungsten alloy containing about 20 weight-percent of tungsten has been deposited under pulsed current conditions consisting of 2 msec T_{on}, 18 msec T_{off}, and

0.5 A/dm^2 j_m, with 90 percent current efficiency and with Co anodes, from the following bath:

Cobalt sulfate	0.5 M
Sodium tungstate	0.5 M
Gluconolactone	1.0 M
Sodium fluoride	100 g/l
Boric acid	50 g/l
pH	4.0
Temperature	55° C

The as-plated alloy had a hardness of 370-420 HV and was crack-free, whereas under similar conditions, but with direct current, deposits of the same thickness were highly cracked (15).

In pulse plating, hydrogen atoms discharged at the cathode combine to form hydrogen molecules during the off-time and are not incorporated into the crystal lattice, whereas in DC plating atomic hydrogen is incorporated into the deposit as a hydride. The effect of pulse plating for the cobalt-tungsten alloy may be similar to that pertaining in chromium plating with pulsed current (see Chapter 8), resulting in crack-free deposits. However, pulse plated deposits showed some cracking after thermal treatment at 500° C. This effect could be eliminated by adding about 8 percent phosphorus by weight into the deposit, through a change in bath formulation consisting of the addition of 50 g/l of sodium hypophosphite. In addition to the complete elimination of cracks, the incorporation of phosphorus into the pulse plated Co-W alloys greatly increased the hardness, to a value of 800 HV after thermal treatment of 500 ° C.

Pulse plated deposits of Co-W-P alloys appeared to be suitable coatings for engineering applications.

11. Nickel-chromium

Electrodeposition of nickel-chromium alloys has been studied by D.S. Lashmore, I. I. Weisshaus and K. Pratt (17) and the part of this study concerned with pulse plating is reproduced hereafter. The electrolyte composition is given in Table 11.4.

The effects of pulsed galvanostatic deposition on chromium content were determined as a function of current density and total cycle time as shown in Fig. 11.10 for a fixed off-time of 2 msec. Generally as the current density increased, the chromium content also increased at each duty cycle from 20 to 90 percent. However, the higher the duty cycle, the higher the chromium content. There is a significant difference in the surface morphology of deposits produced by different waveforms. Scanning electron micrographs (SEMs) compare the surface morphology of a typical DC deposit (Fig. 11.11a) and a coating produced by pulsed current (Fig. 11.11b). The surface of the deposit produced by DC is clearly much rougher.

The effects of increasing duty cycle and current density are shown in Fig. 11.12. The diffraction patterns indicate a single-phase solid solution of chromium in nickel. As the current density increases, the grain size decreases; as the duty cycle increases, the grain size increases.

TABLE 11.4
Electrolyte Composition

Constituent	Amount
$NiCl_2 \cdot 6H_2O$	30-40 g/l
$CrCl_3 \cdot 6H_2O$	100 g/l
Formic acid, 95%	35-40 ml/l
NaBr	15 g/l
NH_4Cl	50 g/l
Boric acid	30-40 g/l
Sodium citrate dihydride	80 g/l
Non-ionic surfactant (non-sulfur-containing)	According to type
pH	3.8
Temperature	30-40° C

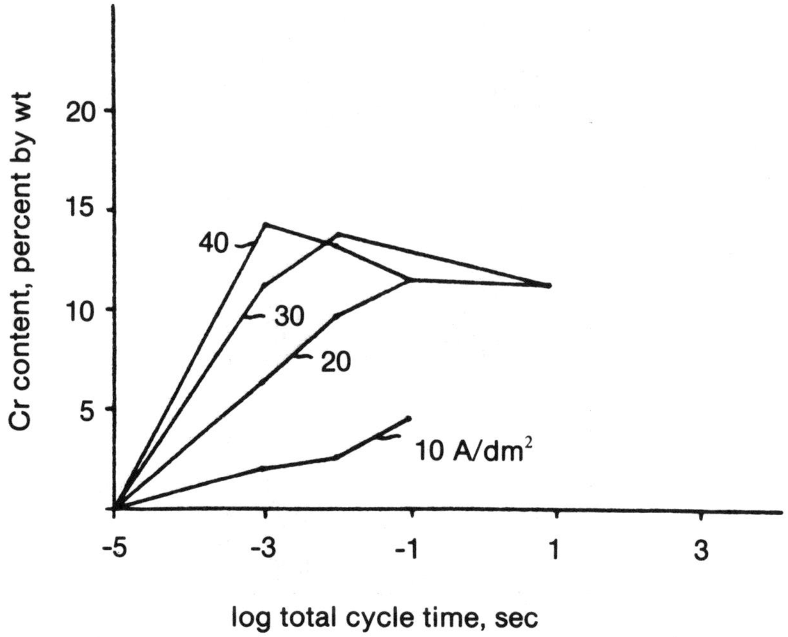

Fig. 11.10—Chromium content as a function of period and peak current density.

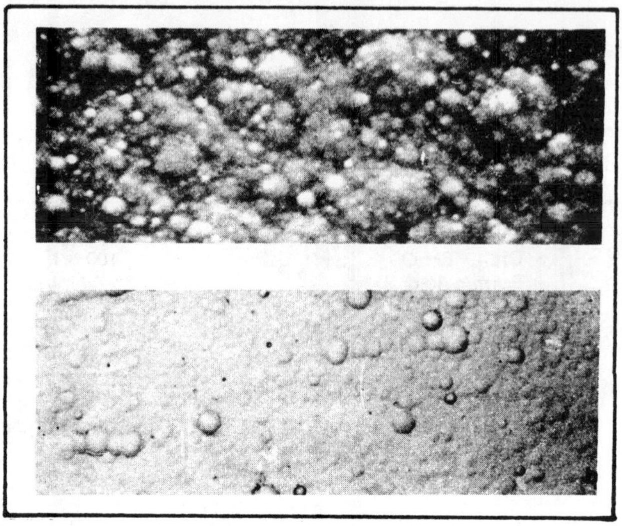

Fig. 11.11—Surface morphology of deposits plated with (a) 1 A/dm^2 DC and (b) 1.25 A/dm^2 pulsed current with a 50 percent duty cycle and T_{on} of 2.5 msec

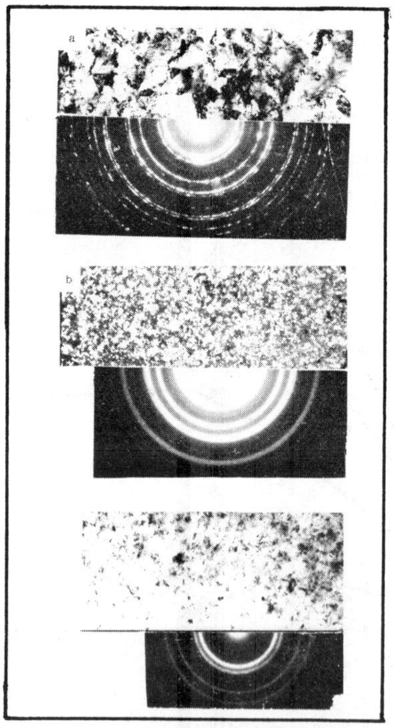

Fig. 11.12—TEMs of (a) an Ni/3 percent Cr alloy produced by pulse plating using a 25 percent duty cycle for 1875 μsec at 5 A/dm^2, (b) an Ni/12 percent Cr alloy produced by pulse plating using a 50 percent duty cycle for 1250 μsec at 12.5 A/dm^2, and (c) an Ni/28 percent Cr alloy produced by pulse plating using a 75 percent duty cycle for 625 μsec at 25 A/dm^2.

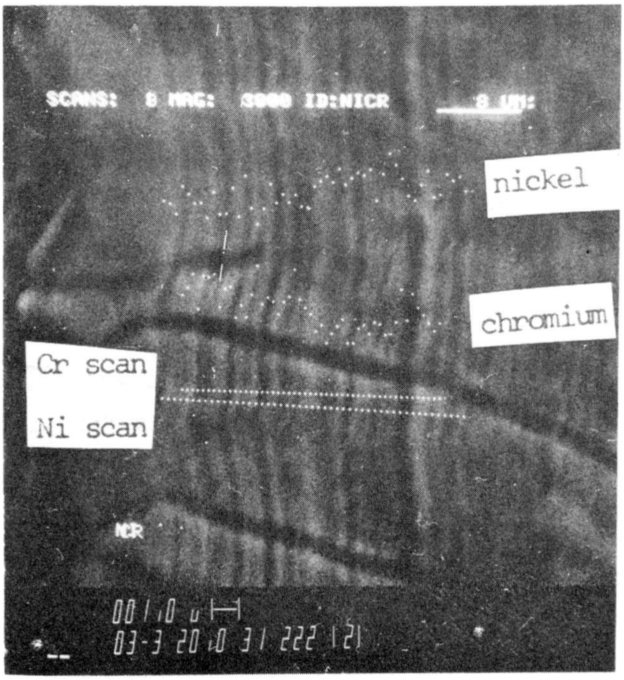

Fig. 11.13—SEM micrograph of an etched cross section of an Ni-Cr electrodeposit with superimposed X-ray line scans on CrKα showing that the layered structure results from concentration variations.

Cross sections of a typical coating show that the deposit grows in a layered structure with the layers exhibiting marked chromium concentration gradients. Figure 11.13 is a SEM of a coating produced at a duty cycle of 33 percent, a current density of 50 A/dm^2, and an on-time of 0.2 msec. It was deposited on copper and etched in aqua regia.

Microhardness measurements made on several coatings as a function of duty cycle are shown in Fig. 11.14 for current densities of 25 and 50 A/dm^2. At both current density values, a minimum microhardness occurred at duty cyle of about 50 percent. Because the chromium content generally increases with the duty cycle, the harder alloys also have high chromium contents.

Corrosion studies were made on brass discs coated by pulse plating under conditions that produced 50-μm-thick deposits with 5, 10 and 15 percent Cr which were mounted on a standard cell holder to expose an area of 3.5 cm^2. The specimens were polarized in a 3.5 percent solution of NaCl in distilled water. The solution was deaerated beforehand for several hours by bubbling high purity argon.

The specimens were first cathodically reduced at -650 mV vs SCE for 5 min, then potentiodynamically scanned at a rate of 0.1 mV/sec to 0.15 mV vs. SCE. The experiments were then repeated for each alloy after soaking in aerated solutions at open circuit for 1 hour. Cathodic and anodic plots were recorded and corrosion reates were assessed by extrapolation of Tafel slopes to the mixed potential. The anodic polarization was extended into the passive range, then reversed to check the tendency

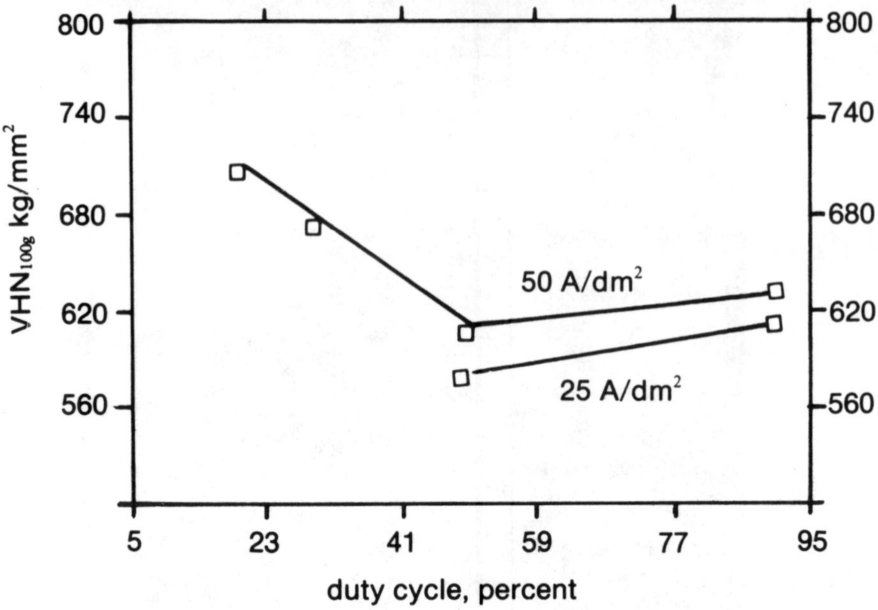

Fig. 11.14—Microhardness as a function of duty cycle for two different current densities. The hardness varies with deposition parameters and can range from 450 to 875 VHN (kg/mm^2).

TABLE 11.5
Corrosion Behavior of Electrodeposited Ni-Cr Alloys in 3 Percent NaCl

			Corrosion rate, μm/yr		
Cr content,	**Nature of polarization curve**	**Width of passivity**	**by b_a**	**by b_b**	**by pol. res.**
5	Primary and secondary passivity	150 mV, primary	3.6	1.04	0.63
10	Primary indication and secondary passivity	20 mV, secondary	0.6	0.8	0.8
15	Wide passive range	400	1.3	4.7	4.1
20*	Wide passive range	1600	—	—	—

*Nichrome.

for pitting. A polarization resistance measurement was also carried out at a span of 50 mV around the open circuit potential.

The mixed potential after exposure at open circuit in an aerated solution was -0.21 V vs. SCE, i.e., in the initial part of the limiting current range. Polarization resistance measurement around this potential was 2.964×10^{-5}, corresponding to a corrosion rate of 0.8 μm/yr. Anodic and cathodic curves around the mixed potential showed Tafel slopes (computed by the system) to be 0.082 and 0.087, respectively. These values corresponded by extrapolation to corrosion reates of between 0.6 μm/yr and 0.8 μm/yr. Reversing the scan in the passive range showed no evidence of passivity breakdown (pitting tendency) at 0.092 V vs. SCE.

Similar measurements were also taken for deposits containing 5 and 15 percent Cr and for commerical Ni-chrome alloy with 20 percent Cr. The results are summarized in Table 11.4.

All three coatings exhibited good corrosion resistance in a relatively corrosive solution with a chloride concentration similar to that of sea water. Also, with passivatable metals, true Tafel slopes are not obtained, and the corrosion rates assessed by extrapolation of tangents to the polarization curves are higher than would be found in reality.

Increasing the chromium content from 10 to 15 percent did not decrease the corrosion rate but considerably stabilized the passive range. Alloys with higher chromium contents should therefore be more immune to changes in the oxidizing conditions in the system. A 50-μm-thick coating can be expected to last for several years in environments such as sea water, based on the assessment of corrosion rates.

The investigations with pulse plating showed distinct differences in surface morphology in the types of crystal phases present between coatings produced by PC vs. DC.

References

1. B. Sturzenegger and J.Cl. Puippe; "Electrodeposition of Palladium-Silver Alloys from Ammoniacal Electrolytes" *Platinum Metals Review,* **28,** (3) (1984)
2. B. Sturzenegger; "Elektrochemische Abscheidung von Silber-Palladium-Legierungen," *Dissertation ETH 7605,* Zurich (1984)
3. B. Sturzenegger, O. Dossenbach and R. Gubelin; "Pulse Current Electrodeposition of a Binary Alloy," *Proceedings of the 35th Meeting of ISE,* Berkeley (August 1984)
4. J. Winkler Jr.; "Electrodeposition of Metal Alloys," US Patent 1,951,893 (March 20, 1934)
5. G.E. Gardam and M.E. Tidswell; "The Electrodeposition of Gold and Other Alloys by a New Method," *Fourth International Conference on Electrodeposition and Metal Finishing,* London (1954)
6. R. Gerard; "Electrodeposition of Thin Magnetic Films," French Atomic Energy Commission, Grenoble Center of Nuclear Studies (June 10, 1965)
7. S.I. Tan, A.F. Mayadas and L.T. Romankiw; "Preparation of Aging-Resistant Magnetic Permalloy Films," *IBM Technical Disclosure Bulletin,* **12,** (3) (August, 1969)
8. Z. Kovac; "The Effect of Superimposed AC and DC in Electrodeposition of Ni-Fe Alloys," *Journal Electrochemical Society : Electrochemical Science,* (January 1971)
9. D.S. Lashmore and J. Weinroth-Aronson; "Pulsed Electrodeposition of Nickel-Phosphorus Metallic Glass Alloys," *AES Second International Pulse Plating Symposium,* Rosemont, IL (October 6-7, 1981)
10. A. Brenner; *Electrodeposition of Alloys;* Vol. II, Academic Press (1963) 465
11. T.A. Eckler, B.A. Manty and P.L. McDaniel; "Pulse Plating of Chromium-Molybdenum Coatings," *Plating and Surface Finishing,* **67,** (9) (1980)
12. M.H. Gelchinski, L. Gal-Or and J. Yahalom; "Pulse Plating of Chromium-Cobalt Alloys Containing a Phase with the A-15 Structure," *Journal Electrochemical Society : Electrochemical Science and Technology* (November 1982)

13. Henry Leidheiser Jr. and A.R.P. Ghuman; "Pulse Electroplating of Silver-Tin Alloys and the Formation of Ag_3Sn," *Journal Electrochemical Society:Electrochemical Science and Technology* (April 1973)
14. V.S. Rachinas; "Cobalt-Tungsten Magnetic Alloys," *Proceedings of the 10th Lithuanian Conference of Electrochemists* (December 1968)
15. C. Barnes; "Electroplating of Cobalt-Tungsten Alloys," *Proceedings of the Institute of Metal Finishing, Technical Conference,* Bournemouth (GB) (1985)
16. A. Knödler, Ch. J. Raub and E. Raub; "Die Galvanische Abscheidung der Nickel-Zink—, Nickel-Zinn-und Blei-Zinn-Legierungen mit Pulse-Plating-Strom," *Metalloberfläche,* **39,** (1) : 21-25 (1985)
17. D.S. Lashmore, I. Weisshaus and K. Pratt; *Plating and Surface Finishing,* **73,** (3) : 48-55 (1986)

Chapter 12

Pulsed Anodic Reactions

H. Konno

1. Introduction to anodic reactions

The anode and the cathode can be simply distinguished by electrons always leaving the anode and flowing to the cathode, that is, an oxidation reaction takes place at the anode and reduction occurs at the cathode. Therefore, when a metal is anodically polarized in an aqueous solution, the following reactions may be proceeding: (a) dissolution of metal, either as aquions or as complex ions combined with electrolyte anions, (b) formation of surface film as oxide, hydroxides and/or other compounds of the metal, (c) oxidation of electrolyte anions and/or (d) anodic gas evolution. Which of the processes predominate depends on the metal and solution characteristics. Electropolishing, electrochemical etching, and electrochemical machining utilize the anodic dissolution of metals, and anodizing is making use of surface oxide formation by anodic polarization.

Electropolishing is generally performed in a concentrated acid solution or a strong alkaline solution with a DC at a high anode potential, where the rate of metal dissolution is controlled by diffusion of dissolved metal ions from the surface to the bulk solution. Electropolishing is widely used in practical applications, but a consistent theory of the complex phenomena at the metal/solution interface has not been established.

In contrast with polishing, the purpose of etching is to roughen the metal surface or to dissolve metal locally to form patterns on the surface. Usually etching is carried out in a neutral or slightly acidic solution containing NaCl, NH_4Cl, $CuCl_2$, etc. In the electrical capacitor production process, however, aluminum is electrochemically etched in a strong acid solution such as HCl and H_2SO_4 using DC, AC, or a superimposed AC-DC. With this process, the effective surface area of aluminum foil can be increased to 80-90 times the apparent surface area. Because of the practical importance and theoretical interests, detailed studies are now proceeding (1-4).

Anodizing, or anodic formation of oxide film, and electrochemical machining will be described in this chapter. Of these, anodizing is the most used anodic process, particularly for aluminum and with AC or PC. Application of pulsed electrolysis to electropolishing and etching is not very common and it will not be discussed further in this chapter.

Most metals form oxide or oxyhydroxide film on the surface when they are anodically polarized in aqueous solutions. When the formed surface film is electron conductive, oxygen gas evolution or oxidation of electrolyte anions, for example, chlorine gas evolution, takes place at high anode potentials by the transport of electrons through the film. The passive oxide films formed on the transition metals are examples of this type. On the other hand, if the formed film has no electronic conductivity, oxide ions and/or hydroxyl ions migrate through the film to the metal/oxide interface, then new oxide layers are formed at both interfaces depending on the transport number of the ions. This type of process is common on the so-called valve metal. Anodizing is performed to form oxide films on the metal surface and gives the metals corrosion resistance, electrical insulation, decorative features, and so on.

There are two types of anodic oxide film, one is the compact barrier type and the other is the porous type. It can be said that the latter type is characteristic to aluminum. On the transition metals such as chromium, iron, molybdenum, nickel, or their alloys, it is practically impossible to form thicker oxide films than a few nanometers, when they are anodized; heating alone, however, makes this possible. On such metals, raising the anode potential, higher than about 0.8V (SHE), causes a breakdown of the oxide film, transpassive dissolution (oxidation to soluble state, e.g., Cr(III)—Cr(IV)), and/or oxygen evolution, even in a nonaggressive neutral solution. On valve metals such as aluminum, niobium, silicon, tantalum, titanium, and zirconium, the formed oxides have poor electronic conductivity and barrier type oxide films having a thickness of several hundred nanometers are readily formed. During anodizing, the relation between anode potential, E_a, and the current passing through the barrier oxide film, I_a, is expressed as

$$I_a = P \exp(Q\, E_a/\epsilon) \qquad [1]$$

where P and Q are constants and ϵ is the thickness of the film. When a metal is anodized in a solution that does not dissolve the formed oxide and a constant current is applied, ϵ becomes larger at a constant rate proportional to I_a and E_a rises linearly with ϵ. On E_a reaching a critical value, arc discharge starts and the film growth ceases, the critical potential is called the dielectric breakdown potential. This potential is dependent on the metal, its purity, and the electrolyte used, but not on the conditions of electrolysis such as current density and temperature. When anodizing is performed at a constant E_a, I_a decreases exponentially with time and finally it attains to a very small value called the leakage current. In this case, the film thickness does not change after an initial short period, but the quality of the film is improved by prolonged anodizing. This is because the imperfections or flaws in the film are repaired during the current decay period.

It should be noted that most of the barrier type oxide films are amorphous in nature and contain small amounts of electrolyte anions. The anodic electric field that can be sustained by the film is around 10^9 V/m and not very different for different metals. This suggests that the electric conductivity of the formed oxide, either ionic or electronic, is determining the growth of the film under an applied anodic electric field.

2. Anodizing of aluminum

The compact barrier type oxide film on aluminum is formed with a DC in a neutral solution with a borate, phosphate, tartarate, or adipate. The main use of this type of film is in the manufacture of electric capacitors.

Porous oxide films are formed in sulfuric, oxalic, phosphoric, chromic, or other acid solutions. Mixed electrolytes are also used, and occasionally alkaline media are adopted. The thickness of the film is usually several tens of micrometers but it can reach more than a few hundreds of micrometers. This is much thicker than the barrier oxide film whose thickness is limited to about one micrometer by the dielectric breakdown potential. The porous film is hard, has abrasion resistance, good insulating ability, and good adhesion to the substrate. Further, it can be dyed and also colored during the anodizing process itself (see also the following sections). Corrosion resistance and weathering resistance are very much improved by the subsequent sealing process which involves treating films with boiling water or high temperature water vapor. The surface finishing of aluminum components with porous oxide films is utilized in a wide variety of fields. For instance, it is applied to structural components, packaging materials, printed circuit boards, electric appliances, wires, and off-set printing plates (PS plate). Recently, there have been attempts to use the porous structure for magnetic memory plates and solar energy collectors, by depositing various metals into pores.

A scheme of the porous oxide film growth is shown in Fig. 12.1. The films are composed of an amorphous oxide with a large number of parallel pores extending through the oxide, nearly to the oxide/metal interface, where there is a thin compact barrier oxide layer. One pore and the region surrounding it is termed a unit cell. During the anodizing, the applied voltage (to be precise, this should be the potential, but the value is so large that it corresponds to the anode potential) is sustained by the barrier layer, and formation and dissolution of the barrier layer occur simultaneously, keeping its thickness constant. With this process pore depth increases at the rate of the retreating barrier layer (Fig. 12.1). The growth rate of the film is proportional to the current density, j_a, and j_a increases with anode voltage V_a. According to recent studies (5), thickness of the barrier layer, ϵ, size of the unit cell 2R, pore diameter, r, and number of pores, N, are dependent only on V_a and not on temperature and concentration of the solution. Increasing V_a, the values of ϵ, R, and r increase so that N decreases. The porosity, which is defined as a ratio of the pore volume to the apparent volume of the film, decreases with increasing V_a, and normally is about 0.1 - 0.2. It should be noted that the porosity is not directly related to film hardness. Density of the oxide, ρ, is about 3-3.5 x 10^6 g/m^3 and higher values are attained with higher solution temperature and lower j_a.

The thickness of the porous layer that can be formed is theoretically unlimited, but due to the chemical dissolution of the pore wall (Fig. 12.1), it is limited. During anodizing, the surface region of the porous oxide is exposed to acid solution much longer than the subsequently formed bottom region, therefore the pore diameter of the surface region becomes larger than that of the bottom region. In the extreme case, thickness of the pore wall at the top of the oxide becomes zero, and film growth ceases. The pore wall in the top part is mechanically weak. Prolonged anodizing can be a cause of 'powdering' in the subsequent treatments. To suppress the chemical dissolution of the oxide, anodizing should be carried out with a high j_a in a solution with low temperature and concentration. These conditions, however, tend to bring about nonuniform distribution of current on the component, and consequently an uneven formation of the oxide film, leading to 'burning.' Generally, high j_a and low temperature (<10° C) result in hard oxide films, thus avoiding burning may conflict with the formation of hard films. The hardness of the film is normally 150-350 HV and it can reach more than 450 HV under certain anodizing conditions.

Pulsed anodizing of aluminum is applied to obtain hard and thick oxide films without powdering or burning, and also to color the oxide film. Further, it enables the formation of uniform oxide film on alloys with considerable amounts of copper, magnesium, and silicon, that are difficult to anodize with DC processes.

Anodizing in an oxalic acid solution is commonly performed with superimposed, AC-DC in Japan (6) while in other countries, DC anodizing is mainly adopted. In Germany, three types, DC, AC and superimposed AC-DC are used, and they are known as Eloxal GX, WX, and WGX respectively. The porous films formed in oxalic

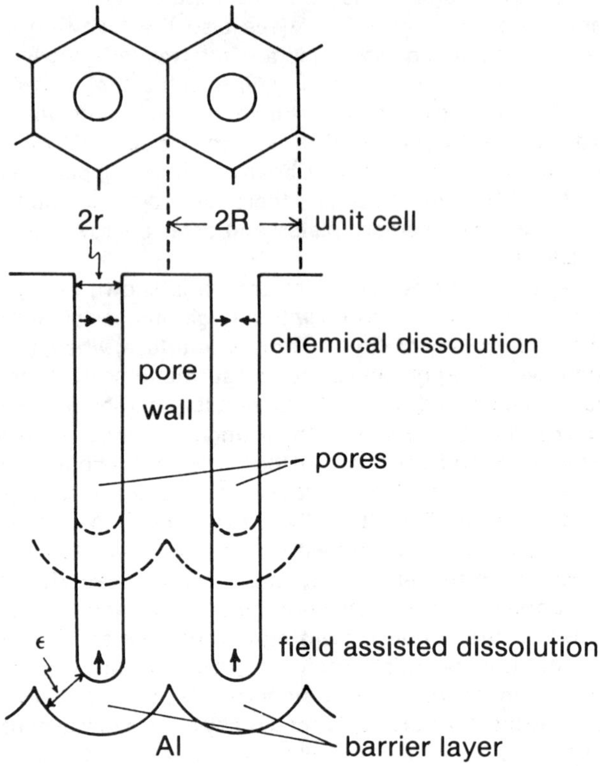

Fig. 12.1—Porous oxide film growth on aluminum.

acid solutions usually have yellowish color, and show better corrosion and abrasion resistance than films formed in sulfuric acid solutions, except for the AC anodized special films. The main disadvantages are the high cost of oxalic acid and difficulties in forming colorless oxide.

With DC the rate of formation is high, but aluminum easily corrodes (pits) in oxalic acid solutions. This is pronounced when the stirring of the solution is not sufficient or when the temperature is high. Anodizing with AC results in a slow formation rate and makes formation of hard and thick films difficult. The AC anodized film is somewhat coarse and flexible, so it is suitable for wires.

Compared with these two processes, the use of superimposed AC-DC is preferable, though the electric power supply system becomes more complicated. With superimposed AC-DC, corrosion of aluminum components is prevented and the rate of formation becomes higher than with DC anodizing. The principal advantage, however, is that the color tone of the formed film can be controlled by changing the superimposed ratio of AC to DC. Increasing the negative component of the current, yellowish and reddish colors become deeper. The mechanism of coloring by this process is not understood completely, but it is believed that fragments of organic compounds, radicals, and/or unpaired electrons formed by electrolysis are included in the oxide and so cause the coloring. When AC or superimposed AC-DC is used, the porous structure of the film is no longer well defined, as shown in Fig. 12.1. There are various sizes of pores and they are many-branched and entwined. In most cases, AC or superimposed AC-DC anodizing in a solution containing organic acids leads to coloring of the oxide film. This type of process is called 'integral color anodizing' or 'self-color anodizing.'

In sulfuric acid solutions, AC anodizing can produce golden colored oxide films. This is attributed to the reduction of sulfate to sulfur during the cathodic half cycle (7), which may be caused by a nascent hydrogen or by a high electrical field that is generated with the momentary alternation of the polarity.

Another process which utilizes superimposed AC-DC is the Hardas process that was developed to form hard oxide films. Solution flow and refrigeration are important in this process. The role of the superimposed AC in this process remains unexplained.

On the continuous anodizing with high current density or high voltage, the surface temperature of components rises locally and this often causes burning. A periodic drop of the voltage can prevent the accumulation of heat and assist it to diffuse evenly. This is one of the merits of pulsed anodizing, and several types of processes have been suggested.

One process, by Takahashi and Saitoh, proposes the following typical conditions: 15 wt% H_2SO_4 solution at 10° C, $V_a = 30$ V, $T_{on} = 50$ ms, $T_{off} = 200$ ms, $j_{m(a)} = 6$ A/dm^2 (8). The formed oxide is brownish without any organic compounds added. They have also designed a pulsed power supply system including wiring for a large scale bath. This system can be easily changed to a normal DC system and is used in some companies in Japan. Another process is patented in the U.S.A. (9), and it is also used by industry. Typically, the power supply system has a capacity of $I_m = 1000$ A and $I_p = 5000$ A. This process is not an on-off type, but pulses are superimposed on a DC. At least six pulses per second and a $j_{p(a)}$ larger than twice the $j_{m(a)}$ are recommended. Normally, sulfuric acid solutions are used, but other solutions such as mixtures of oxalic and sulfuric acids are also suitable. The main advantage of this process is that it can be applied to all aluminum alloys when a uniform oxide film can not be formed by DC processes. This process claims the formation of 25-300 μm thick oxide films in a short period on cast, wrought, or machined componenets. It provides good abrasion and corrosion resistance, and the film shows little cracking, unlike the usual hard films.

The use of a PR current is not very common in anodizing. One such anodizing process has been proposed by Okubo (10). Here a PR current of $\theta = 50$-100 ms with short cathodic pulses is utilized. For films formed in a 15 wt% H_2SO_4 solution at $j_{(a)} = 4$ A/dm^2, with $\theta = 58.5$ ms and duty cycle = 86.7 percent, for 34.8 min, the following hardness data are reported: alloy 1080 (Al $<$ 99.80%), 350 HV at 25° C, 500 HV at 10° C; alloy 5052 (Mg = 2.2-2.8%, Al $<$96.7%), 250 HV at 25° C, 460 HV at 10° C; alloy 2017 (Cu = 3.5-4.5%, Al $<$ 94%), 230 HV at 15° C, 300 HV at 10° C. From these data, one may understand why the refrigeration is important to obtain hard films, and also how difficult it is to form hard oxide films on alloys containing copper and magnesium.

The above processes permit the application of higher anodic current density and reduce the requirement for solution refrigeration to some extent, but refrigeration is still indispensable to obtain hard films. Cutting this cost is one of the important demands.

The current recovery effect, which was first observed by Murphy and Michelson (11), is shown in Fig. 12.2. When an anode voltage of V_{a1} is applied to aluminum in an acid solution, a large anodic current flows in an initially very short period (Fig. 12.2a) and through a minimum value the current reaches a steady value j_{a1} (Fig. 12.2b). When the voltage is decreased abruptly from V_{a1} to V_{a2}, the current drops to a very small value

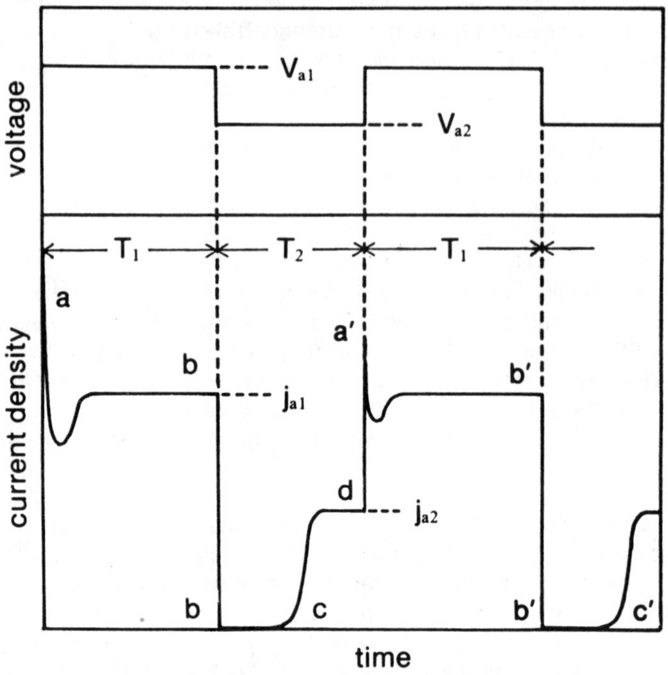

Fig. 12.2—The current recovery effect.

and then gradually increases (c→d) to a steady value, j_{a2}, corresponding to V_{a2}. What is occurring during the recovery period has been observed by electron microscopy and has been explained as follows by Takahashi, Nagayama, Akahori, and Kitahara (12). During anodizing with V_{a1} the oxide film has a large cell size and thick barrier layer corresponding to V_{a1} (a→b); by the sudden drop from V_{a1} to V_{a2}, the electrical field in the barrier layer decreases and the growth of the film almost ceases; then the thickness of barrier layer starts to decrease gradually by chemical dissolution (b→c). In this period, the cell size does not change. When the electrical field has recovered by the thinning of the barrier layer, the current starts to flow again (c→d) and a new smaller cell structure is established corresponding to V_{a2} (Fig. 12.2). Repeated anodizing with V_{a1} and V_{a2} can form a multilayer oxide structure where small cell sized layers are sandwiched in between large cell sized thick layers, if T_2 for V_{a2} is taken long enough (Fig. 12.2).

According to the above basic concepts, a new anodizing process was developed by Yokoyama (13). This process differs from ordinary PC anodizing, which usually adopts a θ shorter than one second: the alternation of the voltage is performed with an interval of several seconds to a few minutes. The results obtained by this process depend on the selection of V_{a1} (j_{a1}), V_{a2} (j_{a2}), T_1, and T_2 for different solutions, temperature, kinds of materials, and the required properties of the oxide films.

TABLE 12.1

Thickness and Hardness of Oxide Films Formed by the New Method under Different Conditions

Solution: 10 g/l oxalic acid, 108 g/l H_2SO_4, Al^{3+} 3.5 g/l

Material	Temperature °C	Time Min	V_{a1}-V_{a2} V	T_1-T_2 sec	Thickness μm	Hardness (surface)
5052S	20	30	18-14	180-10	13.0	
1080S	"	"	"	"	10.0	Hv
5052S	"	60	"	"	26.3	689, 685*
1080S	"	"	"	"	19.4	
5052S	25	30	20-15	"	18.0	399
1080S	"	"	"	"	13.5	
5052S	"	60	"	"	35.0	458, 448†∮
1080S	"	"	"	"	27.0	
5052S	"	30	"	180-60	12.8	
1080S	"	"	"	"	10.0	
5052S	"	60	"	"	26.2	448, 448†
1080S	"	"	"	"	21.0	429†

* Load 300g, † Load 500g
∮ Hardness measured at a section is HV/800 with 500g load.
Thickness was measured by using an optical microscope.

Optimum conditions have to be determined by trial and error, but a number of rules have been determined experimentally (14). In Table 12.1, some properties of the oxide films formed by this process are indicated. Hard and thick oxide films with unevenness of less than 5 percent formed without refrigeration. The suitable operating temperature for hard and flexible oxide film is around 30° C in a mixed solution of sulfuric and oxalic acids, and the maximum obtainable thickness is still as high as 100 μm. With refrigeration this can be increased to a few hundreds of micrometers. The oxide film formed on alloy 5052S with this process showed the electric breakdown voltage of 880 V with 70 μm thick and 1200 V with 100 μm thick films. Another feature of the process is that both colored and colorless oxide films are easily formed by changing the combinations of V_a and T.

The reason why the process enables the formation of thick and dense oxide films at room temperature without burning and powdering can be explained as follows. When anodizing is performed at high V_{a1}, j_{a1} is high and film growth is rapid, leading to a hard oxide film with no burning. Continuous anodizing at V_{a1}, however, results in burning as mentioned before. If the voltage is decreased to V_{a2} before the occurrence of burning and kept there for enough time to disperse the accumulated heat, the temperature of the components becomes even throughout. During this recovery period the barrier layer thickness in each pore becomes even and, on the subsequent anodizing at V_{a1}, the concentration of the current to some pores is prevented.

The oxide films formed with this process have good physical and chemical properties as well as economic advantages, and some 60 companies are using the process in Japan. In Taiwan and South Korea, 30-50 m^3 anodizing baths for architectural components are operated with this process. So far industrial applications are limited, with a few exceptions, to ordinary solutions such as a sulfuric acid solution, an oxalic acid solution, a sulfosalicylic acid solution, or their mixtures. By applying this method to solutions of phosphoric, chromic, or organic acids, it will be possible to increase its application.

The color of oxide films produced by self-color anodizing processes resists fading in light and weathers better than dyed films, but it is very difficult to control coloring at will and sometimes the process is not economical. One of the most widely used coloring techniques, particularly for architectural materials, is electrolytic coloring. For this process, aluminum is first anodized conventionally in acid solutions, usually sulfuric acid solutions, then it electrolyzed again in a solution containing metal salts. With this process, metal or metal oxide particles are deposited in the bottom part of the pores and they develop color by diffusing light. The process is analogous to electroplating, but AC or superimposed AC-DC is most successfully applied, though the use of PC, PR, or simple DC is also possible.

The first electrolytic coloring of porous oxide film on aluminum is referred to by Caboni in Italy in 1936 (15). A commercially useful process, however, was developed by Asada in Japan and only patented in 1963 (16). The process consists of AC electrolysis of porous oxide film in a metal salt solution with a basis of a mineral acid, an organic acid, or its salt with ammonium, amine, or imine. Different metal salts develop different colors; for example light bronze to dark color is developed with nickel, cobalt, or tin salts; a maroon with copper salts; a purple tinge with tin salts; yellow green with silver salts; purple red with gold salts; pale gold with selenate. The world rights for the Asada process outside Japan were transferred to Alcan in 1966 and now the process is known as Anolok. The industrial success of the process set off the development of numerous similar processes, and space limitations do not permit including these here. Two reviews, though they are both somewhat dated, may be helpful to get an outline of this art fundamentally (17) and practically (18).

Relatively few works report on the mechanism of AC electrolytic coloring, in comparison with the number of patents and the reports in connection with these. As mentioned above, the reaction is similar to electroplating, but the details are not so simple. Deposition starts from the bottom of pores and usually a small amount of deposits suffices to develop colors. Deposits are metals when the metal species are relatively noble as copper, tin, and silver, but with nickel or cobalt the deposits are both metal and oxide. It is possible to fill pores with metal, for example with copper or nickel, under certain conditions, but that is not the concern here. If the barrier layer thickness of the porous oxide film is larger than about 40 μm, coloring with this process is not possible. If the voltage is increased, by way of experiment, hydrogen evolution occurs and the oxide film peels off from the substrate. Deposition of metal or oxide takes place during the cathodic half cycle, but it is accepted that the anodic half cycle plays an important role, too. Recently, it was reported that electrolytic coloring is possible with DC but it is still necessary to interrupt the current or apply voltage stepwise. There are several explanations about the function of the anodic half cycle, but so far it can only be said that for successful coloring, the barrier layer has to be activated or stimulated and this is achieved with the anodic half cycle or the discontinuities in the current.

3. Anodizing of metals other than aluminum

Anodizing of metals other than aluminum is not widely used. Anodizing of magnesium, its alloys, and tantalum is performed at an industrial scale, and several processes are known for beryllium, niobium, titanium, tungsten, zinc, and zirconium. For magnesium and zinc, AC or superimposed AC-DC is often applied, but for other metals DC is usually used. If AC or superimposed AC-DC is suitable for anodizing, much better anodizing conditions would theoretically be obtained with PC. Actually, there has been an attempt to anodize magnesium alloys by a similar method to the one described for aluminum under the current recovery effect discussion above. Extensive studies, however, have not been carried out in this field, probably due to the small demand. Following, AC anodizing of magnesium and zinc is briefly introduced and other DC processes are omitted.

Magnesium and its alloys are widely used for aircrafts, automobiles, electronic components, and various machinary. For the purpose of corrosion protection, conversion coating with chromate is sometimes applied, but anodizing provides much better protection. As acidic systems, the following processes are well known: Dow 17 process (solution containing $NH_4F{\cdot}HF$, $Na_2Cr_2O_7$, and H_3PO_4, at 70-80° C), and Cr 22 process (CrO_3, NH_4OH, HF, and H_3PO_4, at 80-90° C). As alkaline systems: HAE process (KOH, KF, Na_3PO_4, $Al(OH)_3$, and $KMnO_4$, at room temperature to 65° C), Dow 12 process (NaOH, oxalic acid, and ethyleneglycol, at 70-80° C), and an alkaline process by P.R. Cutter (phenol, NaOH, and Na_2SiO_3, at room temperature). Most of these recommend AC or superimposed AC-DC. Each process gives different thicknesses, colors, and mechanical and chemical properties of the film. For instance, when alkaline processes are applied, the surface must be neutralized before paint application or further surface finishing. Details are covered in MIL-M-3171A and MIL-M-45202A.

Anodizing of zinc can be applied not only to cast or wrought zinc but also to galvanized steel or thermally-sprayed zinc. Chromate and phosphate based neutral solutions or chromate and silicate based alkaline solutions are used. For both, relatively high terminal voltage, 90-200 V, and high temperature, 70 - 80° C, are required. Each type gives different color and protective ability, but, in many aspects, the surface properties are superior to those obtained by conversion coating with chromate.

4. Pulse electrochemical machining

In electrochemical machining (ECM), or electrolytic machining, a metallic material is set to a counter electrode as a tool with narrow clearance in an electrolyte solution, and it is dissolved anodically with a concentrated and controlled current in the order of 10^3-10^4 A/dm^2 according to set dimensions. The ECM appears similar to electropolishing, but the latter is a simple surface finishing technique. There are a few metal processing techniques other than ECM, utilizing electrolysis. One is electrochemical grinding (ECG), in which metal bonded diamond particles or other compounds are used as a grinder. Metallic materials are ground both mechanically and electrochemically at the same time. In this case, material and the counter electrode (grinder) are constantly in contact, and this distinguishes ECG from ECM. Another is electrochemical etching. The surface of the material is masked with nonconductive and noncorrosive substances, and only the part desired to be dissolved is exposed to the solution. The widely used electric discharge machining (EDM) does not make use of electrochemical reactions. For EDM, a material and a counter electrode are immersed in a nonconductive solution, usually in oil, and the high temperature generated with spark discharges between them removes the metal.

Results with ECM depend on the electrolyte, but industrial materials are so diverse that there is no universal electrolyte solution. It must be selected depending on the purpose and material, and neutral solutions are most commonly used. An important consideration is to avoid metal deposition on the tool during ECM. This can be achieved by using alkaline metal salts as electrolyte and by removing reaction products at high speed. From the beginning of the development of ECM, a NaCl solution has been used extensively. In this solution, most metals dissolve with 100 percent current efficiency over a wide range of j_a and the throwing power of the anodic dissolution is also good. This leads to poor machining precision, as the material far from the tool dissolves together with material near the tool. Recently, $NaNO_3$ solutions are used in ECM, the dissolution efficiency is around 80 percent at $j_a > 5000$ A/dm^2. This leads to good precision because only the parts close to the tool dissolve. It is explained that at low j_a dissolution is prevented by passivation, but transpassive dissolution takes place at high j_a in this solution. Acid solutions are not generally used these days, but the precipitation of the dissolved species does not occur in such solutions, so they are sometimes adopted. For example, a HF, HCl, or HNO_3 based solution is suitable for materials containing titanium, which are difficult to process with ECM in neutral solutions due to the formation of adhesive oxide films on the surface. Alkaline solutions are rarely used, but they are useful for materials containing large amounts of tungsten and molybdenum. These metals are oxidized to tungstate and molybdate and dissolve into alkaline solutions.

The advantages of ECM can be summarized as (a) little stress or strain remains after machining; (b) the hardness of the surface is practically unchanged; (c) there is no burr; (d) a constant and relatively high machining rate is possible independently of hardness and tenacity of the material, and the rate is changeable, controlled by the current density; (e) complicated shapes can be machined in one step; (f) smooth finishing of the surface is obtained; and (g) wear and tear of tools does not occur. On the other hand, the following disadvantages must be taken into consideration: (a) precision of machining is not very good, especially for complicated and delicate shapes; (b) manufacturing of tools is not easy; (c) washing of materials is necessary to remove the electrolyte solution completely; and (d) solutions cannot be disposed simply.

As normal working conditions of ECM, the clearance between material and tool is 0.02-0.7 mm and j_a is 3-20 kA/dm^2. Larger j_2 results in a high machining rate and better precision and finish. Because of these conditions, huge amounts of heat and reaction products such as hydrogen gas and metal ions are accumulated in the gap between material and tool. Unless the heat and the products are removed quickly, ECM does not continue. Therefore, rapid circulation of the solution is indispensable. Usually pressurized solution is passed through the gap at a rate of 6-60 m/s, and the flow must have a constant velocity and uniform distribution in the gap. This also makes the manufacture of tools difficult. To overcome these difficulties, use of PC has been attempted. This is called pulse electrochemical machining (PECM).

The development of PECM started in the 1970s and it is still progressing. There are a number of fundamental and practical reports, in which contributions by Russian and Chinese workers are considerable. It is reported that several types of PECM instruments are developed and used in the U.S.S.R., though details are not sure. The advantages of PECM are: (a) improvement of precision; (b) reduction of surface microroughness; (c) more superficial intergranular attack; (d) larger tolerance of operating conditions; (e) capability of using narrower gap than ECM, and others.

Fundamental aspects of PECM is outlined well in the recent work by Datta and Landolt (20) using simple theoretical treatments for square current pulses. As described above, the main concern during rapid dissolution with a narrow gap is heating of the solution and transport of reaction products. There are two aspects of heating in PECM: one is instantaneous heating by a single high j_p pulse and the other is steady heating. When the solution is flowing at velocity v (i.e.), along the electrode having a length, L, the instantaneous temperature rise, Δt_p, due to ohmic heating is proportional to $j_p^2 T_{on}$, provided that T_{on} is much shorter than the average residence time of the solution in the gap, L/v. If T_{off} is long enough to remove all of the heated solution from the gap, temperature rises do not occur; otherwise there will be a steady temperature rise, Δt_s, in proportion to $j_m^2(L/v) = j_p^2(T_{on}/\theta)^2$ (L/v). Under given hydrodynamic conditions, the ratio of $\Delta t_s/\Delta t_p$ is given by $(T_{on}L/v)/\theta^2$, and this is independent of j_p. Both Δt_p and Δt_s, and also $\Delta t_s/\Delta t_p$ can be minimized by having a short T_{on} and long θ, but this reduces j_m and is not favorable for the machining rate. Usually, steady heating of the solution adversely affects dimensional control of the machining process. The optimal pulse duration is determined by considering balance between the machining rate and the tolerable steady heating.

For describing mass transport in PECM, a duplex diffusion layer model that was originally proposed for pulse plating by Ibl (21) is convenient to use. Details of the original model are covered in another chapter, where only the direction of transport and current is opposite to that used here. Briefly, it is assumed that the diffusion layer consists of two parts, an inner pulsating layer of thickness δ_p, where the dissolved metal ion concentration is a periodic function of time, and an outer diffusion layer of thickness δ_s, where the concentration of dissolved metal ions is independent of time. The total thickness of the diffusion layer, δ is determined only by the hydrodynamic condition of the system. The expression for δ_p at the end of pulse is derived as

$$\delta_p = [2 \, D \, T_{on} \, (1 - T_{on}/\theta)]^{1/2} \qquad [2]$$

For a short T_{on} the value of δ_p becomes much smaller than δ; therefore, a much higher j_p can be applied in PECM than ordinary ECM without reaching j_{gg}. The value of j_{pg} in PECM is given by the following relationship:

$$j_{pg} = j_{gg} \frac{\theta_{gg}}{\theta_{pg}} \left[\frac{\delta_p}{\delta} \left(1 - \frac{T_{on}}{\theta}\right) + \frac{T_{on}}{\theta} \right]^{-1} \qquad [3]$$

where θ_{gg} and θ_{pg} are current efficiencies of dissolution at j_{gg} and j_{pg}, respectively. These expressions, derived from simplified models, allow the selection of experimental parameters to achieve a high rate of mass transport with minimum heating of the solution.

Datta and Landolt (20) have reported that PECM permits the use of a relatively low flow rate, 2.5 m/s. The current efficiency for nickel dissolution in a 6 M $NaNO_3$ solution is higher in PECM than in ordinary ECM at a given j_m. In principle this means that, when metals are passive, a high j_p pulse enables the transpassive dissolution of metals with high efficiency without applying a high j_m and no requirement for a high solution flow rate.

As a variation of PECM, there is wirecut electrochemical machining (WECM). A prototype looks like a scroll saw (22). In this case, the electrolyte solution is flowed along a wire, so that immersion of the material is unnecessary. As a result, the amount of dissolved metal is much smaller than with PECM or ECM and a cut-out part can be used for other purposes. A similar type utilizing electric discharges (WEDM) is already in practical use. Computer controlled WEDM is expanding rapidly. In WEDM, however, wire must be continuously fed due to the consumption by spark discharges. Consequently, only two dimensional cutting is practical. In WECM, three dimensional cutting is possible by using a curved wire or pipe (23). Advantages of WECM compared with WEDM are the absence of work hardening, higher machining rate, and low working cost. A weak point is that machining of fine shapes is almost impossible. WECM is still a developing technique, but it cannot replace WEDM.

References

1. C.G. Dunn, R.B. Bolon, A.S. Alwan, and A.W. Stirling; *Journal Electrochemical Society,* **118** : 381 (1971)
2. N.F. Jackson; *Electrocomponent Science and Technology,* **2** : 33 (1975)
3. H. Fickelscher; *Werkstoffe Korrosion* **33** : 146 (1982)
4. R.S. Alwitt, H. Uchi, T.R. Beck, and R.C. Alkire; *Journal Electrochemical Society,* **131** : 13 (1984)
5. K. Ebihara, H. Takahashi, and M. Nagayama; *Journal Metal Finishing Society of Japan,* **33** : 156 (1982); **34** : 553 (1983); **35** : 205 (1984)
6. JIS H 9510
7. J.M. Kape; *Metal Finishing, (4) : 80 (1974)*
8. T. Takahashi, and J. Saitoh; *Plating and Surface Finishing,* (7) : 36 (1977)
9. US Patent 3,857,766
10. Japanese Official patent Gazzette, 53-131932
11. J.F. Murphy, and C.E. Michelson; *Conference on Anodizing of Al (ADA),* Nottingham, (Sept., 1961)
12. H. Takahashi, M. Nagayama, H. Akahori, and A. Kitahara; *Journal Electron Microscopy,* **22** : 149 (1973)
13. Japanese Patent 735,119
14. K. Yokoyama, H. Konno, H. Takahashi, and M. Nagayama; *Plating and Surface Finishing,* (7): 62 (1982)
15. Italian Patent 339,232
16. Japanese patent 310,401; British Patent 1,022,927; US Patent 3,382,160
17. N. Baba, and T. Sato; *Journal Metal Finishing Society of Japan,* **27** : 491 (1976)
18. E. Hermann; *Galvanotechnik,* **63** : 110 (1972)
19. *Handbook of Aluminum Surface Treatment,* Light Metal Pub. Co. Ltd., Tokyo, p. 171 (1980)
20. M. Datta, and D. Landolt; *Electrochimica Acta,* **26** : 899 (1981); **27** : 385 (1982)
21. N. Ibl; *Surface Technology,* **10** : 81 (1980)
22. R. Maeda, K. Chikamori, and H. Yamamoto; *Journal Japan Society Precision Engineering,* **49** : 709 (1983)
23. K. Chikamori, H. Yamamoto, and R. Maeda, *Journal Mechanical Engineering Laboratory of Japan,* **35** : 318 (1981)

Chapter 13†

Pulse Plating Equipment

N. Osero

1. Characterization of different modulated currents

Many configurations of pulse plating power supplies have been designed and manufactured over the past two decades. The development of high power semiconductor devices has led to the reduction in size and cost of these pulse power supplies.

The term "pulse plating power supply" has been indiscriminately applied to a wide variety of electronic power supplies designed to provide single, multiple or continuous pulses of interrupted current. The two most common types are the square-wave and the modified sine-wave pulse power supplies. To discuss the difference in these two methods, it is important to understand the waveforms involved.

Figure 13.1 shows the relative pulse waveshape of various types of pulse power supplies. The most commonly applied waveshapes include the sine-wave, the half-wave phase delayed, the full-wave phase delayed, the skip-sine, the unipolar pulses, the bipolar pulses (pulse reverse) with or without off-time between the cathodic and anodic, or anodic and cathodic phases, the pulsed pulse reverse, the pulse-on-pulse and duplex pulses.

Various methods of defining a current or voltage pulse have been used in the plating industry. A pulse requires time for the excursion from one amplitude to another amplitude. The exact beginning and ending of this excursion is usually difficult to locate on an oscillogram of the waveform because inductance and capacitance of the power supply and the external load affect the waveshape. Therefore, the pulse rise time (t_r) and fall time (t_f) are specified at the 10 percent and 90 percent points of the pulse amplitude. Pulse duration (T_{on}) is specified at the 50 percent point of the pulse amplitude. This is illustrated in Fig. 13.2.

Square-wave power supplies are the most commonly used type for precious metal applications. Their fast rise time and fall time allow them to be used at frequencies exceeding 10 kHz. Precious metal is often plated at pulse frequencies in the range of 500 Hz to 5 kHz. The pulse frequency is determined by an internal frequency generator which is independent of, and usually much higher than, the power line frequency.

† Editors' note:

The purpose of this chapter is to enhance some relevant points concerned with pulse plating equipment and to discuss some of the typical available units. It was not possible to ask all suppliers to make a contribution to this discussion, but it should to be clear to the reader that pulse plating equipment is available from numerous manufacturers. Opinions may diverge on different points, as for example, on cost comparisons. Equipment related topics are also discussed in Ref. 5 to 7.

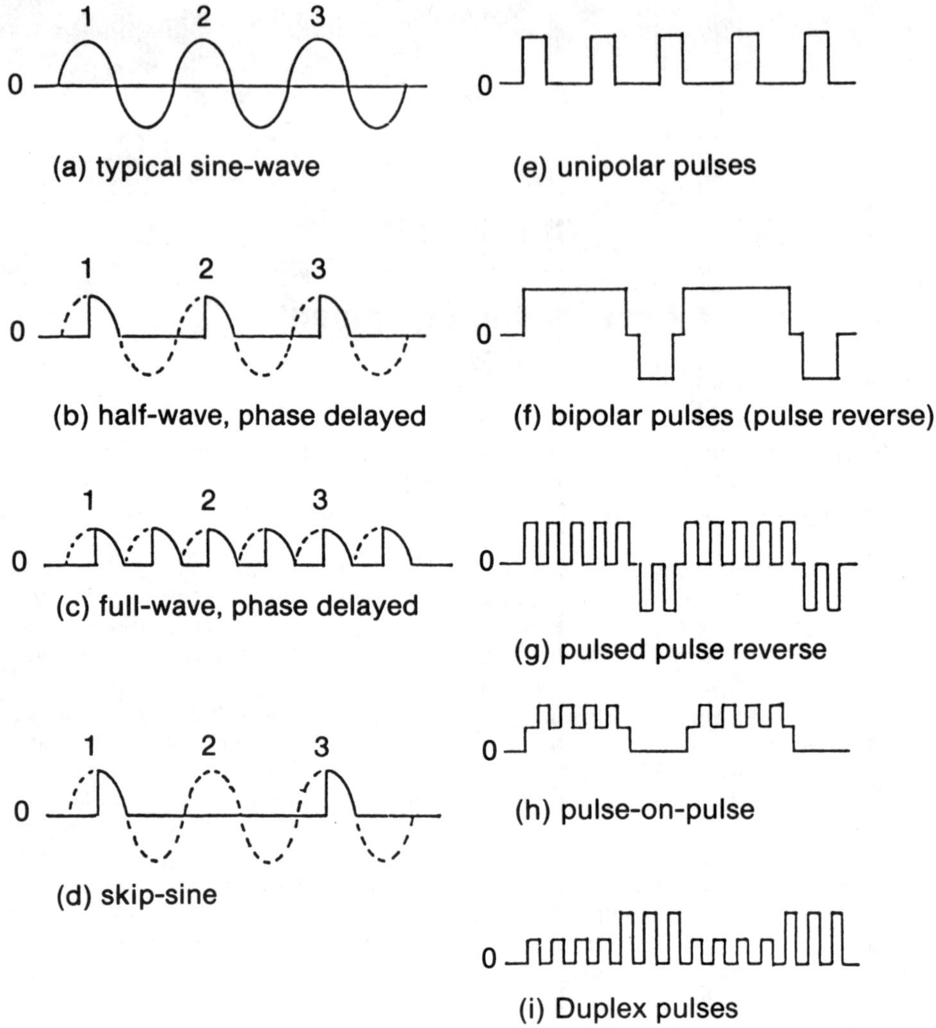

Fig. 13.1—Examples of pulse waveforms.

2. Characterization of different pulse plating equipment

2.1. *Integrated pulse system*

Square-wave power systems can have the DC portion of the power supply as an integral part of the system. They have the advantage of being able to control the amplitude and ripple of the pulse output current. In addition, the DC portion of the system is designed to operate with the pulse circuit, and the user does not have to be concerned with compatibility between the two sections.

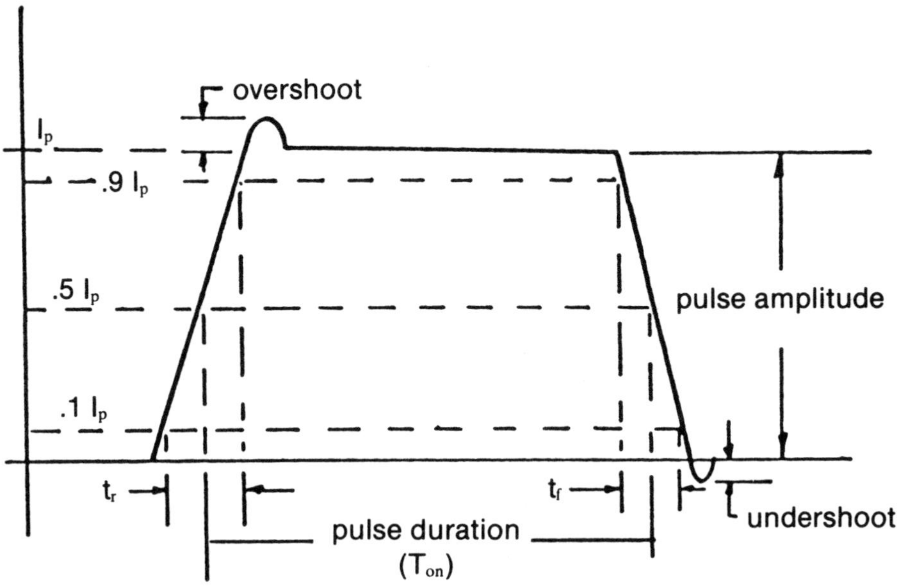

Fig. 13.2—Standard pulse waveform (1).

2.2. *Converter systems*

Square wave add-on-switch or converter systems can be attached to an external DC rectifier.

The linear regulator offers the advantage of being able to regulate the current and minimize the ripple currents present in the output of the DC rectifier. The converter can be either a linear regulator or saturated switch. The saturated switch passes the ripple of the DC rectifier output along to the plating bath.

The DC rectifier must have sufficient output voltage to drive the higher peak amperages used in pulse plating. In addition, to avoid damage to the converter, it is important to stay within the peak current rating of the converter.

2.3. *Modified sine-wave*

The modified sine-wave pulse power supply usually uses the power line as its time base. The output pulse is developed by turning on a semiconductor switch at an adjustable time period after the start of a sine-wave excursion. Thus, a relatively fast pulse rise time is developed, while the turn-off or fall time follows the slow decay of the sine-wave. (See Figure 13.1a, b).

Duty cycles for sine-wave systems typically range from 25-75 percent. Modified sine-wave power supplies can be set to produce an output which skips some of the sine-wave excursions. This is done to obtain duty cycles of less than 25 percent. (See Figure 13.1d).

2.4. *Unipolar vs. bipolar*

The pulse power supplies discussed previously are "unipolar", that is, their output current is unidirectional.

Power supplies have been designed to provide output pulses which alternately change direction. They are refered to as "bipolar" or "pulse reversing" power supplies.

Pulse reversing power supplies are used in applications where it has been determined that reversing the current flow for a short time period results in improvements in the metal deposit, e.g., metal distribution, plating speed and metallurgical properties (2).

Power supplies are available which produce complex waveforms, as shown in Fig. 13.1h and i. These systems tend to be directed towards research applications.

2.5. *Multiple output pulse systems*

The electronics industry continues to reduce the size and increase the complexity of integrated circuits. This technology has resulted in the need to electroplate very small circuit traces with very close tolerances for spacing, line widths, through-hole plating and part-to-part thickness control. Pulse platers, both unipolar and bipolar, have been developed to meet the ever increasing demands on the electroplating processes in the microelectronics industry (3). The most common configuration of this system consists in a common anode and separate cathode current controls. Units with common cathode systems with anode current control are also available.

2.6. *High frequency vs. low frequency*

For purposes of this discussion, high frequency is defined as that greater than 5 kHz and low frequency as that less than 500 Hz. The medium frequencies fall between 500 Hz and 5 kHz.

Many data have been published relative to the use of pulse frequencies in the range of 500-5000 Hz. In terms of repetition periods, this range is 2.0 milliseconds to 0.2 milliseconds. Most square-wave pulse power supplies fall into the medium frequency classification with pulse on- and off-times capable of up to a 0.2 millisecond repetition period. Modified sine-wave pulse power supplies are in the low frequency classification because of their dependence upon the power line for frequency generation.

Pulse plating researchers in recent years have been requesting pulse power supplies capable of frequencies exceeding 5 kHz. This has led to the development of equipment with faster rise and fall times and frequencies in excess of 10 kHz. Their repetition periods are less than 100 microseconds. This places strict requirements on the design of cables between the power supply and the plating cell, as will be discussed later in this chapter.

3. Methods of regulation

Pulse power supplies are offered with various methods of regulation. The method used will depend upon the process requirement or the preference of the user.

3.1. *Current regulation*

The output current of the power supply is regulated, and the resistance or impedance of the load determines the output voltage up to the limit of the power supply maximum rating.

This method is often used for rack plating and continuous strip plating where precise control of the current is required.

3.2. *Voltage regulation*

The output voltage of the power supply is regulated, and the resistance or impedance of the load determines the output current up to the limit of the power supply maximum rating.

This method is usually used where the surface area of the load is constantly changing; for example, barrel plating. This allows the current to fluctuate with the changing surface area, thus attempting to achieve a more uniform current density.

3.3. *Constant current density*

This is a method of power supply control where the output current and voltage are sensed and both are allowed to change as the surface area of the plating load changes. This method is used to maintain the current density constant with an increase or decrease in the surface area.

3.4. *Current/voltage crossover*

This is a combination of current and/or voltage regulation. The power supply can be set to provide current regulation up to a preset voltage limit. Upon reaching the voltage limit, the power supply "crosses over" or changes to voltage regulation. This is useful in those applications where excessive voltage could be detrimental to the process.

The power supply can also be set to provide voltage regulation up to a preset current limit. Upon reaching the current limit, the power supply crosses over to current regulation to limit the maximum current from the power supply.

3.5. *Ramped-start*

Ramped-start is used in conjunction with previously discussed regulation methods where the work load requires a low current at the beginning of the plating process and higher current as the deposition process continues. This is especially useful in the plating of very thin films.

4. Power supply cooling

Electronic equipment develops heat as a normal part of operation. Excess heat is one of the leading causes of component failure. It is essential that pulse power supply manufacturers design to minimize power loss in the equipment and provide for adequate cooling to assure reliability and long operating life of the power supply.

The three common methods of cooling are convection cooling, forced-air cooling and liquid cooling. Each method has its advantages and disadvantages.

4.1. *Convection cooling*

Convection cooling uses the natural movement of air to carry heat away from the components. This method is very inefficiency, but has the advantage of not requiring expensive fans or liquid cooling devices. Also, very little contamination from the plating environment is passed through the equipment.

4.2. *Forced-air cooling*

Forced-air cooling or fan cooling is much more efficient than convection cooling. This permits reduction in heatsink size and overall equipment size.

However, it has the disadvantage of passing contaminated plating room air through the equipment, which increases the chance of components corrosion. Delicate electronic circuits should be enclosed to eliminate their contact with the cooling air.

4.3. *Liquid cooling*

Liquid cooling equipment is offered in several configurations:

Closed-loop cooling uses chilled liquid from a heat exchanger. The heated liquid from the electronic equipment is pumped to the heat exchanger, for example, a cooling tower or refrigeration unit, where the heat is removed from the liquid. The cooled liquid is then recirculated to the power supply.

Open-loop cooling uses the building water supply, which is connected to the liquid cooled heat sinks of the power supply. The supply water must be free of contamination that could plug the cooling lines and cause overheating of the power supply. The exhaust water is either piped to the drain or, if it is not contaminated, is sometimes used as rinse water in the plating process.

Liquid cooling is by far the most efficient method of cooling. It requires very small heat sinks and permits for compact equipment.

5. Important information for the user

5.1. *Anode/cathode conductors*

One of the important considerations in applying pulse plating to production applications is the connecting wires from the pulse plater to the plating bath. Mechanisms and methods familar to the electroplater for conducting conventional direct currents no longer apply. Pulse plating introduces a phenomenon called inductance. Inductance is that property of an electrical circuit that opposes a change of current. When a pulse of current is applied to a wire, this inductance causes the wire to look like an open circuit for a short period of time. The voltage at the pulse plater peaks to a high value trying to overcome this inductance and establish current flow. During this period of voltage overshoot, the current rises slowly to the peak value. Thus the inductive properties of conventional wires or bus bars result in a degradation of the pulse current rise time in the plating bath. This becomes more critical as the pulse frequency becomes higher, requiring faster rise times. See Figure 13.3.

When the pulse is turned off, the inductance again opposes a change in current and causes a reversal of voltage at the output of the power supply. This reverse voltage, if too large, can place excessive stress on semiconductor devices in the power supply.

The inductive effect can be minimized by adding capacitance between the anode and cathode bus wires. This capacitance can be obtained by maintaining tight proximity between the two wires. Twisting the anode and cathode wires together is a good method of increasing the capacitance. An even more effective method is to use several twisted pairs of wire. Figure 13.3 shows typical oscilloscope waveforms for open cables and twisted pair cables. Sheet or ribbon conductors may also be used.

To maintain the integrity of the pulse waveform to the plating bath, it is recommended that the pulse power supply be located as close to the plating bath as practical. This, too, will help to reduce the inductive effect of the bus system.

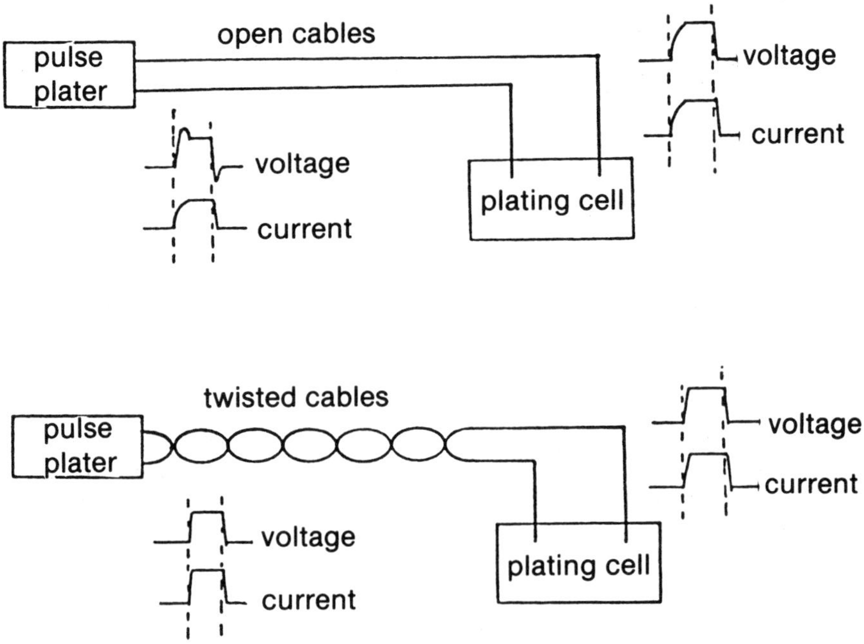

Fig. 13.3—Typical oscilloscope waveforms for open cables and twisted pair cables (4).

5.2. *Conductor ratings*

Another factor to consider is the amperage capacity or current rating of the anode and cathode wires. In order to maintain normal plating rates, high peak currents are used in pulse plating. These peak currents are integrated with on and off timing to arrive at the average current.

The anode and cathode wires must be properly sized to minimize voltage drops. The wire must also be sized properly to allow for the RMS heating effect of the peak currents. Pulse currents generate greater heat than the average current would indicate.

Example: 100 amps peak, 25% duty cycle
Average current 25 amps
RMS current $\sqrt{\text{peak x average}}$
Conductor rating 50 amps

5.3. *Computer control*

Technological developments in the electronics industry are being applied to many industrial processes for improvements in control and operation. The plating industry is no exception. The microprocessor has made profound inroads into all aspects of control of the plating process. Pulse plating power supplies are being offered that will accept commands from process control computers as well as output information to the computer for data logging or for closed-loop control of the power supply.

The computer system can be located away from the harsh environment of the plating area, which could decrease the reliability or life of the control system. Perhaps even greater advantages of computer control are the ease of making process changes for various products being plated, the control and repeatability of the process, and the elimination of operator control.

5.4. *Cost comparisons*

It is difficult to assign costs to each type of pulse power supply because of the immense variations available in the marketplace. In order to give the reader an indication of cost comparison, it can generally be said that sine-wave pulse power supplies are more expensive than DC power supplies. Square-wave power supplies are in the region of two to five times the price of conventional DC power supplies.

6. Summary

This chapter has endeavored to offer the reader some insights into the technical aspects of several types of pulse power supplies. The pros and cons of the various methods of pulsing have been discussed along with the often overlooked, but very important, subject of cables and bussing systems.

Commercial pulse power supplies have been available from several manufacturers since the early 1970s. In its infancy, pulse plating was often considered to be a fad which would soon disappear. Through the perseverance of researchers in chemical and plating labs, production personnel in industry, and equipment manufacturers, pulse plating has developed into a viable and very often required method for metal deposition.

Equipment manufacturers are continually striving to offer pulse power supplies which meet the ever-changing needs of the electroplating industry. Equipment costs will continue to decline as semiconductors are developed that can handle the higher currents of pulse plating. As chemical suppliers begin to offer plating solutions specifically developed for pulse plating, users will be able to apply pulse power supplies directly to their processes without having to engage in their own research programs. This process standardization will have a marked affect on equipment standardization and reduction of equipment costs.

Because pulse power supplies are more expensive than conventional DC power supplies, it is important for the potential user to understand that the pulse power supply is not a replacement for a DC rectifier. Rather, it is a piece of capital equipment that will be purchased to perform a specific function in the manufacturing process.

References

1. MIL-T-21038C, Sect. 6.9, U.S. Department of the Navy.
2. W.F. Hall and A.R. Chaudhuri, *AES 10th Symposium on Plating in the Electronics Industry* American Electroplaters' Society, (1983)
3. L. Missel, P. Duke and T. Montelbano, *Semiconductor International* (February, 1980)
4. N. Osero and A.L. Luckraft; *AES 2nd International Pulse Plating Symposium,* Rosemont, IL (1981)
5. *Proceedings of the AES 8th Symposium on Plating in the Electronics Industry,* (Jan. 20-21, 1981), Phoenix, Arizona
6. *Proceedings of the AES Second International Pulse Plating Symposium,* (Oct. 6-7, 1981), Rosemont, IL
7. *Proceedings of the 11th World Congress on Metal Finishing,* (Oct. 21-26, 1984), Jerusalem, Israel

Chapter 14

Pulsed Current Applied to Surface Area Measurement Techniques

J.Cl. Puippe

1. Introduction

A number of methods for determination of surface area of a body have been proposed, involving geometrical, optical or physicochemical measurements. The latter, for example, include the well-known "BET" method (1), based on measurement of physical or chemical adsorption of gases on the surface. Considering only methods useful in connection with galvanotechnical purposes, the following ones should be pointed out. There is a method (2) in which the surface to be measured is coated with a slow-drying lacquer, followed by immersion in a fluidized bed of uniform glass spheres (diameter 0.2 mm), the surface area being determined gravimetrically from the increase in weight due to adhesion of the spheres in comparison with that obtained for a surface of known area.

Another method utilizes chemical conversion or electroless deposition processes which, in the absence of an applied current, can be assumed to deposit a uniform thickness of coating over the whole of the surface (3). For example, in one process of this type the surface is first electroplated with zinc, followed by chromating of the zinc coating. The chromate layer is then dissolved in a standard volume of a suitable medium, and the chromium concentration is determined photometrically. The surface area is derived by comparison with the value obtained for a specimen of known surface area which is processed together with the test piece. Similar procedures, including electroless copper or nickel plating, may also be based on gravimetric determination.

Electrochemical methods are available based on the proportionality which exists between the cathodic current and the applied voltage (4), or on determination of the anodic current (5). However, methods which utilize information from unstationary current at electrode/electrolyte interfaces, using pulsed current technique, offer unique advantages, particularly in the electrodeposition field. In the following, two methods of this type are described based respectively on capacitance effects and on diffusion phenomena.

2. Surface area measurements using capacity effects

2.1. *Principle of measurement*

As discussed in Chapter 4, an electrolyte/elelectrode interface behaves like a capacitor and a resistance in parallel. Hence at the beginning as well as at the end of a current pulse, the faradic current is accompanied by a capacitive current for charging and discharging the electrical double layer respectively.

For very short pulses, the current is essentially capacitive during the pulse and the pulse charge is used for charging the electrical double layer.

$$Q = C \Delta U S \qquad [1]$$

where C is the specific capacity of the electrical double layer $\mu F/cm^2$, ΔU is the increase in potential difference between the metal and the solution [V], S the surface area of the electrode [cm^2], and Q the pulse charge [coulomb].

$$Q = I_p T_{on} \qquad [2]$$

$$S = I_p T_{on}/C\Delta U \qquad [3]$$

From equation [3] for constant T_{on}, C and ΔU values, S is directly proportional to the intensity of the current pulse.

2.2. *Method of measurement* (6, 7)

A commercial instrument for measurement of surface area on the basis of eqn. [3] is available†. The potential at electrode is measured against a current probe placed in the plating bath, which serves as a reference electrode. Current pulses of a constant duration, which are modulated in amplitude via a feed-back loop, are applied in order that a preset value of ΔU_{a-a} is reached, ΔU_{a-a} being the potential measured at the electrode immediately before and after the pulse. According to eqn. [3], the surface area is proportional to the amplitude of the pulsed current, I_{meas}, and can be calculated by calibration with a standard surface area. In fact, the following proportionality relationship is established experimentally:

$$I_{meas} = k S \qquad [4]$$

where I_{meas} is the measured current, S the surface area and k a proportionality factor. A typical current pulse and the associated voltage response as used in the system are shown schematically in Figure 14.1. The total charge of the pulse ($I_{meas} T_{on}$) should be kept sufficiently low to avoid significant contribution of faradic current during the pulse, which would decrease the sensitivity of the measurement. However, since the measurement is made by comparison with a standard area, significant amounts of faradic current during the pulse duration should not lead to measurement errors so long as no significant concentration polarization occurs. The voltage response in Figure 14.1b shows a sharp increase and decrease at the beginning and the end of the pulse respectively, corresponding to ohmic drops. During the pulse and after the pulse the voltage response follows the charging and discharging curve of the electrical double layer as described in section 2 of Chapter 4. The reference electrode consists of a probe of titanium, stainless steel or any other metal which is inert in the electrolyte. This kind of probe is much less prone to damage than conventional reference electrodes, such as "calomel" electrodes used extensively in laboratory conditions. Since a potential difference is measured, the absolute value of the

† MIM, Egatec SA, Route de Port 20, 2503 Bienne, Switzerland

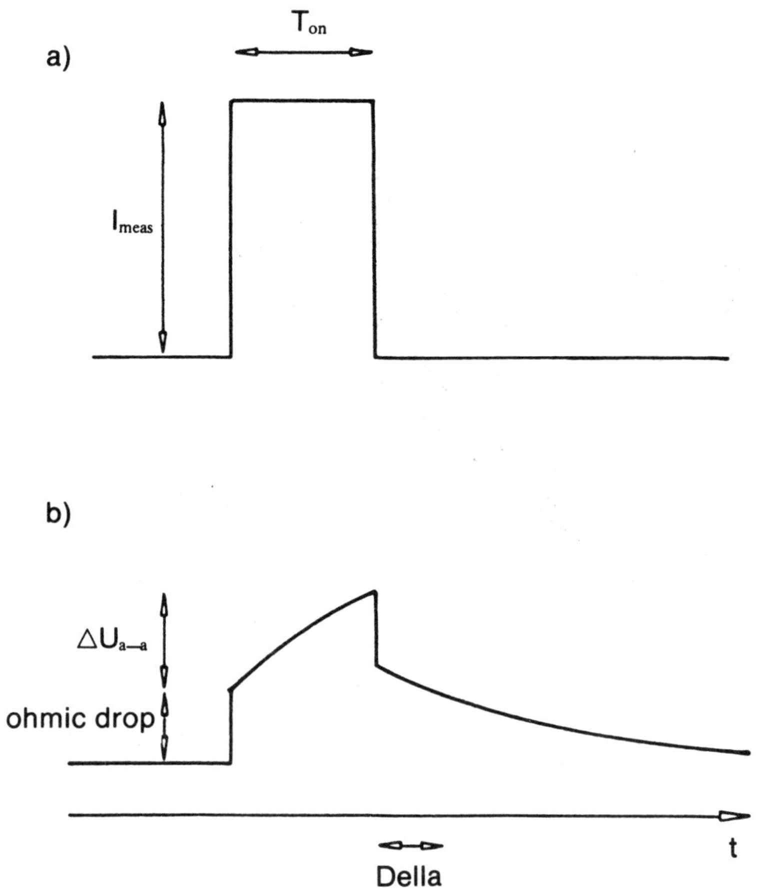

Fig. 14.1—Current pulse (a) and voltage response (b) as used in MIM system for surface area measurement based on capacity effects.

probe potential in the electrolyte is not relevant. To a first approximation, the location of the probe is not important since the voltage measurements are made when the cell is without current and the electrolyte is everywhere at the same potential. In reality, however, this may not be the case since, immediately following the pulse, potental differences may exist between different zones in the electrolyte due to variations in double layer charge between areas of the electrode as a result of nonuniform current density distribution during the pulse. In this situation circulation currents occur between different areas of the electrode, and the electrolyte cannot, therefore, be considered as an equipotential medium in which the location of the probe becomes an important parameter in the measurement system. This difficulty has been minimized by introducing a time, "Della", between the end of the pulse and the potential measurement, designed to allow the charges of the double layer capacity to reach a homogeneous distribution over the electrode area.

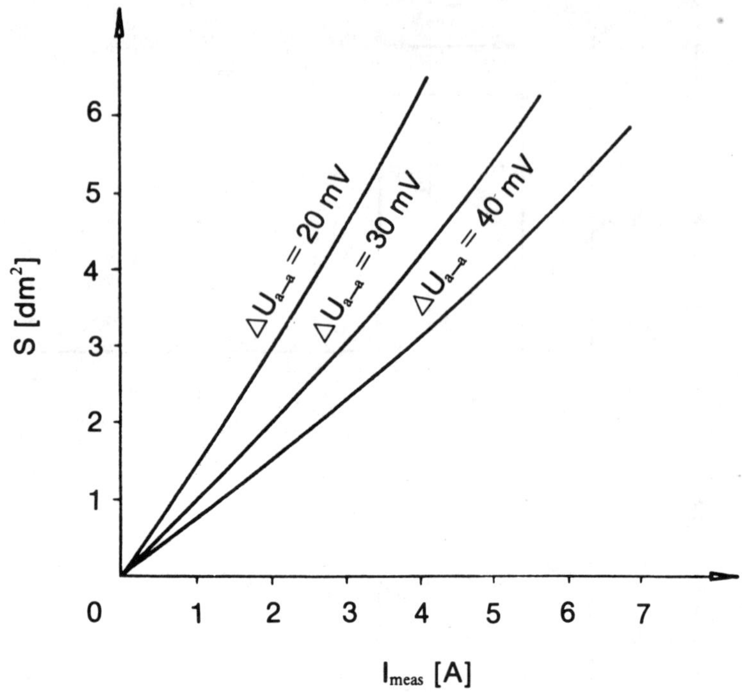

Fig. 14.2—Surface area values S as a function of measured current amplitudes I_{meas} for different preset values of potential differences ΔU_{a-a}.

Typical measurement data are shown in Figure 14.2 where the surface area is represented as a function of measured current amplitudes for different preset values of potential differences ΔU_{a-a}.

The system is not only an area measurement device, but also a pulse generator with which the area measurement is performed during the off-time of the deposition pulse, as seen in Figure 14.3. MIM can be used for electrodeposition and for electrodissolution as well, and operates directly in the corresponding bath. The amplitude of the current pulse for electrodeposition I_{act} is kept proportional to the current I_{meas}, i.e. to the surface area. Typical values assigned to the parameters shown in Figures 14.1 and 14.3 are as follows:

$T_{on\ (meas)}$	50 - 500 μs
Della	50 - 500 μs
$T_{on\ (act)}$	5 - 50 ms
T_{off}	2 - 20 ms
I_{meas}/I_{act}	1 - 1/4
I_{meas}	0 - 60 A
I_{act}	0 - 240 A
ΔU_{a-a}	10 - 50 mV

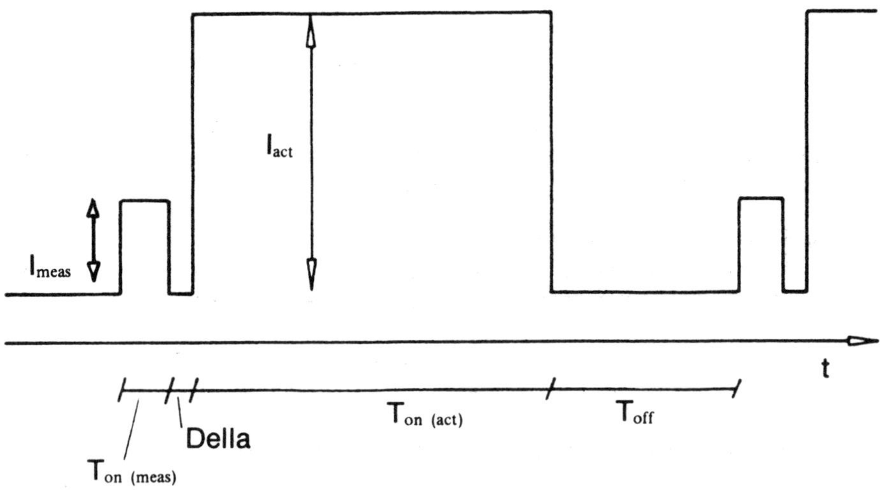

Fig. 14.3—Current profile applied in MIM system, working as pulse generator and surface area measurement equipment.

In practice it is observed that in making measurements with the system, a certain time is required in order to establish reproducible surface area values. This so-called "acquisition time", which depends on the nature of the electrolyte, and ranges from a few seconds to a few minutes, is related to changes in the double layer capacity in the early stages of the deposition process, which may arise from a number of factors:

a. A change in nature of the substrate by deposition of a different metal. This effect is rapid, requiring a very short acquisition time.
b. Changes in surface morphology, involving grain size and roughness factors. A longer time is required for stabilization in this respect.
c. The effect of organic additives, which can change the double layer capacity from the electrolyte side by adsorption on the electrode surface, and on the metal side by incorporation in the deposited layer. Break-down products of these materials may also have a drastic influence.

In addition, attention must be paid to the long-term stability of the electrolyte, since if the deposit deteriorates in terms of roughness, the reproducibility of the system is greatly reduced.

Nevertheless, in spite of the various factors which might change the double layer capacity in an unexpected way, the system has proved quite satisfactory in many practical applications ranging from laboratory to plating shop operations. Practical applications of for nickel, brass, copper and gold deposition are described in Ref. 8.

3. Surface area measurement using diffusion phenomena

3.1. *Principle of measurement* (9)

This method of surface area measurement is based on Fick's second law of diffusion:

$$\frac{\partial c}{\partial t} = D \frac{\partial^2 c}{\partial x^2} \quad [5]$$

as applied to limiting conditions corresponding to the situation when the electrode works at constant potential or at constant current. These two cases are considered separately in the following.

3.1.1. Measurement at constant potential

In electrodeposition processes, constant electrode potential means constant concentration at the electrode, according to the Nernst equation

$$\eta = -\frac{RT}{nF} \ln \frac{c_e}{c_0} \quad [6]$$

In limiting current conditions, c_e tends towards zero with:

$$c = c_0 \text{ for all } x \text{ and } t = 0 \quad [7]$$

$$c = 0 \text{ for } x = 0 \text{ and } t > 0 \quad [8]$$

Applying conditions [7] and [8], the solution of eqn. [5] is as follows:

$$\delta = \sqrt{\pi D t} \quad [9]$$

Equation [9] represents the growing rate of the Nernst diffusion layer δ as defined in Chapter 5, as a function of the time elapsing after the current switch-on. The concentration profiles as a function of time and the corresponding diffusion layer thicknesses are illustrated in Figure 14.4. Equation [9] is applicable only if the corresponding diffusion layer thickness remains thinner than the diffusion layer thickness determined by the hydrodynamics of the system. Therefore, the surface area measurement according to this method is made during a short single potentiostatic pulse. The limiting current density for the discharge of the diffusion-controlled deposited ions is directly proportional to the slope of the concentration profiles represented in Figure 4 and can be written according to the first Fick's law as follows:

$$j = nFD \frac{c_0}{\delta} \quad [10]$$

Replacing δ from equation [9],

$$j = nFD\, c_0 / \sqrt{\pi D t} \quad [11]$$

therefore

$$S = I\sqrt{\pi D t} / nFD\, c_0 \quad [12]$$

The surface area can be calculated from eqn. [12] when I and t are measured with appropriate electronics. The current as a function of time is schematically represented in Figure 14.5. At a given time, for example, t_1, an appropriate

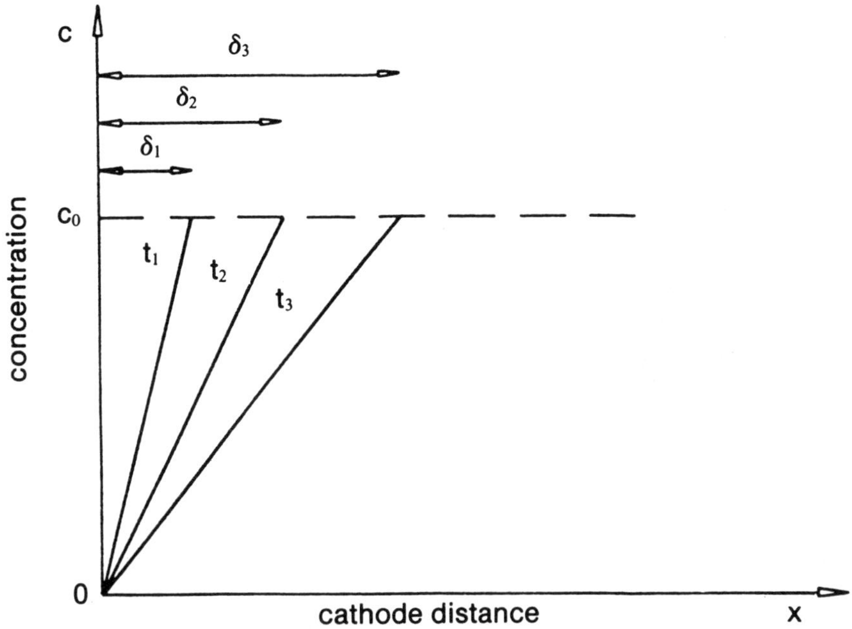

Fig. 14.4—Concentration profiles as a function of time and corresponding δ values for limiting current conditions ($t_1 < t_2 < t_3$).

equipment measures the current I_1, and according to eqn. [12] the surface area is calculated, provided that the measurement is made in an appropriate electrolyte.

3.1.2. Measurement at constant current

In electrodeposition processes, constant current means constant slope of concentration profiles. Figure 14.6 shows the concentration profiles of an electrodeposited species at different times. Under these conditions, the limiting conditions for the solution of eqn. [5] are as follows:

$$c = c_0 \text{ for all x and for } t = 0 \qquad [13]$$

$$D \frac{\delta c}{\delta x} = j/nF \qquad [14]$$

and the solution can be written as:

$$\delta = 2 \sqrt{Dt/\pi} \qquad [15]$$

As in the previous case, eqn. [15] is applicable only if the corresponding diffusion layer thickness remains thinner than the diffusion layer thickness determined by the hydrodynamics of the system. Therefore, the surface area measurement according to this method is performed during a short single

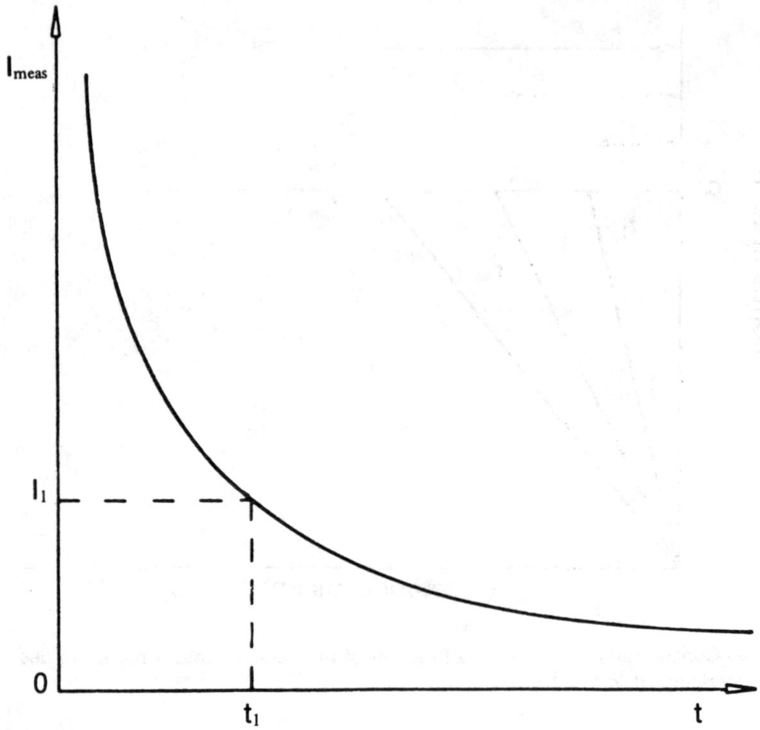

Fig. 14.5—Current as a function of time for a potentiostatic pulse under limiting current conditions. I_1 and t_1: couple of measured values allowing the computation of the surface area.

galvanostatic pulse, the intensity of which has to be higher than the limiting current density in DC.

Combining eqn. [15] with [10],

$$j = nFD\, c_0 / 2\sqrt{D\tau/\pi} \qquad [16]$$

$$S = 2\, I\sqrt{D\tau/\pi}\, /\, nFD\, c_0 \qquad [17]$$

The transition time τ corresponds to the time at which the concentration of the electrodeposited species at the electrode tends toward zero, as for example, at the time t_3 shown in Figure 14.6. In an appropriate electrolyte, a sudden drop in potential occurs at τ, which can be electronically detected. The corresponding current can also be measured, and the surface area computed according to equation [17].

3.1.3. Remarks concerning the constant potential and constant current measurements

Both above types of measurements are performed in an appropriate electrolyte, the essential requirement of which is to show a limiting current plateau on the polarization curve. Since this is not the case for most plating baths, the measurement has to be carried out in a separate tank prior to the plating line.

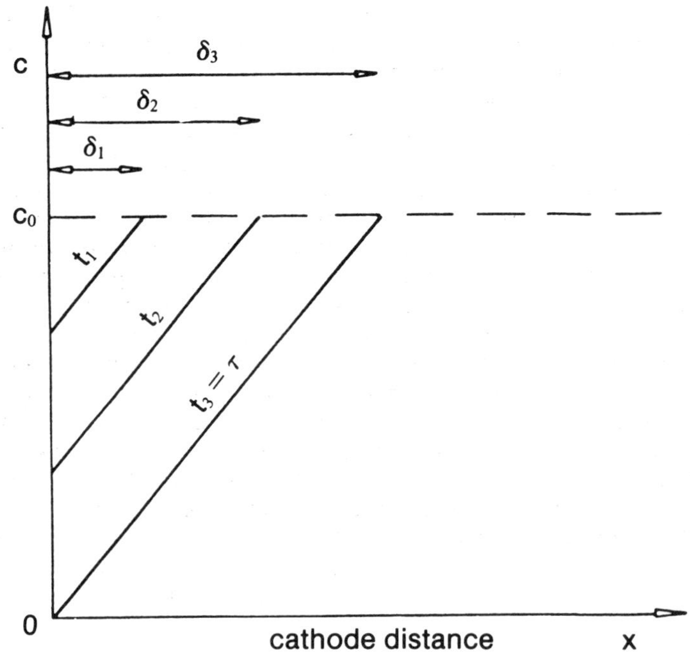

Fig. 14.6—Concentration profiles as a function of time and corresponding δ values ($t_1 < t_2 < \tau$).

The measurement is made during a short pulse, during which practically no metal is deposited.

The mathematical treatment above applies to linear diffusion. However, for areas of a very small characteristic length, like thin lines in printed circuits or small tips or wires, cylindrical or spherical diffusion can occur. In those cases, the mathematics must be modified, or the time at which the current is measured must be reduced in order to keep the thickness of the nonstationary diffusion layer sufficiently small to justify the assumption of a linear diffusion.

According to eqn. [12] and [17] absolute measurements of surface area can be made, provided the diffusion coefficient is known. If not, measurements can be made on a comparative basis.

3.2. Commercial equipment

Equipment for measurement of surface area by the method presented in section 3.1.1 also is commercially available†. The measurement at constant potential is preferred to that at constant current, because with the latter the limiting direct current density and therefore the surface area has to be estimated prior to measuring since, as shown in section 3.1.2, the measurement according to the latter technique involves a current necessarily higher than the limiting DC. For measurement at constant potential, the latter has to be preset, but independently from the surface area, and depending only on the electrolyte and in some cases on the nature of the substrate material.

† DIM, Egatec SA, Route de Port 20, 2503 Bienne, Switzerland

The practical measurement involves the use of a three-electrode system: the working electrode, the counter electrode and a reference electrode. The working electrode is the item to be measured, the counter electrode is an inert anode or a soluble anode compatible with the electrolyte, and the reference electrode can be as simple as a wire, preferably of the same metal as the working electrode. As main parameters, the potential imposed between the working and the reference electrodes during the measurement, the duration of the measurement pulse (typical: 0.1 to 0.5 sec), and a proportionality factor have to be preset on the front panel of the apparatus. The equipment can automatically repeat the measurements for checking its reproducibility. The time between two measurements as well as the number of measurements can also be preset on the front panel, and the results are recorded. The potential corresponds to the middle of the plateau of the polarization curve of the measurement electrolyte. The substrate should not react with the electrolyte, as for example, by an exchange reaction, otherwise the measurement is not feasible. An acid copper electrolyte will cover most of the applications, and substrates reactive to the solution can be covered by a thin copper electrodeposit prior to the measurement. An electrolyte based on the redox couple Fe^{II}/Fe^{III} can also be used, especially when no deposition at all should occur during the measurement. This system covers laboratory and industrial applications from less than 1 cm^2 to several m^2. Practical applications concerned with surface area measurements of screws and different jewelry items are described in Ref. 9 and 10.

4. Concluding remarks

The methods presented in sections 2 and 3 present two main advantages as compared to others mentioned in the Introduction. Firstly, the measurement time is very short, and secondly, the results are obtained in the form of an electrical signal which can be directly treated by computers.

In reality, an automatic plating line can be fully automatic only when the surface area is measured in the line itself. Power supplies are controlled from a central computer, and the latter needs information concerning the surface area in order to supply and maintain the required current density for the electrodeposition process concerned.

The methods presented in sections 2 and 3, based on pulsed current techniques, contribute significantly in practical applications by meeting real needs of the plating industry.

References

1. S. Brunauer, P.H. Emmett, E. Teller; *Journal American Chemical Society*, **60** : 309 (1958)
2. H. Schlegel; *Galvanotechnik*, **59** : 12 (1968)
3. H.W. Lieber; *Jahrbuch Oberflächentechnik*, **35** : 229 (1979)
4. C. Uemura; Japanese Patent 81,160.608 (May 15, 1980)
5. Y. Nishikawa, J. Fukakura; Japanese Patent 80,47,402 (September 30, 1978)
6. E. Robert; *Proceedings of the AES 8th Symposium on Plating in the Electronics Industry*, Phoenix (January 20-21, 1981)
7. E. Robert; *Oberfläche-Surface*, **22**, (8) : 261-266 (1981)
8. E. Robert; *Oberfläche-Surface*, **24**, (9) : 310-313 (1983)
9. J.Cl. Puippe; *Oberfläche-Surface*, **26**, (6) : 222-226 (1985).
10. H. Heuberger, J. Cl. Puippe and Ch. J. Raub; *Galvanotechnik*, **76**, (12) : 1963-1968 (1985)

Glossary

A	proportionality factor between current and potential	[-]
c	concentration	Mol/cm^3
c_{ad}	adatom concentration	Mol/cm^3
c^{o}_{ad}	adatom concentration at a growing step	Mol/cm^3
c_A	concentration of species A, variable	Mol/cm^3
$c_{A,b}$	bulk concentration of species A	Mol/cm^3
$c_{A,e}$	concentration of species A at the electrode	Mol/cm^3
c_e	concentration at the electrode	Mol/cm^3
c_0	bulk concentration	Mol/cm^3
C	capacitance of electric double layer	F/cm^2
D	diffusion coefficient	cm^2/s
D_A	diffusion coefficient of species A	cm^2/s
E	electrode potential	V
E_a	anode potential	V
F	Faraday constant	coul/Mol
I	current	A
I_a	anodic current	A
I_m	average current	A
I_p	pulsed current	A
j	current density	A/dm^2
j_a	anodic current density	A/dm^2
j_A	current density due to charge transfer through species A	A/dm^2
j_b	base current density in superimposed pulse or in pulse-on-pulse	A/dm^2
$j_{b(a)}$	anodic base current density in pulse-on-pulse reverse	A/dm^2
$j_{b(c)}$	cathodic base current in pulse-on-pulse reverse	A/dm^2
j_c	current density at the peak of a profiled electrode	A/dm^2
j_C	capacitive current density	A/dm^2
j_{gg}	limiting current density for DC	A/dm^2
j_F	faradic current density	A/dm^2
j_m	average current density ($= j_p T_{on}/(T_{on} + T_{off})$)	A/dm^2
$j_{m(a)}$	anodic average current density in pulse reverse	A/dm^2

$j_{m(c)}$	cathodic average current density in pulse reverse	A/dm^2
j_{mg}	average current density in plating with limiting current pulses	A/dm^2
j_p	pulsed current density	A/dm^2
$j_{p(a)}$	anodic pulsed current density in pulse reverse	A/dm^2
$j_{p(c)}$	cathodic pulsed current density in pulse reverse	A/dm^2
j_{pg}	pulsed current density of limiting current pulse (a limiting current pulse is a current pulse at the end of which the concentration of electroactive ions at the electrode tends toward zero)	A/dm^2
j_r	current density in the recess of a profiled electrode	A/dm^2
j_t	total current density ($j_C + j_F$)	A/dm^2
j_0	exchange current density	A/dm^2
j^*	dimensionless current density ($= j/j_{gg}$)	[-]
j_∞	current density at $t \rightarrow \infty$	A/dm^2
k	proportionality constant	[-]
L	characteristic length of electrolytic cell	cm
L	length of electrode (in chapter 12)	cm
n	number of electrons transferred per reacting metal ion	[-]
N	number of pores	[-]
Q	charge	A s
r	pore diameter	cm
r_0	radius of rotating disk electrode	cm
R	gas constant ($= 8.317\ 10^3$)	joule/KMol $^\circ K^{-1}$
R	roughness (in chapter 6)	cm
2R	size of unit cell	cm
R_a	activation resistance	Ω
R_e	electrical resistance of electrolyte	Ω
R'_p	polarization resistance of counter electrode	Ω
S	surface area	cm^2
t	time, variable	s
t_A	transport number for species A	[-]
t_c	charging time of double layer (up to 99% of j_p)	s
t_d	discharging time of double layer (from $j_F = 0.999\ j_p$ to $0.01\ j_p$)	s
t_f	falltime	s
t_n	time interval corresponding to the nth increment of potential	s
t_r	risetime	s

t^*	dimensionless time (= $Dt\,\delta^{-2}$)	[-]
t_c^*	charging time of double layer (with $j_F = 0$ during the charge)	s
t_c^{**}	charging time of double layer (up to 98.2 % of $\eta_{a,\infty}$ and constant resistance for electron transfer)	s
T	temperature	°K
or		
T	duration of a group of pulses in pulsed pulse or pulse-on-pulse	s
T'	off-time between two groups of pulses in pulsed pulse or pulse-on-pulse	s
T_a	anodic phase duration in pulsed pulse reverse or pulse-on-pulse reverse	s
T_a'	off-time following a group of anodic pulses in pulsed pulse reverse or pulse-on-pulse reverse	s
T_c	cathodic phase duration in pulsed pulse reverse or pulse-on-pulse reverse	s
T_c'	off-time following a group of cathodic pulses in pulsed pulse reverse or pulse-on-pulse reverse	s
T_{on}	pulse duration (on-time)	s
$T_{on(a)}$	anodic pulse duration in pulse reverse	s
$T_{on(c)}$	cathodic pulse duration in pulse reverse	s
T_{off}	interval between two pulses (off-time)	s
$T_{off(a)}$	off-time following anodic pulses in pulse reverse	s
$T_{off(c)}$	off-time following cathodic pulses in pulse reverse	s
v	flow velocity	cm/s
V_a	anode voltage	V
α	transfer coefficient	[-]
γ	duty cycle	[-]
δ	thickness of Nernst diffusion layer (linearized concentration profile)	cm
δ'	thickness of diffusion layer (real concentration profile)	cm
δ_p	thickness of pulsating diffusion layer	cm
δ_s	thickness of stationary diffusion layer (= δ- δ_p)	cm

δ_v	thickness of hydrodynamic boundary layer	cm
Δt_p	instantaneous temperature rise	°C
Δt_s	steady temperature rise	°C
ΔU_{a-a}	potential difference between beginning and end of a capacitive pulse	V
ϵ	thickness of passive film on anode	cm
η	overpotential	V
η_a	activation overpotential	V
$\eta_{a,\infty}$	activation overpotential by $j_F = j_p$	V
η_c	concentration overpotential	V
η_k	crystallization overpotential	V
η_s	electrode surface overpotential	V
θ	cycle time ($= T_{on} + T_{off}$)	s
θ^*	dimensionless cycle time ($= D\theta\, \delta^{-2}$)	[-]
θ_A	current efficiency for the deposition of species A	[-]
θ_{gg}	current efficiency at j_{gg}	[-]
θ_{pg}	current efficiency at j_{pg}	[-]
κ	specific conductivity of electrolyte	$\Omega^{-1}\, M^{-1}$
ν	kinematic viscosity	cm^2/s
τ	transition time	s
τ^*	dimensionless transition time ($= D\tau\, \delta^{-2}$)	[-]

INDEX

General and Theoretical Aspects

Practical Aspects

Metal and Alloy Deposition

*Gold-cobalt and gold-nickel alloys are dealt with together to avoid duplication of headings, since these systems figure in parallel in much reported work — but note that the references above do not necessarily apply to both systems.

*Diammino salts would form tetrammine complexes in ammoniacal solution as commonly used.